Molecules in the Galactic Environment

Molecules in the Galactic Environment

Edited by

M. A. GORDON

National Radio Astronomy Observatory

LEWIS E. SNYDER

University of Virginia

A WILEY-INTERSCIENCE PUBLICATION

JOHN WILEY & SONS, New York · **London** · **Sydney** · **Toronto**

Library of Congress Cataloging in Publication Data
Main entry under title:

Molecules in the galactic environment.

   Proceedings of a symposium sponsored by the National Radio Astronomy Observatory and the University of Virginia, held Nov. 4-7, 1971.
      1.   Interstellar matter—Congresses.   I.   Gordon, Mark A., ed.   II.   Snyder, Lewis E., ed.   III.   United States.   National Radio Astronomy Observatory, Green Bank, W. Va.   IV.   Virginia.   University.

QB790.M64   523.1'12   73-6555
ISBN 0-471-31608-3

Printed in the United States of America

10 9 8 7 6 5 4 3 2 1

FOREWORD

Optical astronomers have studied the spectra of molecules in the interstellar medium and in stellar envelopes for many years. With the possible exception of the interstellar diffuse bands, the interstellar molecules have been diatomic: CH, CH^+, and CN. Recently, the list has expanded to include H_2 and, following an earlier radio detection, CO in the ultraviolet. The limited number of these molecules has made it difficult to study their origin by optical techniques.

Beginning with the detection of OH in 1963, radio astronomers have expanded the list of interstellar molecules manyfold. Tables 1

TABLE 1

INORGANIC INTERSTELLAR MOLECULES

Diatomic Molecules	Triatomic	Four-Atomic
H_2 – molecular hydrogen	H_2O – water	NH_3 – ammonia
OH – hydroxyl	H_2S – hydrogen sulfide	
SiO – silicon monoxide		

and 2 list the molecules which had been detected in the interstellar

TABLE 2

ORGANIC INTERSTELLAR MOLECULES

Diatomic Molecules	Triatomic	Four-Atomic
CH^+	HCN – hydrogen cyanide	H_2CO – formaldehyde
CH	OCS – carbonyl sulfide	HNCO – isocyanic acid
CN – cyanogen radical		H_2CS – thioformaldehyde
CO – carbon monoxide		
CS – carbon monosulfide		

Five-Atomic	Six-Atomic	Seven-Atomic	Unidentified
H_2CNH – methyleneamine	CH_3OH – methyl alcohol	CH_3C_2H – methylacetylene	X-ogen (CCH ?)
HCOOH – formic acid	CH_3CN – methyl cyanide	$HCOCH_3$ – acetaldehyde	"HNC" (hydrogen isocyanide)
HC_3N – cyanoacetylene	$HCONH_2$ – formamide		

v

medium as of February 1973. The expanded list contains not only diatomics, but polyatomics containing as many as 7 atoms total and as many as 3 heavy atoms. The complexity of these molecules required great changes in old tenets regarding formation of molecules in the tenuous interstellar medium.

This sharp increase in the number of molecules detected in the interstellar medium also brought challenging problems: the seemingly disproportionate ratio of organic to inorganic molecules, indications of non-terrestrial abundances of isotopes, molecular spectra unidentified in the laboratory, and non-equilibrium excitation mechanisms. Solutions to these problems require not only the radiation transfer familiar to all astronomers, but a detailed knowledge of a number of scientific disciplines usually not associated with the main stream of astronomy: chemical reactions on surfaces, laboratory and theoretical spectroscopy in the microwave range, the physics of saturated and unsaturated masers, and the formation of primitive biological systems.

The extreme diversity of the observations in the rapidly expanding field of molecular astronomy required a joint assault by a variety of specialists. Therefore, the National Radio Astronomy Observatory and the Astronomy Department of the University of Virginia organized a Symposium to bring together astronomers, chemists, physicists, and biologists to consider the observations then available and to ascertain future directions to investigate. The organizing committee consisted of M. A. Gordon (chairman) and W. E. Howard III of the National Radio Astronomy Observatory, H. E. Radford of the Smithsonian Institution (Astrophysical Observatory), and L. E. Snyder of the University of Virginia. Our plan was to divide the Symposium into sessions dealing with a review of the interstellar medium, observations of molecules, techniques of molecular spectroscopy, physics of molecular excitation, chemistry relevant to the astronomical environment, and possible biological relevance. Each session began with a general introduction to that specialty and was followed by shorter, more specific contribution.

This volume contains the proceedings of this comprehensive Symposium. The general—occasionally tutorial—nature of many of the articles should make it a useful reference work for those beginning work in this field. It does not contain reviews of observations, however. The changing nature of the basic observations dates such reviews, and the interested reader should consult current literature. We list below several excellent reviews available at the time of this printing.

Many people have contributed much to this Symposium and to this book in a general sense. In particular, we thank Dave Buhl, Bert Donn, Frank Drake, George Herbig, Gerhard Herzberg, Carl Sagan, Phil Solomon, Larry Slobodkin, Barry Turner, and Gart Westerhout for suggesting appropriate topics and speakers and, in some cases, for serving as session chairmen. Our secretaries, Phyllis Jackson and Mary King Frazier, were the real power behind our executive decisions. Our printer, George Fitch, together with Deborah Stoughton and Shirley Baker, oversaw the preparation of this book. The National Radio Astronomy Observatory and the University of Virginia sponsored the Symposium and book. We thank the California Institute of Technology and the Carnegie Institution of Washington for permission to use the copyrighted Hale Observatories photographs.

M. A. Gordon

L. E. Snyder

Charlottesville, Virginia
February, 1973

SUPPLEMENTARY READING

Bok, B. J. 1972, "The Birth of Stars", *Sci. Amer.*, **227**, 48 (August)—description of theories of star formation for the non-specialist.

Buhl, D. 1971, "Chemical Constituents of Interstellar Clouds", *Nature*, **234**, 331—an introductory article of general interest.

Heiles, C. E. 1971, "Physical Conditions and Chemical Constitution of Dark Clouds", *Ann. Rev. Astron. Ap.*, **9**, 293—a detailed discussion of the physics of dark clouds.

McNally, D. 1968, "Interstellar Molecules", *Adv. Astron. Ap.*, **6**, 173—a detailed discussion of interstellar molecules (primarily optical) prior to the recent expansion of the field.

Rank, D. M.; Townes, C. H. and Welch, W. J. 1971, "Interstellar Molecules and Dense Clouds", *Science*, **174**, 1083—a review of observations emphasizing the physics of excitation.

Snyder, L. E. 1972, "Chemistry Between the Stars", in *The Emerging Universe*, edited by W. C. Saslaw and Kenneth C. Jacobs (Charlottesville: University of Virginia Press)—a brief overview of astrochemistry for the astronomer.

Snyder, L. E. 1972, "Molecules in Space", in *Spectroscopy, Physical Chemistry Series 1*, **3**, *MTP Int. Rev. Sci.*, (London: Butterworth)—a comprehensive review article, particularly thorough in describing the molecular lines observed.

Snyder, L. E. and Buhl, D. 1970, "Molecules in the Interstellar Medium", *Sky and Tel.,* **40,** 267 and 345—a general survey article for the advanced amateur astronomer or non-specialist.

Solomon, P. M. 1973, "Interstellar Molecules", *Physics Today,* **26,** 32—an excellent overall discussion of the observations, physics and chemistry of interstellar molecules.

Turner, B. E. 1973, "Interstellar Molecules", in *Galactic and Extragalactic Radio Astronomy,* edited by G. L. Verschuur and K. E. Kellermann (New York: Springer Verlag)—a comprehensive review for the specialist, emphasizing possible formation mechanisms.

TABLE OF CONTENTS

Molecules in the Galactic Environment

Session 1

The Interstellar Medium:
A Review

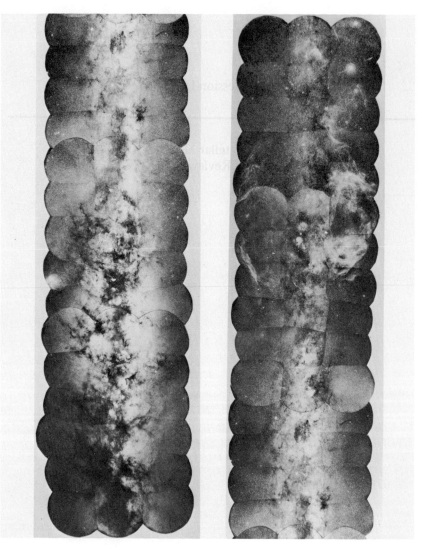

Composite photograph of the Southern Milky Way made by B. J. Bok from photographs in the light of hydrogen taken from the Mount Stromlo Atlas. Dark clouds of all shapes and sizes are seen superimposed upon the background nebuosity; these contain molecules described in this book.

Stellar Life Cycles and Cosmic Abundances

Icko Iben, Jr.

*Joint Institute for Laboratory
Astrophysics,
University of Colorado, Boulder, Colorado
and
Massachusetts Institute of Technology,
Cambridge, Massachusetts*

Edge-on view of the spiral galaxy NGC 4565 in the constellation Coma Berenices. This galaxy may have structure similar to that of our own. Note the extensive absorption caused by the dark nebulae confined to the plane of the galaxy. Photographed by the 200-inch in red light. (Hale Observatories photograph)

I. MOTIVATION

What bearing do interstellar molecules have on stellar structure and stellar evolution? Even more to the point, what do stars have to do with molecules in space?

After some thought, several technical answers to these questions become apparent. (1) Using molecular lines as probes, the spectroscopist can tell us something about the nature of condensations that actually form in nature as necessary preludes to star formation. How a star forms is certainly in the domain of stellar evolution. (2) The nature of condensations, the distribution and excitation state of the emitting and absorbing molecules, are in turn influenced by the spectral energy distribution produced by stars, both during quiet stages — in the form of Planckian distributions of photons and power law distributions of stellar wind particles — and during spectacular phases — in the form of high energy X- and γ-rays and cosmic rays. Thus, it is important to know which stars provide what sort of input into the interstellar medium at what stages in their lives. (3) The immediate origin of many interstellar molecules may well be in the extended, expanding atmospheres of certain types of cool stars. Which stars in what stages of evolution can act as direct sources of interstellar molecules is therefore an important concern.

The relationships I have cited are certainly valid reasons for being interested simultaneously in stellar evolution and in interstellar molecules. However, to me they are not very exciting reasons. This is partially because the enumerated relationships can be explored much more competently by the person well versed in the physics of the interstellar medium.

I suspect that the real reason a discourse on stellar life cycles has been commissioned for this symposium is a concern with and a curiosity about our ultimate origins and perhaps our ultimate ends. This concern is shared, perhaps, by many of us. A partial answer to the question of origins is provided by the stars.

The initial excitement engendered by the discovery of polyatomic molecules in space centered about technical points. First, not many thought such molecules could be present — "Collisions in interstellar space are too infrequent," we said. Second, the intensity of several emitting regions suggested the phenomenon of stimulated emission — "Think of that," we raptured, "There are giant water 'masers' in the sky!"

But, then, more and more polyatomic molecules, including several simple organic molecules that occur in living things, were found in space — "Life began in space!", we shouted. So it is that, recently, much of the interdisciplinary interest that centers about the discoveries of the interstellar molecular spectroscopist is connected with the thought that much of the chemical buildup that precedes the development of the most elementary life forms may have taken place in space (on grains in dirty interstellar clouds, or at the surfaces of stars?) in the absence of the benevolent environment provided by the earth and its energy source, the Sun.

Currently, the chemist and the radio astronomer combine forces to find more and more exotic molecules in space. The physicist and interstellar medium theorist try to figure out how H, C, N, O, etc., can manage to bind together to form NH_3, H_2O, CH_3OH, H_2CO_2, CO, CH_3CN, OCS, CS, HCN, NH_2CHO, etc. The biologist and the chemist speculate on biospheres and whether or not it is necessary to postulate lightning and thunder in great abundance to initiate and promote the early stages of animate evolution on earth.

In summary, what makes molecules in space exciting for many of us is the role that they may play in the story of life. The question that unites us is the question of our origin. If, then, we share a curiosity about how H, C, N, O, etc. bind together in space to form simple organic molecules; about how some of these molecules bond together (perhaps elsewhere than on earth) into complex chains that act as coding templates for replication and for the transmission of inherited characteristics; about how the simplest living cells are formed — must we not also share an abiding curiosity about where and how the elements H, C, N, O, etc. themselves were born? This is the story that a study of stellar structure and evolution can perhaps eventually provide. The details of the story are far from being understood, but we know almost with a certainty that the stars hold the answer to the question "Whence the elements," In the memorable words of Willy Fowler, "All of the elements that make up our bodies (with the exception perhaps of hydrogen) have been formed in stars. Thus, each and everyone of us is, in a very real and literal sense, a little bit of stardust."

II. THE WHEN AND WHERE OF ELEMENT FORMATION

A. Hydrogen and Helium

Spectroscopic observations indicate that H and He are overwhelmingly the most abundant species at the surfaces of most stars and in the interstellar medium. Perhaps the most remarkable feature of the galactic distribution of H and He is the apparent constancy of the abundance ratio He/H. In almost every situation where it can be measured or estimated, this ratio turns out to be ~ 0.1 by number (or 0.4 by mass). Comparison of lines emitted by hydrogen and by ionized helium in H II regions places a lower limit of $Y \sim 0.25$ - 0.27 on the abundance by mass of He in these regions. A model atmosphere mediated analysis of He and H lines (in absorption) at the surfaces of hot O and B stars reveals, for a large sample of disk stars, a Gaussian distribution of estimated Y values peaking at $Y \sim 0.25$ - 0.3. Theoretical models of the sun give $Y \sim 0.25$ -

0.27. Theoretical models of globular cluster stars give Y $\sim$ 0.3 ± 0.05.

Thus, all indications speak for a nearly universal abundance ratio of He to H that is independent of the abundance of elements heavier than He. A currently fashionable explanation for the apparently universal He/H ratio is provided by an examination of element synthesis in Big Bang models of the universe. In the framework of these models one may extrapolate backward in time to reach conditions ($T > 10^{9\circ}$K) under which all matter and energy are in the form of protons, neutrons, electrons, neutrinos, gamma rays, and gravitons. The exact form of the extrapolation depends, of course, on estimates of the current matter density and radiation density in the universe when averaged over dimensions large compared with intergalactic separations as well as on estimates of the current expansion rate of the universe as measured by the Hubble time. There are also differences related to the degree of homogeneity and isotropy, the degree and spectrum of fluctuations, etc. that one can postulate. Nevertheless, the net result with regard to element synthesis is remarkably insensitive to the model details and to the uncertainties in estimates of current properties of the universe.

Having extrapolated backward to conditions when matter exists only in the form of neutrons, protons, and electrons, we ask what happens if we again look forward in time. As the model universe expands and temperatures drop, the rate at which fusion reactions turn nucleons into alpha particles exceeds the rate at which photodisintegrations undo nuclei formed in intermediate steps. As the expansion proceeds still further, the rate of nucleosynthesis continues to decrease and, eventually, the conversion of nucleons into nuclei effectively ceases. The net result is a final He/H ratio of about 0.1 and an abundance of elements heavier than helium that is many orders of magnitude smaller than that found in even the most metal deficient stars.

Big Bang models of the universe thus provide a simple explanation for the apparently universal abundance ratio of He to H in the galaxy. They suggest at the same time that elements heavier than helium were made within galaxies (the current picture is that galaxies are products of the smallest condensations that survive radiative dissipation just before hydrogen recombines in the early, post fireball expansion of the universe). The proximity of Big Bang estimates of Y (that apply to pregalactic matter) to values of Y estimated for objects now in our galaxy suggests that the formation of the heavier elements in the galaxy was accompanied by very little additional helium production.

In leaving this stage of our story, we should recognize that we have not answered the question of our ultimate origin. To this the best answer remains "And God said, 'Let there be light.'" Today we need add only: "And God said, 'Let there also be matter and laws governing the behavior of matter and light.'" All else follows if we but have the patience and intelligence to read and properly interpret the signs that the Heavens provide.

B. *The Heavy Elements*

Spectroscopic and model atmosphere analysis reveals that, at the surface of most stars, the distribution of elements heavier than helium is, to a first approximation, independent of the total abundance of the heavy elements, even though this later quantity varies by three orders of magnitude as one ranges from the most metal rich stars in the galactic disk to the most metal poor stars in the outer fringes of the halo.

If we assume that element formation occurs at singularities (e.g., cooking in stars and regurgitation from same), the existence of an approximately universal distribution for the heavy elements implies one of two things: either (1) the distribution of elements emerging from each singularity is the same, or (2) products expelled from a multitude of singularities mix at a rate that is fast relative to the rate at which new singularities are born out of the steadily diminishing supply of interstellar gas. Our current partial understanding of the late stages of stellar evolution suggests that the second alternative is the more likely one.

The total abundance of heavy elements at a star's surface seems to be roughly correlated with the star's age and with its mean (rms) distance above the galactic plane. In general, the older the star, or the greater its mean distance above the plane, the lower is its metal abundance. There also appears to be a distinct dichotomy between stars in the halo and stars in the disk of the galaxy. Most of the stars in the halo are distinctly metal poor and are, within a spread of perhaps a billion years or so, all about 10^{10} years old. Most of the stars in the disk are comparable to the Sun in metal richness and have ages that run the entire gamut from 10^{10} years to 0 years.

The simplest way to correlate these general features is by means of a collapse model of the galaxy. We assume that, at some stage in the expansion of the universe, our galaxy separated out as a distinct element in which internal gravitational forces finally overwhelmed the cosmologically induced tendency to expand. Gravitational contraction accelerated until, eventually, when inward gravitational forces reached values that would maintain the overall rotation rate at a steady value, our galaxy stabilized at its present dimensions. It is not unreasonable to suppose that conditions favorable for star formation first occurred near the galactic center where densities are highest. The e-folding time for a mean density increase during the final stages of our galaxy's collapse may be estimated to be only a few hundred million years. One might expect that a rapid upsurge in the rate of star formation occurred during this last e-folding time and that the birth rate reached a maximum when the galaxy approached its present dimensions and thereafter declined.

If the rate of heavy element enrichment of the galactic gas is proportional to the rate of star formation, one might guess that the last stages of galactic collapse also saw a great upsurge in the rate at which heavy elements spewed forth from production singularities. These liberated heavy elements perhaps flew out from the central parts of the galaxy to contaminate the gas in regions further from the center where star formation was proceeding less rapidly. Since many heavy elements can act as efficient cooling agents, one might even guess that the contamination from the center of the galaxy promoted or facilitated star formation in regions far from the center. Since the flux of heavy elements outward is attenuated by capture in prestellar matter, we have, with this picture, neatly accounted for the anticorrelation between a halo star's distance from the galactic center and its metal abundance. We have also made plausible that much of the halo gas was catalyzed into stars at about the time our galaxy reached its present dimensions some 10^{10} years ago (over a time interval that is comparable to the time it takes elements ejected from stars near the galactic center to reach the periphery of the galaxy or comparable to the time it takes matter at the periphery of the galaxy to free fall to the center). Finally, the near constancy (within a factor of 3) of the total heavy element abundance in disk stars and the approximate universality of the distribution of heavy element abundances in all stars is accounted for if the bulk of star formation and heavy element formation indeed occurred over a short interval of a few hundred million years and if stirring of matter in the galactic gas was efficient over regions large enough to contain many heavy element sources.

III. STELLAR STIRRING MECHANISMS AND MASS EJECTION MECHANISMS

We are persuaded that the heavy elements in the interstellar medium were made by stars not just because the anticorrelations between surface metal abundance and stellar age and distance from the galactic plane suggest production by galactic objects. We are persuaded also because heavy nuclei are synthesized during successive nuclear burning stages in the interiors of the spherically symmetric models of stars that we construct. The sad truth is, however, that we do not yet know how to transfer elements from regions in stars where they are synthesized back into the interstellar medium. Therefore, any current attempt to construct a theory of heavy element enrichment by stars must be of a speculative and fanciful nature.

Several instabilities have been proposed as triggers for the explosive ejection of synthesized material. In only one case, however, has it been demonstrated that the distribution of elements and state variables requisite for explosive ejection is a likely end product of previous stages of quiet nuclear burning. Embarrassingly enough, the one model with an acceptable pedigree suffers from fatal deficiencies: (a) it spews out almost nothing but iron peak elements (whence then, C, N, and O, the most abundant of the heavy elements?); (b) it shatters completely, leaving no remnant (whence, then, the pulsars?) and floods the galaxy with a superabundance of iron; (c) the energy of the explosion is comparable to the energy that supernovae are inferred to emit and yet the birthrate of the precursor stars is somewhat larger than the rate at which supernovae are inferred to occur in our galaxy (why, then, so few supernovae?).

Other less spectacular processes lead to the ejection of synthesized material from the surfaces of stars. Again, unfortunately, we do not yet know how to calculate from first principles the rate at which they operate. The observations, however, tell us unequivocally that these processes do occur.

In order that the interstellar medium be enriched by matter ejected from the stellar surface, synthesized matter must first be brought to the surface from regions deep in the interior. Several processes are possible candidates. When a star becomes sufficiently cool at the surface, convection can extend down into the interior as far as the location of active nuclear-burning shells, bringing up products of burning left behind by now-defunct shells. A demonstration of the occurrence of this phenomenon is provided by the two components of a-Aurigae. In one component, Li remains in a thin shell near the surface. In the other, which is in a slightly more advanced stage of evolution, Li has been spread out over a large convective region and is therefore considerably reduced in abundance at the surface relative to the abundance at the surface of the less evolved component.

Convective regions are also thought to arise within and above thermally unstable nuclear burning shells. The associated mixing may lead to the production of several exotic species of elements and also to the transportation of these exotic elements closer to the surface from whence they might be later brought all the way to the surface by envelope convection or some other stirring mechanism. One likely site for mixing of this sort is in the long period Cepheids that belong to the halo population. Calculations suggest that elements built up by neutron capture reactions that occur slowly relative to the rate of subsequent beta decays (s-process elements) may be synthesized between the He-burning and H-burning shells of these stars. Additional calculations suggest that these newly produced s-process elements may be carried to the surface by convection during one phase of the relaxation that follows each thermal pulse in the He-burning shell. Sad to say, the halo population stars which these models are supposed to describe show, if anything, a deficiency of s-process elements at their surfaces!

Additional stirring processes are important in rapidly rotating stars. The more rapidly a star rotates, the greater is the role that meridional and/or spin-down currents play. Magnetic field configurations may also strongly influence the nature of these currents. Perhaps the best demonstration that rotation-related currents can bring nuclear processed matter to the surface is the distribution of elements at the surfaces of carbon stars. The high abundance of nitrogen relative to both carbon and oxygen in carbon stars can be understood easily if we postulate that, in the main sequence progenitors of these stars, large scale mixing currents occurred with turnover times short compared to main sequence lifetimes.

Other processes that can alter surface abundances relative to the initial abundance distribution involve gravitational settling in slowly rotating stars (halo blue stars that show an abnormal deficiency of He are perhaps examples) and resonance absorption that can selectively drive elements to the surface of slowly rotating stars (the abundance distributions in Am and Ap stars may be examples).

Once elements have been brought to the surface, our task is to then expel them into space. Several mechanisms may be important. Stars with a well developed subphotospheric convective zone and a low surface gravity are likely to produce strong stellar winds. The turbulence in the convective zone generates energy that is then fed by acoustic waves into a corona whose limiting temperature is roughly proportional to the surface gravitational potential. Matter then evaporates from the hot corona. A stellar wind of this type is most likely to be effective as a rapid mass loss agent during the red giant phase of a star's evolution (a star becomes a red giant after each successive fuel is exhausted at the stellar center).

Another mass loss mechanism may become effective on the giant branch if the flux of energy generated in the interior becomes large enough that the radiation pressure force outward on a particle near the surface exceeds the inward gravitational force on that particle. This occurs when the ratio of luminosity to mass exceeds a large, critical value which depends on the major opacity source near the surface. In very cool, metal-rich red giants, carbon grains may form and be forced out very effectively by radiation pressure, carrying surrounding matter with them.

Surface mass loss may also be driven by a mechanism that involves favorable coupling between rotationally induced currents and magnetic fields. Flares are another possibility.

Yet another possible mass loss mechanism is associated with the thermal instability that occurs in the helium burning shell of a star that proceeds up the giant branch after He has been exhausted at the center. Calculations indicate that the amplitude of each successive instability increases and it is not an outlandish speculation to suppose that the amplitude eventually grows until all of the matter above the He-burning shell is ejected from the star in one final pulse, leaving a remnant that contracts toward the white dwarf state.

Examples of all of these types of mass loss occur in nature. Red giants such as a-Orionis are notorious emitters of mass. The Sun produces a steady wind and, during periods of high sunspot activity, emits low energy cosmic rays. Planetary nebulae (found everywhere in the sky) are shells of matter expanding out from central stars which are contracting toward white-dwarf dimensions. In the Hyades cluster, white dwarfs of mass no larger than $\sim M_\odot$ (certainly less than 1.4 $M_\odot$) are found side by side with main sequence stars of mass 2-3 $M_\odot$; since initially more massive stars evolve more rapidly than initially less massive stars, the implication is that the progenitors of the white dwarf were more massive than 2-3 $M_\odot$ and that they must have lost considerable mass at some stage between being main sequence stars and being white dwarfs. Some of this inferred mass loss undoubtedly occurred on the giant branch in the form of a steady wind, some of it may have been a consequence of a final thermal pulse that led to the ejection of a large shell of matter.

IV. FURTHER OBSTACLES TO A QUANTITATIVE THEORY OF ELEMENT ENRICHMENT DUE TO STARS

The observations tell us that convective and rotationally induced mixing processes are frequently important in bringing heavy elements to the surface from whence they may be expelled by various ejection mechanisms. They tell us also that instabilities occur (novae, supernovae, planetary nebulae) that lead to the sudden ejection of a large fraction of a star in which there are perhaps newly synthesized elements. We do not, however, know how to calculate any of these processes from first principles (with the possible exception of convective mixing in certain limiting situations). How then can we hope to produce a believable, quantitative theory of heavy element enrichment by stars?

Suppose that, by good fortune, we eventually learn how to handle rotation-related and magnetic field-related mixing processes and how to follow stellar evolution from start to finish, including relevant mixing processes and including mass loss by stellar winds and expulsion by explosion. Will we then be in a position to construct a satisfactory description of galactic nucleosynthesis? The history of a star depends on its initial mass, as well as on rotation rate and magnetic energy content. What shall we choose as the birth rate function (the mass spectrum) and how shall we distribute angular momentum and magnetic field energy among the stars for the first few generations of stars that appeared during the late collapse phase and early post collapse phase of the galaxy and were therefore the major source (we think) of heavy elements in the galaxy? The birth rate function must be a (possibly strong) function of the gaseous environment. Since the environment 10 billion years ago, when probably most of the elements were made, was undoubtedly considerably different than it is today and since a birth rate function constructed from current observations is strictly applicable only to the present environment, we must invent some all inclusive theory of star formation that will permit us to extrapolate back to conditions 10 billion years ago. Such a theory does not yet exist.

The list of obstacles to a satisfactory theoretical reproduction of galactic element synthesis could be continued almost indefinitely. I will mention only two more features of the problem that are possibly crucial but as yet have not received adequate attention in the context of a theory of galactic nucleosynthesis.

Many stars (perhaps even the majority) are formed as binary pairs. If the stars in a binary system are close enough they will interact. If and when the radius of either component exceeds the Roche limit for the pair, mass will be transferred to the other component. The evolutionary histories of both the recipient and the donor are altered. Mildly "cataclysmic" variables and blue stragglers are examples of the effects of close proximity. So too are at least some novae and perhaps also some supernovae of type I. During mass transfer, the "secondary" component may turn "inside out," depositing products of element synthesis on the "primary" component. The primary component may eventually explode or perhaps just lose mass from its surface and from the binary system in a less violent way. The net result is an additional, perhaps important way of getting elements out into space following production in the deep stellar interior. If binary evolution plays a crucial role in galactic nucleosynthesis, we are then faced with an additional set of difficult and, as yet, unsolved problems: What is the probability of binary formation? What is the probability for any given choice of initial orbital elements and for any initial distribution of mass between the two components? What are the details of mass transfer?

A final example of a possible obstacle to a satisfactory theory of galactic nucleosynthesis has to do with element processing that may occur in and above the stellar surface. Discussions of galactic element production usually assume that most of the synthesis and processing of the heavier elements takes place in stellar interiors. However, in order to account for the occurrence of light elements such as Li, Be, and B, spallation reactions at or near the surfaces of stars have sometimes been invoked. The necessary high energy protons are presumably accelerated in flare type activity that we know is present at the surfaces of many stars (the Sun and T-Tauri stars, for example). Are there possibly other situations in which spallation reactions may be important? Some models of pulsars suggest that particles may be accelerated to cosmic ray energies near the surfaces of neutron stars. What do these energetic particles do to the matter that was ejected in the supernova explosion that accompanies the formation of a neutron star core? Is perhaps a large part of the ejected iron reconverted to C, O, and a particles?

V. ELEMENT SYNTHESIS IN STARS

Perhaps I have been too pessimistic. Pessimistic views are sometimes the consequence of faulty glasses that focus clearly only when we look at those parts of the picture that we do not yet fully understand and fail to focus when we look at those parts of the picture that we do understand. And some things we really understand quite well. We do know where some of the lighter elements are made in stars, despite the fact that we don't know how to get them out at the right time.

A. Core Hydrogen Burning Phase

Most of any single star's life is spent converting hydrogen into helium in a fairly small volume near its center. The core hydrogen burning lifetime of any star with mass less than roughly the Sun's mass is in excess of 10^{10} year. Thus, single stars initially somewhat less massive than the Sun cannot have contributed to the observable abundance of heavy elements in the galaxy.

Stars with initial mass in excess of 50 - 100 $M_\odot$ appear to be exceedingly rare. If their rate of formation relative to the rate at which less massive stars are formed was similar 10^{10} years ago to what it seems to be now, we might expect that stars more massive than 50 - 100 $M_\odot$ have not played a significant role in the history of galactic nucleosynthesis. Thus we are led to consider stars with initial, main sequence masses in the range $\sim 1\ M_\odot \rightarrow \sim 50\ M_\odot$.

Stars in this mass range remain near the main sequence until they have exhausted their hydrogen fuel over at least ten percent of their entire mass. Wherever hydrogen has been converted completely into helium any initial C and O will have been converted almost completely into N. Since nitrogen synthesis is bypassed in later stages of nuclear burning, it is possible that the main source of N in the galaxy is synthesis in hydrogen burning regions by stars of successive generations that were formed from gas that was enriched by the C and O produced and ejected by stars of prior generations.

Parenthetically we should note that, since stellar lifetime is a very sensitive function of stellar mass, the word generation has an ambiguous meaning when applied to stars of different mass. A thousand generations of 10 $M_\odot$ stars come and go for every generation of 1 $M_\odot$ stars. Thus 20 or so generations of 10 $M_\odot$ stars could have formed during the final 2×10^8 yr of the collapse phase of the galaxy while the 1 $M_\odot$ stars formed during this phase are just now reaching the end of their nuclear-burning lives.

Carbon stars provide rather convincing evidence for the suggestion that N results from the conversion of C and O during H-burning phases. At the surfaces of these stars — massive red giants — the ratio N/C is in excess of 30. This is in contrast to a ratio N/C ⟨ 1 that is appropriate for the Sun, where surface abundances are thought to be relatively unaffected by nucleosynthesis occurring in the deep interior. The immediate progenitors of carbon stars are probably massive main sequence stars that are sufficiently rapid rotators that significant mixing by meridional currents is expected (such currents are thought to be minor in the slowly rotating Sun). In a rapidly rotating star near the main sequences, products of CN cycle burning may be continuously mixed up to near the surface by rotation-induced currents. When the star becomes a red giant and a convective envelope extends downward from the surface into the deep interior, these products are then brought right up to the photosphere. A ratio of N to C such as is observed in carbon stars may be achieved if approximately ten complete turnovers (center to surface mixing) occur over the main sequence lifetime of the progenitor star.

B. Central Helium Burning Phase

Once hydrogen is exhausted over about 10 percent of its mass, a light star ($M/M_\odot < 2.5$) rapidly develops a large envelope and a dense core in which electrons are degenerate. We say that the star is on the red giant branch. Hydrogen burning in a thin shell adds mass to the rapidly condensing helium core and the luminosity and radius of the star increase rapidly. Stellar radius can become several hundred times larger than the Sun's radius and luminosity can reach several thousand times the Sun's luminosity. Once the mass in the core exceeds about 0.5 $M_\odot$, helium burning reactions are ignited. After a brief interval, during which nuclear energy is converted into sufficient electron kinetic energy to relieve the condition of degeneracy, the star enters into another stable phase of nuclear burning. Helium is converted into C and O (possibly in comparable amounts) near the center of the hydrogen exhausted core and hydrogen burning continues in a thin shell. Perhaps only a percent of the light star's entire lifetime is spent in the central helium burning phase. Comparison between theoretical models and the observations suggests that as much as 20-40 percent of a light star's mass may be lost during the giant branch stage. Those stars that are either initially light enough ($\sim$0.5 - 0.8 $M_\odot$) or lose sufficient mass as giants become pulsating RR Lyrae stars for a portion of the central helium burning phase.

Stars initially more massive than $\sim$2.5 $M_\odot$ have a somewhat different history. In them, densities are sufficiently small and temperatures are sufficiently large that, when hydrogen is exhausted over at least the inner ten percent of the stellar mass, electron degeneracy does not play the role of delaying the onset of helium burning. They ignite helium almost immediately on reaching the giant branch and then swing back into a region of the Hertzsprung-Russell diagram intermediate between the main sequence band and the red giant branch. Their luminosity during the ensuing phase of core helium burning and shell hydrogen burning is not much different than during the main sequence phase. As a consequence, roughly 10-20 percent of their entire lifetime is spent as core helium burners. Pulsating stars known as Cepheids are in this phase. Comparison between models and the observations suggests that Cepheids have not lost a very substantial fraction of their mass during their previous sojourn on the giant branch, consistent with the short time theory assigns to this sojourn and with the large surface gravity such stars have relative to that of low mass stars that, owing to the presence of an electron-degenerate core, evolve up to similar luminosities on the giant branch.

The mass of the hydrogen exhausted core at the onset of central helium burning can be quite large in massive stars, thanks to the occurrence of a central convective region during the main sequence stage. In a large convective core, hydrogen from regions that are quiescent with regard to the synthesis of helium is stirred into the hottest, central regions where burning can occur and helium produced in the central regions is stirred out to quiescent regions. Near the end of the central hydrogen burning phase of massive stars, the convective region can contain mass far in excess of ten percent of the star's entire mass. For example, at the end of the central hydrogen burning phase, the mass of the convective region in a metal-rich 15 $M_\odot$ star is about 3 $M_\odot$. This is also the mass in the helium core at the commencement of the central helium burning phase.

C. Shell Helium Burning Phase

After helium is converted completely into carbon and oxygen at the stellar center, nuclear burning continues in two shells encasing a rapidly contracting C-O core. The star again becomes a red giant, if it is not already one. Burning in the two shells is, however, not necessarily monotonic. A thermal instability can develop in the helium burning shell, each thermal burst in the shell being followed by a relaxation phase. The hydrogen burning shell is intermittently extinguished and then reignited. During phases of extinction, the convective envelope can extend down past the hydrogen-helium discontinuity, carrying it closer to the helium burning shell. The net result is that, at least for stars less massive than about 8 $M_\odot$, the mean locations of the hydrogen and helium burning shells approach each other, leaving only a few hundredths of a solar mass of helium between the two shells. Thus, all of the helium built up so carefully during the main sequence and subsequent phases is destroyed during the double-shell source phase. This is a rather encouraging result since the observational fact of a nearly universal He/H ratio that is independent of the heavy element abundance suggests that stars have not contributed much new helium in the process of adding heavy elements to the interstellar gas.

In stars more massive than about 8 $M_\odot$ there may still be a considerable amount of helium between the two burning shells when carbon burning commences in the core, but indications are that this store may diminish to a negligible amount during the phase of central carbon burning.

There are two interesting phenomena that may be associated with the occurrence of a thermal instability in the helium burning shell. One of these has to do with element production; the other has to do with mass loss.

During the maximum-luminosity phase of a thermal pulse, conditions become favorable for convective mixing over a region extending from the discontinuity between C-O and He to beyond the He-H discontinuity. Thus, hydrogen may be convected into a hot region containing lots of carbon and helium. A sequence of reactions ensues that has the net effect of liberating neutrons. If the appropriate seed nuclei are present, the so-called s-process elements (neutron capture rates slow compared to intermediate beta decay rates) may be built up.

As the mass left between the helium burning shell and the surface diminishes and as the mass in the C-O core grows nearer the critical mass for carbon burning, the amplitude of the thermal instability grows. It is quite conceivable that, when this shell-to-surface mass become sufficiently small and/or when the C-O core mass becomes sufficiently large, a last violent swing might carry off the entire region above the helium burning shell. So far, no calculations have demonstrated directly that this will be the case.

D. Core Carbon Burning

The mass 1.4 $M_\odot$ plays a critical role in the life of a star. First, if there is one electron for every two nucleons (as is the case for any mixture of He^4, C^{12}, or O^{16}), 1.4 $M_\odot$ is the maximum mass for which a cold object can exist in hydrostatic equilibrium supported by electron pressure. Second, conditions in the C-O core of a double shell source star do not become ripe for carbon burning until this core reaches approximately 1.4 $M_\odot$.

We conclude that stars initially less massive than 1.4 $M_\odot$ will not get beyond the synthesis of C and O. If such stars make any contribution to galactic nucleosynthesis, this contribution is confined to newly formed C and O, to N converted from C and O (produced primarily in stars of earlier generations), and to s-process elements in an abundance proportional to the abundance of appropriate seed nuclei (produced in stars of earlier generations).

Stars initially more massive than 1.4 $M_\odot$ may well lose mass in sufficient quantity during giant phases that they are less massive than 1.4 $M_\odot$ before shell helium burning has gone to completion. The emitted mass may contain N that was converted from old C and O in H-burning regions and stirred out by rotation induced processes.

If we trust all of the model details currently in the literature, those stars that do survive with a mass larger than 1.4 $M_\odot$ but less than about 8 $M_\odot$ will suffer a most spectacular fate when the C-O core reaches a mass of 1.4 $M_\odot$. In models of such stars, carbon burning begins at the center, initiating a detonation wave that sweeps through the core, turning all C and O into Fe-peak elements as it passes through the core, leaving processed matter with a velocity greater than escape velocity.

We have enumerated in a previous section some of the undesirable consequences of this model. There are several ways in which the difficulties may eventually be removed. First, there is a factor of ten uncertainty in the effective rates of the reactions that produce C relative to the rate of the reaction that produces O. The uncertainty is such that, over a large central region, the final product of helium burning may be almost entirely O. If this is true, then the helium exhausted core must be increased ever so slightly in mass (by further action of the helium burning shell) until densities and temperatures are high enough to initiate oxygen burning. Electron captures on the Fe-peak elements formed following the detonation of oxygen may produce a sufficient under-pressure that a portion of the core begins to collapse toward neutron star dimensions. The detonation wave proceeding outward may be sufficiently modified that burning goes all the way to equilibrium for Fe-peak elements over only a portion of the core. The ejected matter may therefore contain O and C produced during prior stages of shell helium burning (the final ratio of C to O left behind by an advancing He-burning shell increases as this shell moves outward in mass).

Another possibility depends on whether or not the temperature profile in the C-O core has been correctly calculated (it requires a knowledge of the conductivity of a degenerate, relativistic electron gas and of neutrino loss rates) and whether or not the neglect of the thermal instability in the helium shell is justified. The course of the detonation wave might be significantly altered if it were initiated at some distance from the center rather than at the center. Among other things, a remnant might result from a detonation wave propagating toward the center, raising densities and temperatures in such a way that an implosion induced by electron capture occurs.

In any case, current models for stars initially less massive than about 8 $M_\odot$ when carbon burning commences suggest a universe flooded with Fe-peak elements and a deficiency of the most abundant elements C, N, and O. Evidently, the models require further study.

Stars initially more massive than 12 $M_\odot$ develop a nearly pure C-O core that is larger than 1.4 $M_\odot$ by the time He is exhausted at the center. The electrons in the C-O core do not become degenerate before C-burning is initiated. Hence, carbon burning will not lead to an explosion. After carbon is exhausted at the center, the core (composed mostly of O, Mg, and Ne) will contract and heat rapidly until electron degeneracy sets in. Thereafter, as mass is added to the core by the C-burning shell, core degeneracy continues to increase. When the central density reaches about 4×10^9 gm/cm^3, electron captures on Mg^{24} may initiate a collapse of the central region. Subsequent contraction and heating may lead to explosive O-burning that may result in an expulsion of the outer portions of the core. Hopefully, what comes out will have lots of C, O, Ne, Mg, Si, Fe, etc. in the right proportions. Only time and patience will tell if this, in fact, will be the case.

VI. POSTSCRIPT

How will the story turn out? So far we've concluded that God made electrons, nucleons, and other assorted forms of energy; a Big Bang left us with H and He and incipient condensations that became galaxies; galaxies made stars and stars made the elements. But how the elements got out of the stars, so far God only knows.

During the rest of this symposium we may occasionally be treated to guesses as to how molecules are made out of God-given atoms (on grains in clouds, near surfaces of stars, in solar nebulae) and as to how chains are made out of molecules in dark globules or under more theatrical circumstances in hot vapors sparked by electrical discharges. We will probably wind up learning that, as yet, only God knows how to make life and that He may need a very special set of circumstances that does not prevail frequently in the Universe.

Perhaps we should contemplate the possibility that life may, after all, be a unique event confined to the earth and that the form of life that tortures itself with questions about beginnings and ends may cease to exist, thanks to the evolution of the Sun, long before it has reached a satisfactory answer. This much we know: 4½ billion years ago the average temperature on the earth would have been only 22°F if the earth had not relied on internal sources to keep it warm and if atmospheric properties were roughly similar to what they are now. Average temperatures would not have reached 32°F until one to two billion years had elapsed. In another two billion years from now, temperatures in the shade, if there is any shade, will average 86°F. In yet another 3 billion years the oceans and the blood in our veins will boil and our progeny, if any, will have been forced to evolve into quite different beings. In still another billion years our modified descendents, if they exist, will have to develop cities that will float on molten lava. Perhaps, then, with energies being devoted entirely to survival, the question of whence life or whence the elements will no longer frequently be raised and suggestions for symposia such as this will be considered unconscionably frivolous. On the other hand, given man's peculiar nature, speculations as to how and why may occur with accelerating frequency as his final end draws nearer.

It is a pleasure to thank Carl Hansen, John Castor, John Cox, and Hollis Johnson for conversations that contributed considerably to my understanding of topics herein discussed. Support by the NSF (GP-11277) is gratefully acknowledged.

Thermodynamic Structure

George B. Field

University of California, Berkeley

I. EQUILIBRIUM

A typical parcel of interstellar gas is buffeted by many energy fluxes. It falls in the galactic gravitational field; it oscillates in the galactic magnetic field; it is ionized by starlight and cosmic rays; it is accelerated by shock waves, and heated by X-rays. In such a situation one cannot easily generalize about thermal properties; rather one must discuss specific phenomena.

It is instructive to consider the situation in HII regions as a prototype which is reasonably well understood. Close to a hot star, photons with < 912Å burn their way through and rapidly photoionize any H atoms present at the rate Γn_a cm^{-3} sec^{-1}. Re-combinations occur at the rate $a(T)n_e n_i$ cm^{-3} sec^{-1}, where the recombination coefficient a depends on the kinetic temperature T. With

$$\frac{dn_e}{dt} = \Gamma n_a - a n_e n_i \tag{1}$$

we see that equilibrium holds if the left hand side is much smaller than either of the terms on the right. If we define a dynamical time scale for the change of n_e by

$$\tau_d = \frac{n_e}{\overset{.}{n_e}} \tag{2}$$

this condition is expressed by

$$\tau_i \ll \tau_d \tag{3}$$

where

$$\tau_i = \frac{1}{a n_i} \tag{4}$$

is an ionization time scale defined by equation (1). Typically $\tau_i \sim 10^4$ years for $n_i = 10$ cm^{-3}, while $\tau_d \sim 10$ pc $\div$ 10 km/sec $\sim 10^6$ y. Thus, equation (3) is fulfilled, and we have ionization equilibrium:

$$\frac{n_e n_i}{n_a} = \frac{\Gamma}{a} \tag{5}$$

Similarly, the balance of energy gains and losses establishes the equilibrium temperature of the gas. In an HII region, the gains are due to the kinetic energy of the electrons photoejected from atoms by stellar photons:

$$G = \Gamma n_a \Delta E \text{ erg cm}^{-3} \text{ sec}^{-1} \tag{6}$$

where $\Delta E (\approx$ few eV) depends upon the blackbody temperature of the star. The losses in this case are primarily due to excitation of fine-structure and forbidden electronic levels of impurity ions such as O$^+$ by thermal electrons. Such a process occurs at the rate

$$L = n_e n_t \beta E_{ex} \exp (-E_{ex}/kT) \quad \text{erg cm}^{-3} \text{ sec}^{-1} \qquad (7)$$

where n_t is the density of target ions, $\beta(T) = \langle av \rangle$ is a rate constant which varies slowly with T, and the exponential factor signifies that only the tail of the electron energy distribution above the threshold, E_{ex}, is effective. If

$$\tau_t = \frac{\frac{3}{2} nkt}{L} = \frac{3 \ nkT \exp (E_{ex}/kT)}{2 \ n_e n_t \beta} \qquad (8)$$

is small compared to τ_d, (τ_t = thermal time scale), there is equilibrium, and the temperature is established by

$$n_e n_t \beta E_{ex} \exp (-E_{ex} kT) = \Gamma n_a \Delta E \qquad (9)$$

This can be combined with the ionization equilibrium equation (5) to yield

$$\frac{\beta(T)}{a(T)} \exp (-E_{ex}/kT) = \frac{\Delta E n_i}{E_{ex} n_t} \qquad (10)$$

From this simple equation we learn some important results. Note that the right hand side tends to be independent of density. This is because close to a hot star $n_a \ll n_i$ for hydrogen, so n_i approaches the total abundance hydrogen n_H; also, for ions like oxygen, the same tends to be true - $n_t \sim n_O$. Hence the right hand side is a constant determined by the relative abundance of oxygen, n_O/n_H. Therefore, the equilibrium temperature tends to be independent of density.

Furthermore, we note that β/a varies only slowly with T compared to exp $(-E_{ex}/kT)$, and is a large number ($\sim 10^6$). Although $\Delta E \ n_i/E_{ex} \ n_t$ is a large number also ($\sim 10^3$), exp $(-E_{ex}/kt)$ is small ($\sim 10^{-3}$), and only the Maxwell tail participates in the cooling.

Therefore
$$T = \frac{5040 \ E_{ex}(eV)}{\log \left(\dfrac{n_t}{n_i} \dfrac{E_{ex}}{\Delta E} \dfrac{\beta}{a} \right)} \qquad (11)$$

depends only logarithmically on abundances, energy levels and rates. For cooling by O^{++} ($E_{ex} = 2.5$ eV), $T \sim 4000°$. Thus, we expect the temperature to be relatively constant from place to place, near $T \sim 4000°$.

The thermal equilibrium of HII regions can be tested observationally. By studying the ratios of different emission line strengths of the same element (which have different E_{ex}), one can infer the value of T. By summing all the emission lines, one can check whether the energy emitted is in fact consistent with the theoretical heat loss at the observed temperature, and whether this in turn is in balance with the theoretical heat gain from the observed exciting star. By and large the thermodynamics of normal diffuse nebulae is well understood along these lines. The result is often summarized in this rule of thumb: the temperature of HII regions is of the order of 10^4 °K under most conditions. Thus, we can picture galactic ionized hydrogen as lying close to a single isotherm in the p - n diagram.

Matters are not that simple for regions where molecules are found. Most molecules would be disassociated by the $\lambda < 912$Å photons one finds in HII regions, so that by and large, we don't expect to find molecules there. Therefore molecules will exist mostly in HI regions, where they are shielded from ultraviolet; this appears to be consistent with the observations. The unfortunate fact is that our knowledge of the thermal properties of such regions is far more limited than that of HII regions. This is because (i) the heat source cannot be unambiguously identified (unlike HII regions, where the exciting stars are seen); (ii) the cooling mechanisms are uncertain, because the cooling radiation involved is presumably in the as yet unobserved far infrared (unlike that from HII regions, largely in the visible).

This situation is most vexing, because most of the mass of the interstellar medium is HI, and the thermal processes taking place there are of great interest, because they determine the gas pressure, which is one of the forces resisting gravitational collapse to form stars.

From the point of view of interstellar gas dynamics, the molecular phenomena which are the subject of this Symposium are of great interest. First, the observed molecular transitions offer an opportunity to infer the temperature of HI regions from line ratios. Second, there is a suspicion that molecules form particularly well in dusty regions of high density − which are just the regions most likely to undergo gravitational collapse, so that observations of them are of particular interest for star formation. Third, by emitting millimeter photons (as recently detected from CN, CO, etc.) they may provide important cooling mechanisms for lowering the gas temperature to the 2.7°K minimum allowed by the cosmic blackbody background. Thus, they may play an active role in promoting star formation.

II. HEATING AND COOLING OF HI REGIONS

Spitzer pointed out in the 1940's that HI regions will be much cooler than HII regions. Oversimplified, his argument was that some of the same electronic forbidden transitions are still available for cooling (e.g., excitation of the rather abundant ion S^+, which is the dominant form of sulfur even in an HI region because S° is ionized by $\lambda < 1200$Å, which penetrates HI regions), while the great heat input from the photoionization of hydrogen which occurs in HII regions is missing in HI regions. He concluded that the major heat inputs are (i) the photoionization by starlight of easily-ionized elements like carbon (ii) the impact ionization of hydrogen and helium by penetrating cosmic radiation. Mechanism (i) is qualitatively like the heating of HII regions, but much weaker because of the low abundance of carbon ($\sim 3 \times 10^{-4}$ H). Thus, the right hand side of equation (10) is ~ 3000 x smaller, so T must be smaller as well. It appears from what we have said so far that it would fall to about $1000°$, but at this low temperature, fine-structure levels such as that of C^+ at 0.013 eV begin to dominate the cooling because the electronic levels have a strong negative exponential factor. These levels (which are less important in HII regions because there the larger heating forces the temperature up to where the more effective electronic levels can be excited), dominate in an HI region. Since β/α is of the order of 10^6, $n_t/n_i \sim 1$, and $E_{ex}/\Delta E \sim 10^{-2}$, we see from eq. (11) that T $\sim$ 5000 $E_{ex}/4 \sim 16°$K. Mechanism (ii) was evaluated on the basis of the then-known cosmic ray spectrum down to 1 GeV, and found to be substantially smaller. Spitzer therefore predicted that HI regions should be at $\approx 20°$K or so.

When the 21-cm line was discovered in the early 1950's, it was found that HI regions generally have $\langle T^{-1} \rangle^{-1} \sim 100°$. Qualitatively, this vindicated Spitzer's idea that T(HI) $\ll$ T(HII), but as C^+ cooling at $100°$ is 5000 x more effective than the same process at $20°$, it is clear that the postulated mechanisms cannot be correct in detail. The answer to this problem is still not certain for the reasons given previously. We list the possible mechanisms in Table 1 and comment on them in what follows.

Table 1

HEATING AND COOLING MECHANISMS IN HI REGIONS

Heating (H)
1. Photoionization of C, Fe, Mg, Si, S
2. Collisional ionization by observed cosmic rays, E$>$1 GeV
3. Shock waves generated by cloud collisions and expansion of HII regions
4. Collisional ionization by low-energy cosmic rays, E = few MeV
5. Photoionization by soft X-rays, E$<$1000 eV
6. Photoionization by bursts of ultraviolet from supernovae, E $\sim$ 30 eV
7. Effective heating caused by depletion of coolants by, e.g., sticking to grains

Cooling (C)
1. Electron collisional excitation of fine-structure levels of ions of C, Fe, Mg, Si
2. H-atom collisional excitation of fine-structure levels of the above, and of O atoms
3. Electron and H-atom collisional excitation of forbidden electronic levels of O, N, S

4. H-atom collisional excitation of H_2 rotational levels
5. Electron and H-atom collisional excitation of polar molecules (e.g., CO)
6. Electron collisional excitation of Lyman lines of H.

We have already commented upon mechanisms H1 and H2 above. H3 seemed like a promising possibility until it was shown that the postshock cooling is too rapid (if coolant abundances are normal). Figure 1 shows the results of a recent hydrodynamic code calculation at Berkeley (Meszaros, 1971). A shock at a typical velocity ($\sim$ 10 km/sec) heats the gas to $\sim 3000°K$, but cooling

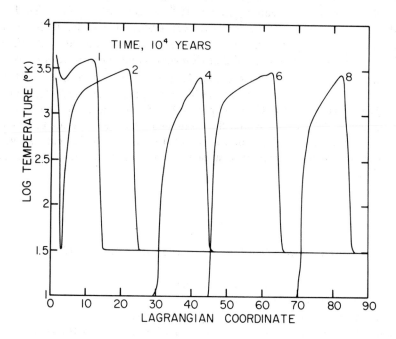

1. Cooling behind a shock wave proceeding into a cloud of n = 10 cm^{-3}. A piston moves from the left at 10 km/sec, heating the gas from $\sim 30°K$ to $\sim 3000°K$ immediately behind the shock. Radiative cooling reduces T below $10°K$ in $\sim 2 \times 10^4$ years (e.g., coordinate 45 between 4×10^4 and 6×10^4 years). From Meszaros (1971).

mechanisms C1 and C2 bring it below $100°K$ in $\sim 10^4$ years. Since the time for a shock to cross a 10-pc cloud is $\sim 10^6$ years, at any one time only 1 percent of the cloud is hot, and the harmonic mean temperature is raised only 1 percent. Earlier results (Field *et al.,* 1968) showed that introduction of a magnetic field decreases the heating by the shock wave, but prolongs the cooling period by inhibiting the density increase. The resulting cooling times are not very different. (However, if coolants are depleted by a factor 10-100 (H7), the results might be very different, and shocks might well be important in heating HI regions).

Mechanisms H4 and H5 have been extensively discussed recently. They are qualitatively similar in that (a) energetic particles or photons which have much

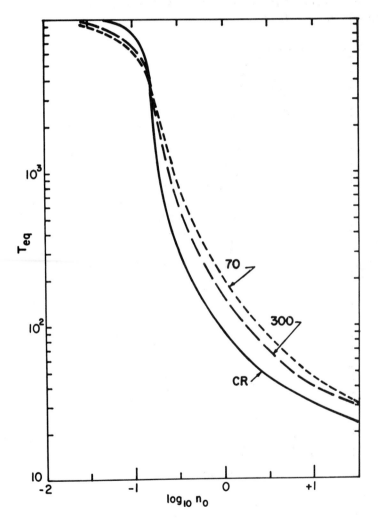

2. The density-temperature relation for heating by cosmic rays and X-rays (Habing and Goldsmith 1971). Ionization rates were adjusted to yield maximum pressure at $n = 0.2$ cm^{-3}; the dashed curves correspond to different X-ray spectra (see text).

larger mean free paths than stellar photons near the Lyman limit penetrate HI regions, weakly ionizing hydrogen and helium, and giving their excess energy to the ejected electrons and (b) in neither case (H4 or H5) is there direct conclusive evidence for the presence of the required particles or the photons, but only indirect arguments in their favor. Habing and Goldsmith (1971) have shown that the temperature-density relation in equilibrium is qualitatively similar for the mechanisms H4 and H5 (Figure 2), where T is plotted against n, the total density of nuclei. In Figure 2, Habing

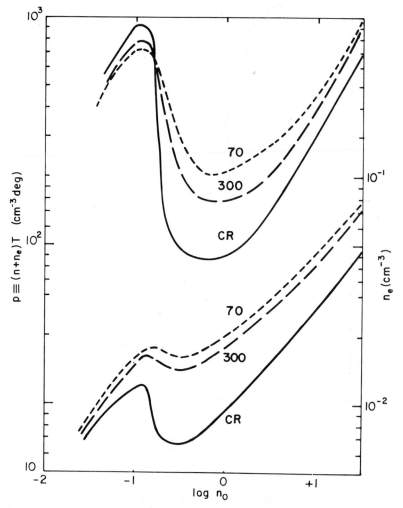

3. The pressure-density relation for heating by cosmic rays and X-rays. (Habing and Goldsmith 1971). See Figure 2. The electron density is also shown.

and Goldsmith took the primary ionization rate by cosmic rays to be 4 x 10⁻¹⁶ sec⁻¹, and that by X-rays to be 1.4 x 10⁻¹⁶ sec⁻¹ (curve 70) and 3.5 x 10⁻¹⁷ sec⁻¹ (curve 300) in cases where an assumed bremsstrahlung X-ray spectrum has a low energy cut-off at 70 eV and 300 eV, respectively. This insures that the corresponding p - n relation shown in Figure 3

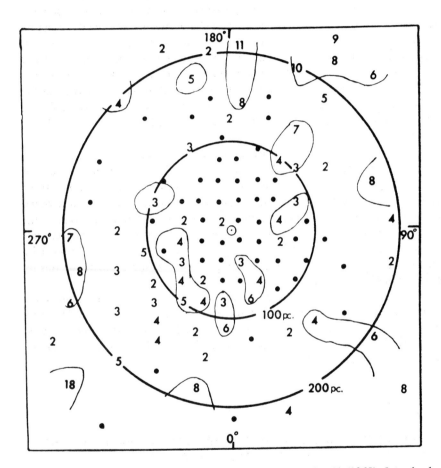

4. Patchiness in color excesses near the Sun according to FitzGerald (1968). Intercloud regions have color excesses (noted on the diagram in units of 0.01 mag) less than 0.01 mag out to 200pc, so $A_v < 0.15$ mag/kpc there. About a dozen clouds are evident, each with about 0.05 mag color excess, or 0.15 mag, extinction. Since the mean line of sight in a cloud is ~ 25pc, the dust in clouds is at least 40 x as dense as in the intercloud region. (As the data provide only upper limits on cloud dimensions, we conclude the density contrast may be 100 x or greater.)

(where $p = [n + n_e]$ kT) all have a maximum pressure at the same density (0.2 cm^{-3}). The curious p - n relation in all cases resembles that of a Van de Waals gas, in that as n increases, p increases roughly in proportion to n (T $\sim$ const $\sim$ 10^4 °K). It falls rapidly at n ~ 0.2 cm^{-3}, only to rise again roughly in proportion to n (T $\sim$ const $\sim 10^2$ °K) at higher densities. These characteristics, so different from the HII situation discussed above, accompany thermal equilibrium based on an input proportional to the first power of the density (as is the case here, because the target atoms—hydrogen and helium, are the dominant ionization state, unlike mechanism H1, where the density of target atoms, C°, is proportional to n_e $n(C^+) \sim n^2$, together with a cooling law which rises rapidly at low temperature because of fine-structure cooling (C1 and C2) and again at high temperatures because of excitation of electronic levels (C3 and C6). Because the cooling is proportional to n^2, it gains relative to heating as n rises, so T falls, finally to the point that fine-structure transitions can take over at a much lower temperature.

The curious shape of the p - n relation means that in principle there can be both a hot (intercloud) and cold (cloud) phase each in thermal equilibrium, at dramatically different temperatures but at the same pressure. This at one stroke accounts for discrete interstellar HI clouds, a prominent feature first discovered in the distribution of optical extinction (see Figure 4) and in optical sodium-line profiles, but later verified by high-resolution 21-cm studies. Normal clouds exert a finite pressure (p/k $\sim$ 10^3 cm^{-3} deg) but are apparently not confined by self-gravitation. The postulated intercloud medium would account for this by exerting an equal inward pressure.

If we assume that the intercloud pressure adjusts to the maximum value permitted for the hot phase, and if we revise ζ upward to 1.2×10^{-15} sec^{-1} (assumed to be cosmic rays, but X-rays would yield qualitatively similar results) we derive the data in Table 2, a revision of Field, et al. (1969) based on more recent cross-sections and abundances.

Table 2

PREDICTED AND OBSERVED PROPERTIES OF INTERSTELLAR HI NEAR THE GALACTIC PLANE

Quantity	Predicted	Observed
ζ, primary ionization rate	1.2×10^{-15} sec^{-1} *	1.2×10^{-15} sec^{-1}, from observations of 21-cm and low-frequency absorption (1)

Table 2 continued

Quantity	Predicted	Observed
$\bar{n}$, mean atomic density	1 cm^{-3}*	1 cm^{-3}, from 21-cm emission profiles near b = 0 (2)
n_i, intercloud density	0.24 cm^{-3}†	0.26 cm^{-3}, from intermediate b observations of broad 21-cm component (3)
T_i, intercloud temperature	7500°K†	$\sim$ 8000°K, from intermediate b observations of broad 21-cm component (3)
p_i, intercloud pressure	1800 cm^{-3} deg†	
n_c, cloud density	24 cm^{-3}	
T_c, mean cloud temperature	75°K*	72°K (1), 67°K (3) mean values from absorption lines at 21-cm
ζ, coolant depletion factor in clouds	0.16**	
η, cloud filling factor	0.032	
n_e, mean electron density	0.026 cm^{-3}	$\sim$ 0.03 cm^{-3} from pulsar dispersions
$\eta n_{ec}^2 \, T_{ec}^{-3/2}$, absorption parameter in clouds	4.1 x 10^{-7} cm^{-6} deg$^{-3/2}$	1.4-2.4 x 10^7 cm^{-6} deg$^{-3/2}$ (4,5) from low frequency absorption of nonthermal background and of point sources

* Assumed equal to observed value.
† Assumed to be that at the maximum pressure in the hot phase.
** Taken to give agreement with cloud temperature.

Notes to Table 2
   (1) Hughes, Thompson, and Colvin, 1971.
   (2) Kerr and Westerhout, 1965.
   (3) Radhakrishnan, Murray, Lockhart, and Whittle, 1971.
   (4) Ellis and Hamilton, 1966.
   (5) Bridle, 1969.

Since the earlier work, there has been considerable observational progress, some of which is compared with prediction in Table 2. Radhakrishnan, *et al.* (1971), in studying the 21-cm absorption of 35 sources, noted a wide emission component whose intensity rises smoothly in proportion to csc b. Noting that

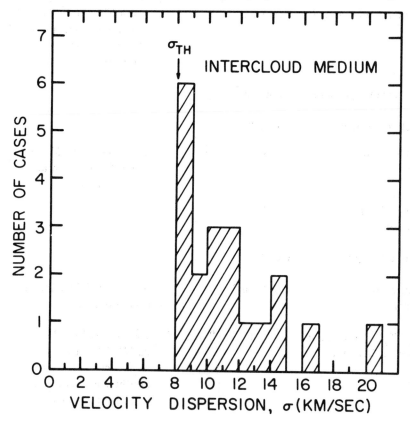

5. Root-mean-square velocity dispersions of the wide 21-cm emission component observed by Radhakrishnan, *et al.* (1971) for 20 different directions. Note the sharp lower limit to σ. If this is identified as thermal broadening, T = 8200°K, close to theoretical prediction.

this component showed no absorption features, as expected for hot HI gas, they attributed it to the intercloud medium, and derived an HI density, which if augmented by 10 percent ions and 10 percent helium, agrees well with prediction. Observations of extinction within 200 pc (Figure 4) show that the density of dust in the intercloud medium is $\leqslant 1$ percent of that in clouds. If the same thing is true of the gas, and if $n_c = 24$ cm^{-3}, this implies $n_i \leqslant 0.2$ cm^{-3} (Table 2), also in rough agreement. The histogram of the velocity dispersions noted for the hot component appears in Figure 5. It is striking that in no case out of 20 was σ observed to be <8.3 km/sec, even though the instrumental width was only 2.1 km/sec. We interpret this as due to a hot medium with superimposed turbulence. Then the temperature of the medium is given by the minimum velocity dispersion:

$$T_i = \frac{m_H \sigma_{min}^2}{k} = 8200°K \qquad (12)$$

We see in Figure 6 that the Mach number of the turbulence, defined by varies

$$M = \frac{\sigma_{tu}}{\sigma_{th}} = \left(\frac{\sigma^2 - \sigma_{min}^2}{\sigma_{min}} \right)^{1/2} \qquad (13)$$

varies from 0 to 1.01 (except for one case) with a mean value of 0.67. Thus, the intercloud medium would be mildly supersonically turbulent in this interpretation, a not unexpected result in view of the dynamical processes going on there, and the rapid dissipation which sets in for $M < 1$. This result agrees qualitatively with a recent 21-cm study at Berkeley by Baker (1971), who found evidence for comparable turbulent and thermal broadening in the broad component seen at intermediate latitudes.

The temperature deduced from equation (12), $\sim 8200°K$, agrees well with prediction (Table 2). Note that the product of this temperature and the intercloud density derived from the same observations, yields a pressure of $\sim$ 2200, also in fair agreement with theory. The predicted cloud parameters do not agree with observation, however. To achieve agreement with the cloud temperatures observed one must postulate further that cooling elements are depleted (Field, et al., 1969) by a factor of about 6. This ad hoc assumption has some basis in the observation that with the larger electron density predicted by cosmic- or X-ray heating, the abundance of Na^0 and Ca^+ should be larger than is observed (equation 5); perhaps they are depleted also. Note that while depletion can increase T from $\sim 20°$ (predicted for cosmic-ray heating in the absence of depletion) to $\sim 75°$, it cannot do the same thing in Spitzer's original picture

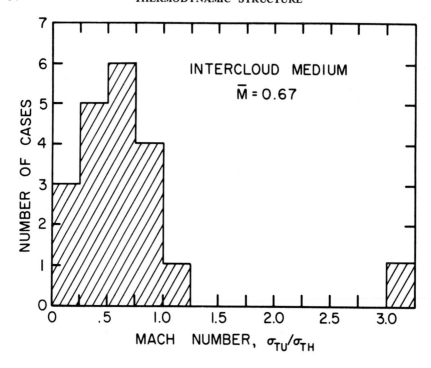

6. Mach numbers computed from the velocity dispersions of Figure 5, assuming T = 8200°K. Note the fairly sharp cut off at M = 1, suggesting rapid dissipation of supersonic motions.

based on H1 and C1. The reason is that in that case the efficiency of H1 would also be reduced by the depletion factor, leaving T the same. Further consequences of depletion will be discussed below.

We have now discussed all heating mechanisms but H6, which was developed by Bottcher, McCray, Jura and Dalgarno at Harvard (1971). Here the idea is that the UV output from a supernova explosion may be so enormous that a vast region will be suddenly and completely ionized (as in an HII region) and that during its long drawn-out recombination phase will mimic the patchy, partially ionized HI distribution we actually observe. The process is aided by the growth of thermal instabilities to form clouds. In order to fit the data, every parcel of interstellar gas must be exposed to UV every 10^{13} sec = 3 x 10^5y ($\zeta = 10^{-13}$ sec^{-1}). Ordinary hot stars fail to do this by a factor of >100 (Gould, Gold, and Salpeter, 1963) so that a dramatic new source is needed. If supernovae occur every 30 years (10^9 sec), each must ionize about 10^{-4} of the galactic disk (which contains 5 x 10^{66} cm^3, or 5 x 10^{66} H atoms). Thus, some 5 x 10^{62} ionizing photons are needed from each supernova. At 30 eV = 5 x 10^{-11} erg per photon, this amounts to 3 x 10^{52} ergs. While this is far larger (>10^3 x) than the visible light, our knowledge of supernovae is not adequate to decide whether it is possible. The idea has gained impetus from the interpretation of the Gum Nebula as a fossil HII region of enormous extent (Brandt, Stecher, Crawford, and Maran, 1971) with some 2 x 10^{62} free electrons. These authors point to the presence of a supernova remnant near the center of the nebula as evidence for the supposition that the nebula was ionized by a pulse of ultraviolet from a supernova.

This is not the place to make a detailed comparison of various advantages of H4, H5, and H6. However, in Table 3 we set forth some considerations which may ultimately decide the situation.

Table 3

ADVANTAGES AND DISADVANTAGES OF HEATING MECHANISMS

Consideration	Remarks		
	H4	H5	H6
Energy required	$10^{50.5}$ erg per supernovae; reasonable.	10^{41} erg/sec from discrete X-ray sources; possible.	$10^{52.5}$ erg per supernova; questionable.
Transfer of energy to interstellar medium	Instability limits speed to Alfven speed; much too slow.	No problem.	Localized to source; should look for large spatial variations.
Observational tests	Upper limit on flux >5 MeV from abundances of spallation-produced light elements.	Direct observation of relevant sources.	Young pulsars should have large dispersions.

III. EVOLUTION OF CLOUDS IN THE TWO-PHASE MODEL

In this section we describe some of the consequences of the two-phase model which have been studied at Berkeley, bearing in mind that such a model is by no means observationally proven. We have seen that in this model it is necessary to postulate depletion of cooling elements. This happens in any event as a result of collisions of the relevant atoms and ions with interstellar dust grains. Calculations indicate that C^+ ions will stick to either graphite or silicate surfaces, with a mean free time of $8 \times 10^8/n$ years, if the dust density is proportional to the gas density n. However, there is a question as to what happens when the C^+ reacts on the surface of the grain with trapped H atoms to form CH^+. The energy released in the reaction is certainly sufficient to eject the molecule into the gas phase. The probability s that this will not happen is the effective sticking probability, since the atoms almost always stick to the grain. If s = 1, a mantle builds up rapidly on the grain, as first described by van de Hulst. If s is small, molecules are ejected into the gas phase (see below). Actually, the situation differs somewhat for each atomic species, depending on its mass and charge. We define ξ as the depletion factor = number of atoms in gas phase ÷ total number of atoms (gas plus grains).

Field and Meszaros (1971) have made calculations of the evolution of clouds on the assumption s = 1 for all species but H and He. This assumption does not permit molecules to form in the gas phase (see below). It gives the maximum

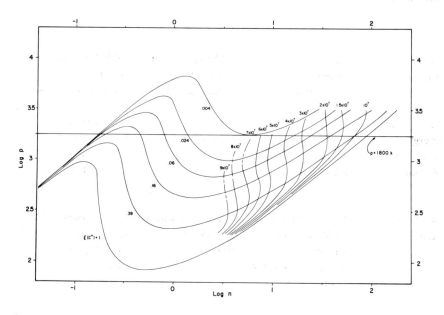

7. Evolution of a cloud at constant pressure by accretion of cooling agents on grains. If p = 1800 k, constant-pressure evolution is possible to log n = 0.7, ξ (C^+) = 0.004, but not beyond, as ξ must continue to drop (Field and Meszaros 1971). Time scales apply only to an undisturbed cloud.

rate of cloud evolution; the rate for other assumptions can be obtained simply by dividing the time scale by s. These authors adapted $\zeta = 4 \times 10^{-16}$ sec^{-1} — significant changes would occur for the larger ζ considered previously.

Figure 7 shows the results. For each value of n and ξ, we solved the equations of thermal equilibrium and ionization equilibrium to yield T and n_e. This permits one to plot p(n) for fixed ξ. Then we assumed that a cloud starting with $\xi = 1$ would evolve slowly enough so that constant pressure would be maintained by the intercloud medium. (This assumption is discussed further below.) This permits one to solve the time-dependent differential equation for depletion, deriving t(ξ) for each fixed pressure. Thus, the isochrones in Figure 7 are labelled with the time t to reach the associated density n and depletion ξ at the constant pressure which specifies the point on the isochrone. Thus, a cloud evolving at p = 1800 k expands from an initial n = 90, T = 20 to n = 5.5, T = 330 in 7×10^7 years, ξ (C$^+$) dropping to 0.004.

At 7×10^7 years, the cloud developes a critical condition. The expansion, induced by the ever-increasing temperatures accompanying the depletion of coolant species, cannot proceed at constant pressure if ξ continues to decrease (as it must). The dynamical behavior has now been studied using a 1-D hydrodynamic code (Meszaros, 1971).

As assumed above, the slow evolution before the critical condition permits p to remain constant (Figure 8). What happens then is rather surprising. Because

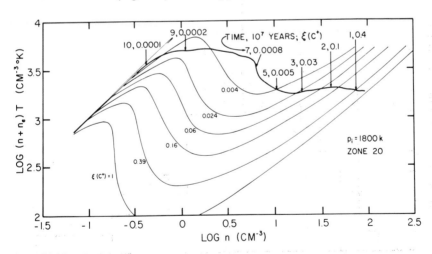

8. Hydrodynamic calculation of cloud evolution. Cloud is subjected to an intercloud pressure p_i = 1800k. As a typical interior zone approaches the critical condition ξ (C$^+$) = 0.004, dynamical evolution commences, causing the internal pressure to rise above p_i and the cloud to expand, finally approaching equilibrium at intercloud densities (Meszaros 1971).

the cloud is near the unstable branch of the p - n relation, it heats up rapidly, time scale is smaller than the dynamic one, the pressure rises above the external value and a rapid expansion ensues. From the evolutionary track of a typical mass element in the p - n diagram shown in Figure 8, we see that in effect the cloud evaporates back into the intercloud medium. This effect may be significant astronomically as a mechanism for the destruction of clouds initially formed by thermal instability from the intercloud medium.

All of this must be taken with a grain of salt, however. After all, in the conventional theory at least, clouds collide on a time scale of $\sim 10^7$ years, causing the shock waves we spoke of earlier. These shocks, because they heat the gas and cause the grains to move supersonically through the gas, can momentarily reverse the development described above by sputtering volatile materials off the grains (Aannestad, 1971). On the other hand, they can also cause an enhancement of the accretion rate by compressing the cloud gas to a high density in the cool region following the shock. The net effect of these events is still under investigation, but it presently looks as if the sputtering is probably a minor effect, and that the depletion can be fairly complete after one or two cloud collisions.

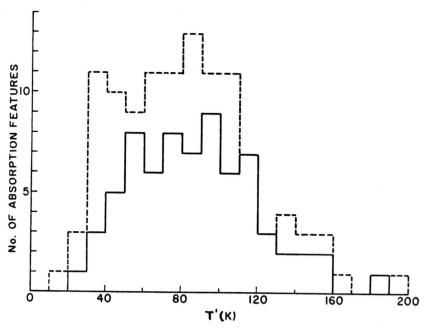

9. Temperatures of 119 clouds according to Hughes, Thompson, and Colvin (1971). This can be interpreted as the result of clouds found in various stages of depletion; see the range of temperatures in Figure 10.

It is interesting to see whether this model can explain the known facts about clouds, at least qualitatively. One fact which has been emerging in the last few years is the wide variety of cloud temperatures. Figure 9 is a histogram of the temperatures of 119 clouds observed at 21-cm by Hughes, Thompson, and Colvin (1971). We see that there is a factor of 10 spread, from 20 to 200°K, with a mean value of 89°K. (When corrected for the intercloud contribution, this mean falls to the 72°K quoted in Table 1.) This is qualitatively in agreement with the predictions of the constant-pressure evolution of Figure 7, if we assume that clouds are being observed at various random points along the evolutionary tracks. According to our picture, there should be some high-temperature clouds ($> 300°K$) in the process of evaporating back into the intercloud medium. The fact that such are not observed may be a consequence of the fact that the 21-cm absorption coefficient is inversely proportional to T, so that such clouds, if they existed, would give absorption lines too weak to detect. Indeed, the experiment of Hughes, *et al.* appears to be unable to detect absorption by gas hotter than 600°K, and is only marginally able to detect gas in the 200-600° range.

IV. MOLECULE FORMATION IN NORMAL CLOUDS

Aannestad (1971) is completing a study of molecule formation in normal clouds (typically, densities < 100 cm^{-3}), using a number of mechanisms: gas-phase reactions, catalysis on grain surfaces, and sputtering of grain mantles. The latter occurs only under conditions when a high atom-grain velocity is present, and we shall not discuss it here (although it may be important in molecular sources where a large spread of velocities is evident). Here we consider only gas-phase and catalytic reactions in quiescent clouds.

The former have been studied by Solomon and Klemperer (1971). One of the most rapid gas-phase reactions is $C^+ + H \rightarrow CH^+ + \gamma$. The CH^+ can then form CH via $CH^+ + e \rightarrow CH + \gamma$, and CO via $CH^+ + O \rightarrow CO + H^+$; CH can form CN via CH + N $\rightarrow$ CN + H. Other reactions, including photodissociation, are of course included.

This scheme permits one to form the molecules CH^+, CH, CO and CN, but not OH (which cannot radiate to the ground state in an atomic collision). However, as we have mentioned above, gas-phase OH can be formed by the reaction of O atoms and H atoms trapped on the grain surface, and subsequent escape from the surface (a process we previously assigned the probability 1 - s). Clearly to get any OH, we must take s <1; Aannestad has arbitrarily taken s = 0.5. This means that half of the oxygen atoms hitting the grain leave as OH, building up its density in the gas phase, while the other half remain on the grain, presumably to react further to form an H$_2$O mantle.

If one selects a cloud of a given mass, one can follow the evolution of its molecular abundances as well as the other parameters discussed previously, as the depletion process proceeds. It is convenient to discuss the results as a function of the depletion parameter ξ (C^+), as the time dependence of this parameter may be greatly changed by the intervention of cloud collisions.

The results for a cloud of 500 $M_\odot$ and p = 1800 k are shown in Figure 10. We note that the cloud heats and expands at constant pressure as in the previous discussion. The time scale is about twice the previous one because in order to generate OH we have taken s = 0.5 rather than unity. The visual extinction A_v through the cloud rapidly increases from 0.1 mag to 0.5 mag as the grains grow mantles and their cross-sections increase. However, for $\xi < 0.3$, A_v decreases

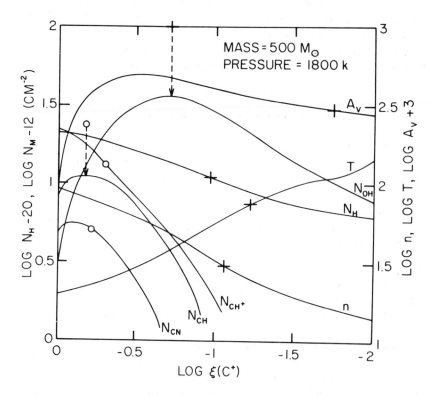

10. Evolution of a cloud of 500 $M_\odot$ at p = 1800 k. The sticking probability is s = 0.5. The column densities predicted for CH^+, CH, and CN are comparable to these observed (open circles) if ξ (C^+) is near unity. On the other hand, clouds where OH is seen can be explained if they have smaller ξ (C^+); the crosses indicate the observations of these clouds. Note that clouds of the OH type would not be expected to have strong CH^+, CH, or CN. (Aannestad 1971).

slowly because of the expansion of the cloud and the consequent decrease of N_H (which is proportional to the inverse square of the cloud radius).

The column densities of CH^+, CH, and CN are highest at the start, where n is large, because their formation rate is proportional to n^2. OH, on the other hand, because it forms at a rate proportional to nv times the cross-sectional area of the grains per cm^3, would be expected to have a column density proportional to $nT^{1/2}A_v$. This reaches a peak later because the increase in $T^{1/2}A_v$ more than compensates the decrease in n.

Aannestad has compared these predictions for a 500 solar mass cloud with recent observational data given by Frisch (1971) and Davies (1971). Frisch has reanalyzed the Adams' optical data on CH^+, CH, and CN, using the most recent f-values. One sees that both the CH^+ and CN observations can be satisfied with ξ >0.3, although the theory fails by a factor of 2 to account for the CH observations.

Davies (1971) has discussed the properties of clouds in which OH is found in absorption. Again, we find fair agreement between the mean of the observations with a theoretical 500 solar mass cloud having ξ<0.1, except that OH fails by a factor of somewhat more than 3. Considering the uncertainties in the rate coefficients and the spread in the observational data, the agreement is acceptable.

The following point emerges from a study of Figure 10. Roughly speaking, we see that CH clouds are "young" – they have relatively large ξ, while OH clouds are "old," having small ξ. The theoretical reason for this is that OH (as a product of grain catalysis) persists over a much larger period of cloud evolution, while CH^+, CH, and CN (as products of gas-phase reaction) can be found only in "young" clouds where depletion has not yet occurred, so that the gas is cool, and the intercloud pressure is able to compress the cloud to favorably high densities. This may explain why OH has not been seen in radio observations of regions where CH^+, CH, and CN have been seen optically.

Obviously such studies are still in their infancy, but they give a hint of the correlations between cloud properties which may soon become much more meaningful.

Dark Clouds

As it is well known, molecules are prominent in dark clouds for which A_v>1 (Heiles, 1971). This is natural, because extinction reduces the penetration of the cloud by UV, and thereby reduces the photodissociation rate of molecules. It is believed that these clouds represent a qualitatively different phenomenon from ordinary clouds with A_v <1. The reason is this: Heiles (1971) states a typical dust cloud has a total column density $N_H + 2N_{H_2} \rangle 8 \times 10^{21}$ cm^{-2}, mostly in the form of H_2, a density (including helium) > 2200 cm^{-3}, a gas temperature of $\sim 5°$, and a radius ~ 0.7 pc. The pressure is therefore $> 10^4$ k, at least 5 times the intercloud pressure. Confinement by the latter alone is therefore not possible. One therefore considers confinement by the joint effect of external pressure and self-gravitation, which is negligible for ordinary clouds. This problem has been solved subject to the approximation of isothermality by Ebert (1955) and Bonner (1956) (see Spitzer 1968 for a review). For a given intercloud pressure p_i and cloud temperature T_c there exists a critical mass

$$M_c = 1.2 \frac{(RT_c/\mu)^2}{(G^3 p_i)^{1/2}},$$

which for $T_c = 5°$ and $p_i = 1800$ k is 2.2 $M_\odot$ (we have assumed $\mu = 2.33$, appropriate for a mixture of H_2 and He). For masses much less than the critical one, confinement is largely by the intercloud medium, with $\rho_c = \rho_* \equiv \mu \, p_i/RT_c$. As the mass rises, gravitation is increasingly important, the central pressure

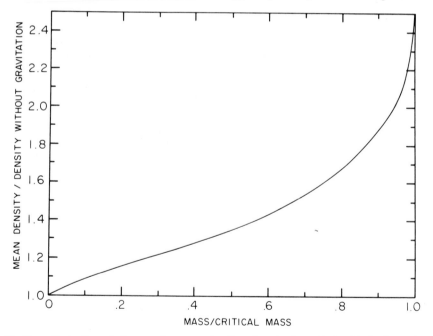

11. The effect of gravitation on the mean density of an isothermal cloud in hydrostatic equilibrium. When $M \ll M_c$, the density is just $\rho_* = \mu P_i/RT_c$. As $M \rightarrow M_c, \rho \rightarrow 2.48\rho_*$. Larger clouds are unstable.

exceeding the external one by the gravitational pressure $\int \rho g dr$. At the critical mass, the mean density ρ (Figure 11) has risen to the critical value 2.48 ρ_*; further increase in mass reduces the external pressure required for equilibrium. The actual pressure, which is greater, would cause the cloud to collapse. Hence only clouds with mass less than the critical one can exist in stable equilibrium and the mean density is therefore $\rho \leqslant 2.48 \rho_*$. In the present case this means $\langle n_H + n_{H_2} \rangle < 750$ cm^{-3}, too small to account for the observations. At $M = M_c$, R = 0.15 pc (too small) and the mean column density is $\langle N_H + 2N_{H_2} \rangle = 2.7$ x 10^{21} cm^{-2} (also too small). What are the chances that a cloud will become a dark cloud on this picture? We may consider a dark cloud to be one for which $\rho > 2\rho_*$ (or ρ center $> 6 \rho_*$). From Figure 11, this requires that 0.95 $M_c < M <$ M_c — a circumstance which seems improbable.

(We note that $M_c \propto (T_c/\mu)^2$. In normal clouds, $T_c = 75°$ and $\mu = 1.27$, compared to $T_c = 5°$ and $\mu = 2.33$ in dark clouds. Hence M_c for normal clouds is much larger, about 10^3 $M_\odot$. Gravity is important for dust clouds because T_c is small and μ is large (H_2).)

Observed dust clouds do not appear to have the properties required by the model based on binding by gravitational forces and external pressure. For $M = M_c$, the quantities in the virial theorem

$$-\Omega = 2T - 3p_i V = 3RTM/\mu - 4\pi R^2 p_i \qquad (14)$$

are in the proportion 1 : 1.67 : -0.67. For the cloud parameters cited by Heiles (1967), these values are in the proportion 1 : 0.11 : -0.03, and the internal pressure is inadequate to resist the compression of gravity (external pressure is negligible, but goes in the wrong direction to resolve this problem). We conclude that a balance between gravity and pressure cannot explain these configurations. This conclusion is strengthened by the improbability of finding clouds in the required narrow mass interval.

This leads us to investigate other mechanisms which can oppose gravity. Heiles (1971) points out that the OH line widths (usually $\sigma \sim 0.6$ km/sec) are much larger than would be expected from thermal motions alone (0.05 km/sec at 5°K). He therefore suggests that turbulence is present. In cloud 2, the line is split with $\Delta V = 0.8$ km/sec, which is consistent with large-scale turbulence. If there is small-scale turbulence, it will act as an additional stress in the virial theorem. Its approximate effect is to act like an effective temperature $T_{eff} = \mu \sigma^2/R = 100°K$–20x the actual value. This happens to be roughly what is required to stabilize the cloud. Unfortunately, this turbulence would have a Mach number $M = (T_{eff}/T)^{1/2} = 4.5$. Such turbulence would ordinarily form shocks within a time $\tau \sim L/\sigma$, where L is the turbulent scale. Since small-scale turbulence should have $L < R \sim 0.7$ pc, $\tau < 0.7$ pc/0.6 km/sec $\sim 10^6$ years. Since these shocks will be highly dissipative (see above), the turbulence damps in a time short compared to the lifetime of the cloud assumed to be ≫ free-fall time = 10^6 years. Thus a model constructed along these lines must solve the problem of how to regenerate this turbulence in the rather short time available.

A similar problem exists for ordinary clouds. Spitzer (1968) gives $\sigma(Ca+) = 3$ km/sec, while Hobbs (1969) gives $\sigma(Na^\circ) = 1$ km/sec, both far larger than the thermal width at, say, $75^\circ K$, implying turbulence Mach numbers in the ranges 1.4 to 4. In the case of the 21-cm absorption features measured by Radhakrishnan *et al.* (1971), both the velocity width and the temperature are measured directly, permitting one to plot a histogram of Mach numbers (Figure 12), where again we see that M = 1 - 4 is most frequent.

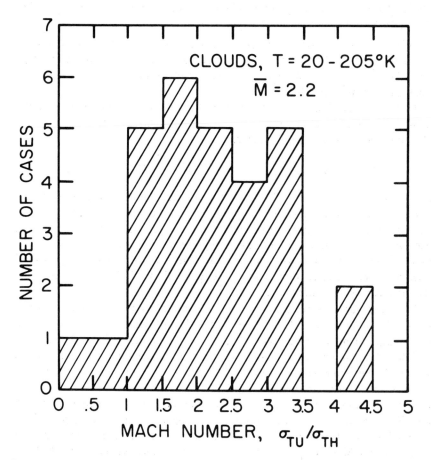

12. Turbulent Mach numbers inside clouds, from data by Radhakrishnan *et al.* (1971). Contrast this histogram with Figure 6, which probably pertains to the intercloud medium.

This histogram should be compared with that for the intercloud medium (Figure 6), where the interpretation in terms of weak supersonic turbulence is plausible. Inside clouds, however, the turbulence appears to be strongly supersonic, contributing an effective pressure M^2 (up to 20) times the thermal pressure. It has been suggested that in ordinary clouds, the large widths are in fact due to systematic expansion of the cloud following cloud collisions (Stone, 1970). In a calculation of this effect, Stone found expansion velocities $\simeq$ sound speed $\simeq$ 1

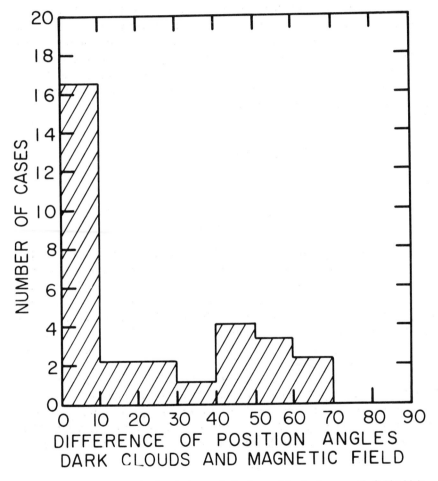

13. Position angles of dark clouds relative to that of a model galactic magnetic field which explains stellar polarization data (Verschuur 1970a). The large number of cases less then 10° suggests that the clouds may be stabilized by magnetic forces, but there are difficulties (see text).

km/sec, which probably is not enough to account for Figure 12. The origin and maintenance of such high velocities remains an outstanding problem.

Another possibility is that the clouds are stabilized by magnetic forces. This is strongly suggested by the observation (Shajn, 1955) that dust clouds tend to be aligned with the direction of the interstellar magnetic field deduced from stellar polarization measurements. Verschuur (1970a) has studied this problem using more modern data for 31 clouds. The histograms of position angles seen in Figure 13 seems to confirm this idea.

It is sometimes believed that a magnetic filament can form with the radial acceleration of gravity being countered by the j x B force associated with a field along the filament as suggested by the Shajn effect. Indeed, for a infinitely long filament, this is possible because then there are no gravitational forces along the direction of B which must be compensated. If the undisturbed field is B_o and the density before gravitational forces act is ρ_o, if $B^2/8\pi \gg p$, as is appropriate here, the density profile is approximately

$$\rho = \rho_{center} \, J_o(2.4 \, \frac{r}{R}) \tag{15}$$

where

$$R = 0.19 \, \frac{B_o}{\rho_o \sqrt{G}} \tag{16}$$

For $B_o = 3\mu G$ and $\rho_o = 1.8 \times 10^{-22}$ gcm^{-3} (corresponding to $p_i = 1800$ k and $\xi = 1$), R = 1.3pc, in fair agreement with observations. The above equation is based on flux-freezing, which implies that B has increased over its original value by a factor of $2200/90 = 24$, to a value of 72 μG. An upper limit from the Zeeman splitting of the OH line (Verschuur, 1970b) gives $B < 170 \, \mu G$, while a much smaller limit $B < 5 \, \mu G$ from the 21-cm line in cloud 2 (Turner and Verschuur 1970) may not be pertinent because the H atoms are believed to be in the outer part of the cloud (see below).

Unfortunately the magnetic filament model suffers a grave defect if one considers a finite length along the magnetic field, because it turns out that then the gravitational forces along the filament are even larger than those across it, and now only thermal pressure is available to counter them. The result is that the configuration shrinks along the field (Field, 1970), the maximum length attainable being of the order of

$$L = \frac{RT_c}{\mu(2\pi Gp_i)^{1/2}} \tag{17}$$

which for $T_c = 5°$, $\mu = 2.33$, $p_i = 1800k$ is 0.7pc — even less than the diameter of the cloud! Hence the cloud will not be a filament at all, but a pancake, with B parallel to the minor axis, in contradiction to the Shajn effect. The dynamics of dark clouds for the moment are obscure.

Chemical and Thermal Equilibrium of
Dark Clouds

One of the most conspicuous observations about dust clouds is their lack of 21-cm emission or absorption (Heiles, 1969a) in spite of large column densities of grains. If one assumes that the dust-to-gas ratio is constant, one obtains the column densities of gas quoted earlier. The upper limits on atomic hydrogen alone are one to three orders of magnitude lower (Solomon and Werner, 1971). Most workers believe that this is best interpreted by postulating that the gas is present in the expected amount, but is in the form of H_2 molecules rather than H atoms. That this is true in at least some clouds is known experimentally from the work of Carruthers (1970), who found $H_2(N = 1.3 \times 10^{20}$ cm$^{-2})$ in the direction of ξ Per ($A_v = 1$) but not in ϵ Per ($A_v = 0.15$) by rocket ultraviolet spectroscopy. Carruthers' observations are consistent with the notion that interiors of clouds with $A_v > 1$ (dark clouds) will be shielded from the 960-1010Å UV radiation which destroys the H_2 catalyzed on grain surfaces, with the result that H_2 builds up in such clouds.

This problem has been studied theoretically by Hollenbach and Salpeter (1971) and by Hollenbach, Werner, and Salpeter (1971), who find that the formation of H_2 on grains is 10^{-17} n n_H cm^{-3} sec^{-1}, where $n = n_H + 2 n_2$ (n_2 is the density of H_2). The destruction rate by UV is 10^{-10} n_2 a cm^{-3} sec^{-1}, where a ($\leqslant 1$) is the attenuation factor due to absorption in the cloud. Thus,

$$f = \frac{n_2}{n_H + n_2} = \frac{1}{1 + 10^7 \, a/n} \tag{18}$$

which is small ($\approx 10^{-7}$ n) for $a = 1$, but which is large for $a < 10^{-7}$ n. Thus, H_2 may occur in dust clouds, where $a \ll 1$ and n is large. It turns out that the major attenuation effect is by the H_2 molecules themselves in the outer parts of the clouds (rather than by the grains). Figure 14 shows the results for clouds of three masses (200, 500, 2000 $M_\odot$) as a function of $\tau_v \approx 1/2$ A_v, the optical depth to the center in the visible (a measure of column density). This figure shows how f at the center of the cloud varies as the cloud contracts to higher density and τ_v increases. An ordinary cloud (500 $M_\odot$, n = 25 cm^{-3}, $\tau_v = 0.2$) is only 1% H_2, while a dark cloud of the same mass (n = 100 cm^{-3}, $\tau_v = 0.5$ or $A_v = 1$) is 50% H_2. Dark clouds with n $\approx 10^3$ cm^{-3}, $A_v = 10$ are almost completely H_2 according to this theory.

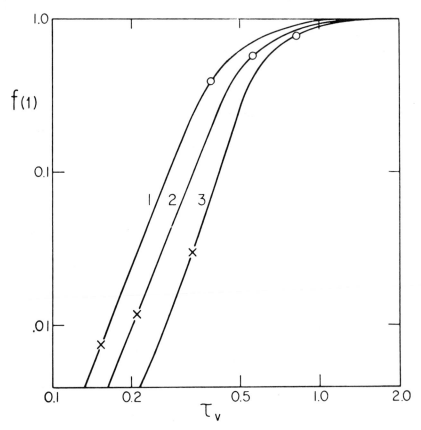

14. Fractional abundance of H_2 at the center of a cloud versus τ_v ($\simeq 1/2\ A_v$, the visible extinction through the cloud). Clouds of mass 200, 500, and 2000 $M_\odot$ are labelled 1, 2, 3, respectively. Crosses denote a density of 25 cm^{-3}; open circles, 100 cm^{-3}. Dark clouds are almost completely H_2 (Hollenbach, *et al.* 1971).

Within this framework, one may consider the formation of molecules like OH. The question is, if OH can form by a similar process, why is OH/O only $\approx$ 10^{-3} as observed (Heiles, 1971)? We find that the formation rate is $\approx 3 \times 10^{-18}$ (1 - s) n_o at 5°K, where s is the probability that O sticks permanently to the grain. The lifetime of OH against photodissociation is unknown, but it is probably at least 100 years. Inside a cloud with $A_v = 10$, $\tau_v = 5$ and τ_{uv} is doubtless >10. Hence the destruction rate is less than $e^{-10} n_{OH}/3 \times 10^9$ sec, so

$$\frac{n_{OH}}{n_O} \geqslant 10^{-4}\ n(1 - s) \qquad (19)$$

and for n = 10^3 $n_{OH}/n_O \geqslant 10^{-1}$ (1 - s). To agree with observations, $n_{OH}/n_O \approx$ 10^{-3}, either there must be some other method of destruction, or 1 - s $\leqslant 10^{-2}$ — only 1% of the O atoms come off the grains as OH. This is a weak argument that the depletion process we have discussed above occurs in dark clouds. If so, it will cover the grains with a mantle, increase their cross-section a factor ~ 10, and decrease the column densities of gas inferred from A_v, with important implifications for the dynamics of the clouds.

The gas temperatures of dust clouds are very low — about $5°K$. This is not surprising, since even if we ignore the additional cooling effects to be discussed below, normal clouds with high densities would be expected to have temperatures $\approx 10°K$ as a result of the increased cooling at higher densities (Figure 2). Moreover, there are several processes which occur in dark clouds and not in ordinary clouds, most of which tend to lower the temperature even further:

(i) The high column densities of gas provide a shield against both low-energy cosmic rays and X-rays, decreasing the heating by H4 and H5 (but not H2).

(ii) The high column density of carbon in the gas phase shields carbon atoms against ionizing radiation, with the result that $C^+ \rightarrow C°$. The latter has an effective cooling level at $23°K$ (compared to $92°K$ for C^+), so that the effectiveness of cooling is increased. At the same time, heating mechanism H 1 is decreased. (This is somewhat offset by the fact that the resulting high column density of $C°$ re-absorbs the emerging IR photons, and the high density means some of the cooling efficiency is lost because the collisions become saturated.)

(iii) A variety of polar molecules such as CO can be formed, which have rotational transitions in the millimeter region. These will be helpful in cooling (C5).

(iv) Because A_v is large, the grains are shielded from the galactic radiation field. Their temperature drops, and atoms or molecules which return to the gas after sticking to their surfaces (e.g., H_2) will be cool.

(v) The only mechanism which tends to raise the temperature is H7 — the depletion of cooling elements by sticking to grains. This will proceed on a time scale of $\approx (10^{10}/n)$ years at $5°K$ for $C°$ atoms, or 5 x 10^6 years for n = 2000. This is short enough so that one may question whether there are significant numbers of impurities — particularly $C°$ — in the gas phase. This could raise the temperature significantly.

These problems are only now being investigated. Solomon and Werner (1971) have studied the effect of low-energy cosmic rays, and have concluded that at high densities every interaction of a cosmic ray with a molecule (total rate 1.1ζ) finally gives 2H (directly, or via H_2^+ or H + H^+). Equating this to the destruction of H via grains gives

$$n_H = \frac{1.1\zeta}{2 \times 10^{-17}} \tag{20}$$

or for ζ 1.2 x 10^{-15} sec^{-1} (see above),

$$n_H = 66 \text{ cm}^{-3} \qquad (21)$$

larger than the 21-cm upper limits in the five cases (≈ 10 cm^{-3}). They conclude that if low-energy cosmic rays exist, they do not penetrate the dust clouds. Indeed, this is what one expects, because a 2 MeV proton stops in a total column density of 2 x 10^{21} cm^{-2} along its trajectory. Allowing for the fact that only 1/3 of the distance travelled in a magnetic field is along a given direction (i.e., into a cloud), this corresponds to a column density of only 7 x 10^{20} cm^{-2} — only 10% of that of a typical cloud. Thus 2 MeV cosmic rays penetrate only 10% of the cloud diameter, or 4 x 10^{17} cm. If n_H = 66 cm^{-3} in this region, N_H = 3 x 10^{19} cm^{-2}, consistent with the upper limits in four cases, and only twice it in another.

Werner (1970) has studied the question of the $C^+ \rightarrow C^\circ$ equilibrium in dark clouds. He finds that for normal clouds (n $\leqslant$ 100 cm^{-3}), C^+ dominates as assumed earlier in this report, but that for n = 10^3 cm^{-3} dominates for all but the outer 0.8 pc of the cloud, while for n = 10^4 cm^{-3}, in only the outer 0.02 pc is carbon ionized. Thus, a cloud with R = 0.7 pc and n = 2000 cm^{-3} is in an intermediate case, and one cannot always assume carbon is neutral throughout a dust cloud.

The effect of shielding on the grain temperature has been studied by Werner and Salpeter (1969). They included heating by the cosmic blackbody radiation, starlight, and absorption of infrared emitted by the hotter grains in the outer layers of the cloud. They conclude that the temperature of a graphite core-ice mantle grain (which is 17°K in normal clouds) will fall to 10°K in a cloud with A_v = 10. Impurity cooling is capable of lowering it still further, to 4.8°K. Thus, grains may be a coolant for the gas as well, although their temperature cannot fall to the extremely low value required to freeze out H$_2$ ($<3^\circ$K).

Heiles (1969b) has computed temperatures of dust clouds having various densities. He assumed a grain temperature of 3°K, and included their cooling effect via gas-kinetic collisions. He included the cooling due to C° (as well as other species) and assumed that UV is attenuated 8.7 mag. Self absorption in the cooling lines was ignored. His results for ζ = 7 x 10^{16} sec^{-1} were 8°K at 100 cm^{-3} and 6°K at 10^3 cm^{-3}, the cooling being due mostly to C°. More precise calculations, taking into account the exclusion of cosmic rays, the effect of polar molecules, more precise grain temperatures, and possible depletion, are under way. At present the only problem we see in this area is the effect of depletion. The low abundance of OH suggests that depletion is effective, and this will tend to counter the very strong cooling effects we have listed above.

REFERENCES

Aannestad, P. A. 1971, Private Communication.
Baker, P. 1971, Ph. D. Dissertation, University of California, Berkeley.
Bonnor, W. B. 1956, *M. N.*, **116**, 351.
Bottcher, C.; McCray, R. A.; Jura, M. and Dalgarno, A. 1970, *Ap. Letters*, **6**, 237.
Brandt, J. C.; Stecher, T. P.; Crawford, D. L. and Maran, S. P. 1971, *Ap. J.*, **163**, L99.
Bridle, A. H. 1969, *Nature*, **221**, 648.
Carruthers, G. R. 1970, *Ap. J.*, **161**, L81.
Davies, R. 1971, Liege Symposium 18.
Ebert, R. 1955, *Z. f. Ap.*, **37**, 217.
Ellis, G. R. and Hamilton, P. A. 1966, *Ap. J.*, **146**, 78.
Field, G. B. 1970, Liege Symposium 16.
————. 1971, *Dark Nebulae, Globules, and Protostats*, ed. B. T. Lynds (Tucson: University of Arizona Press), chapter 12.
Field, G. B.; Goldsmith, D. W. and Habing, H. J. 1969, *Ap. J.*, **155**, L149.
Field, G. B. and Meszaros, P. 1971, in preparation.
Field, G. B.; Rather, J. D. G.; Aannestad, P. A. and Orszag, S. A. 1968, *Ap. J.*, **151**, 953.
FitzGerald, M. P. 1968, *Ap. J.*, **73**, 983.
Frisch, P. 1971, in preparation.
Gould, R. J.; Gold, T. and Salpeter, E. E. 1963, *Ap. J.*, **138**, 408.
Heiles, C. 1969a, *Ap. J.*, **156**, 493.
————. 1969b, *Ap. J.*, **157**, 123.
————. 1971, *Ann. Rev. Astron. and Astroph.*, **9**, 293.
Hobbs, L. M. 1969, *Ap. J.*, **157**, 165.
Hollenbach, D. J. and Salpeter, E. E. 1971, *Ap. J.*, **163**, 155.
Hollenbach, D. J.; Werner, M. W. and Salpeter, E. E. 1971, *Ap. J.*, **163**, 165.
Hughes, M. P.; Thompson, A. R. and Colvin, R. S. 1971, *Ap. J. Suppl.*, in press.
Kerr, F. J. and Westerhout, G. 1965, *Stars and Stellar Systems*, **5**, ed. M. Schmidt (Chicago: University of Chicago Press), p. 167.
Meszaros, P. 1971, Private Communication.
Radhakrishnan, V.; Murray, J. D.; Lockhart, P. and Whittle, R. P. J. 1971, *Ap. J. Suppl.*, in press.
Shajn, G. A. 1955, *Astron. Zh.*, **32**, 381.
Solomon, P. M. and Klemperer, W. 1971, *Ap. J.*
Solomon, P. M. and Werner, M. W. 1971, *Ap. J.*, **165**, 41.
Spitzer, L., Jr. 1968, *Stars and Stellar Systems*, **7**, ed. B. M. Middlehurst and L. H. Aller (Chicago: University of Chicago Press), chapter 1.
Stone, M. E. 1970, *Ap. J.*, **159**, 277 and 293.
Turner, B. E. and Verschuur, A. L. 1970, *Ap. J.*, **162**, 341.
Verschuur, G. L. 1970a, *Interstellar Gas Dynamics: I. A. U. Symposium No. 39* (Reidel Publishing Co., The Netherlands), chapter 9.
Werner, M. W. 1970, *Ap. Letters*, **6**, 81.
Werner, M. W. and Salpeter, E. E. 1969, *M. N.*, **145**, 249.

Jacobs: Please comment upon the effect of shock waves, resulting from spiral density waves cycling through the galaxy every 10^8 to 10^9 years, upon the heating and cooling of the interstellar medium.

Field: The initial effect of the shock wave is to cause condensation of clouds. But as the wave passes, the pressure is reduced below the critical pressure required to support such clouds, and the clouds tend to evaporate (see Figure 3). Clouds may also be destroyed by the depletion-induced evolution which I've mentioned. This mechanism predicts cloud lifetimes of 10^8 years (see Figure 8), comparable to the period of the density wave. Collisions between clouds may accelerate their depletion by as much as an order of magnitude.

A Study of Dark Nebulae

Bart J. Bok and Carolyn S. Cordwell

Steward Observatory
University of Arizona
Tucson, Arizona

I. PREAMBLE

For many years dark nebulae were not considered to be clouds of cosmic grains, but they were rather thought of as "holes in the sky." This idea was discarded in the late 19th century, when Barnard began his photography of the Milky Way. Photographs made with high resolution telescopes showed bright nebulous structure in many of the obscured areas.

Dark nebulae are very likely closely linked with star formation, and it is of increasing importance to discover the physical processes taking place in these areas to further our understanding of pre-protostar conditions. Radio observations of molecular lines are revealing some of the chemical constituents of dark clouds, as well as indicating temperatures. Optical studies furnish data on absorptions and distances of nebulae, which in turn yield minimum values of densities and masses, provided we make certain fundamental assumptions concerning grain characteristics. Radio and optical data, combined with theoretical interpretation, are leading to reasonably definite conclusions regarding the physical conditions and composition of unit dark nebulae in the pre-protostar stages.

We present first a brief summary of different types of dark nebulae together with representative properties for each variety we discuss. Next, we discuss the methods for finding approximate absorptions and distances of dark clouds through the use of star count data and from results obtained with modern color techniques. A summary of available catalogs and photographic atlases of dark nebulae is given. In the concluding sections of the paper we present three tables with lists of positions and properties of some known dark nebulae.

II. VARIETIES OF DARK NEBULAE

Clumpiness is a basic characteristic of the interstellar medium. Interstellar dust is especially found in concentrations such as clouds and lanes of dark nebulae. Although many area of the sky show filaments of absorption, there are also many unit clouds with well-defined boundaries and dimensions. These single structures can be roughly categorized into three types, according to size.

The smallest unit clouds that we can observe directly are seen in relief against bright emission nebulae. These clouds are the small globules, first postulated as possible regions of star formation by Bok and Reilly (1947). They are frequently observed in clumps or chains; see for example, photographs of NGC 2244 (the Rosette Nebula) and IC 2944. These globules appear as opaque specks, with shapes varying from extremely round, compact spheres, as seen in NGC 2244, to more windswept filaments such as those seen in Messier 8. Bok, Cordwell and

Cromwell (1971) have shown that diameters for these objects range between 0.01 to 0.1 parsec, and that their masses are somewhere between 0.1 and 1.0 solar mass.

The second category of dark nebulae includes isolated large globules with angular sizes between about 2' and 20', which are seen as voids against rich star fields. Many of the nebulae cataloged by Barnard (1927) fall in this category, hence these globules are generally referred to as "Barnard objects."

Two large, semi-transparent globules, Barnard 361 and Barnard 34, were found by Bok *et al.* (1971) to have photographic absorptions of about 2.5 mag., masses around 60 solar masses, and radii in the neighborhood of 1 parsec. Many of the large globules appear opaque to the limit of the plate, and only minimum absorptions can be assigned to them. Examples of these are Barnard 227 and Barnard 335. Unlike the tiny globules, which appear in turbulent surroundings, the Barnard objects seem to be isolated objects in quiescent regions.

The third category of dark nebulae encompasses large dark complexes of dust, such as the ρ Ophiuchi Nebula, several clouds in Taurus, and the Southern Coalsack. The ρ Ophiuchi Nebula and three regions in Taurus were studied extensively by Bok (1956). Star counts were made from the original Palomar Sky Survey plates, and absorptions were mapped for each region. The ρ Ophiuchi Cloud shows a distinct radial density gradient, with a total absorption amounting to 8 magnitudes in the center, decreasing to about 3 magnitudes near the edge. This cloud has a radius of about 4 parsecs, and a mass of the order of 2100 solar masses. It is of interest that Heiles (1969) finds a temperature for this cloud near $10°K$ or less, as indicated by OH absorption lines.

The Taurus region displays several large dark lanes, with absorptions running as high as 7 magnitudes. Young T Tauri stars have been found in this area, and radio astronomers have searched these clouds for interstellar molecules.

The Southern Coalsack, as seen in the older Atlases, appears as a large unit cloud, but upon close inspection of higher resolution plates, it looks as though it is divided into many knots of absorption. It is possible that fragmentation is taking place in this cloud, leading possibly to the formation of a star cluster. Rodgers (1960) finds absorptions ranging from 0.7 to 2.4 magnitudes.

Each category of dark nebulae is studied in a somewhat different manner. The following sections summarize the current methods of analysis which lead to values of absorptions and distances for these objects.

III. STAR COUNTS

Most determinations of distances and absorptions of dark nebulae involve a statistical analysis of star counts. Star counting is still a laborious process, but through the application of modern techniques it is becoming both simpler and more accurate.

A fully automatic star counting instrument, called the GALAXY machine, has been developed in Edinburgh (Walker, 1971 and Pratt, 1971). For a given photographic plate, this machine can measure automatically the x, y coordinates and image sizes of 900 stars per hour, 24 hours per day. Some one million stars have already been measured in this fashion. With the aid of a GALAXY machine, the process of counting stars to successive limits of magnitude for well-defined areas on a photographic plate becomes fully automatic and large-scale counting of stars is possible. If required, data from B, V, and R photographs can be combined to yield statistical data on the distribution of colors and magnitudes.

The object of star counts is generally to find the numbers of stars per magnitude interval to as faint a limit as possible. Stellar images of different sizes are counted, and their sizes are then converted to magnitudes by means of a standard magnitude sequence. For the study of a dark nebula, star counts are made both in the obscured region of interest, and in one or two comparison fields. The choice of a good comparison field (or fields) requires much care, since a comparison field must be a field little affected by the dark nebula under investigation, yet as near to it as is practicable.

An iris photometer provides the most readily available accurate means for measuring image sizes. The aperture readings are converted into magnitudes by means of a standard photoelectric sequence. The stars can then be counted in successive magnitude intervals.

As an illustration, consider B and V stellar plates which have photoelectrically determined standard sequences on them. By utilizing the method of photographic photometry, we can obtain B and V magnitudes for any star on our photographic plates. Through measurement of B and V magnitudes for all stars in a designated area, we can obtain the numbers of stars per square degree to successive limits of apparent magnitude, B and V.

Effective methods of counting were developed by Lindsay and Bok (1936) and Miller (1936). A magnitude sequence of image sizes is imprinted at the top and bottom of a transparent reseau, which is then placed on the plate to be counted. A specific scale image is selected, and all stars which have sizes less than or equal to this calibration image are counted column by column in the manner shown in Figure 1. The process is repeated for successive scale images.

If no sequence of standard magnitudes is available, then one can get a rough total photographic absorption by counting stars to the limiting magnitude on a blue plate. The areas counted are reduced to an area of one square degree, and counts are compared directly with van Rhijn's (1929) tabulation of $\log N(m)$ versus m for old galactic latitude and longitude. Here, $N(m)$ is the number of stars per square degree of magnitude m or brighter. Conversion from equatorial (1950) coordinates to new galactic coordinates may be made by means of a graph which can be found in the book by Kraus (1966). To convert from new to

1							
2							
3							
4							
5	etc.						
6	11						
7	10						
8	9						

1. Illustration of a reseau which may be used for star counts. The star magnitudes are shown at the top and bottom of the reseau. The observer mentally fixes a particular image size, and counts down a column at a time in the order shown. He can refer to either image set when necessary. After the total area has been counted down to a particular image size, he goes to the next image, and counts all stars with sizes greater than or equal to that size considered. This gives a series of N(m) values.

old galactic coordinates, one may subtract 33° from ℓ^{II} and leave b^{II} approximately the same.

As an example of an application involving the van Rhijn tables, suppose the counts in the comparison field give a $\log N(m)$ which corresponds to m = 18 in van Rhijn's tables, and that the star counts in the dark nebula correspond to van Rhijn's counts for m = 15. We can then conclude that the nebula absorbs about 3 magnitudes. For this method to be applicable, we require that the counts must reach sufficiently faint limits so that most of the counted stars lie beyond the dark nebula.

There are no tables such as van Rhijn's for star counts in red colors. To get an approximate red absorption for the nebula one can look at the ratio of numbers of red to blue counts in the comparison fields, and use this ratio as a multiplying constant to go from red to equivalent blue counts in the dark nebula. Van Rhijn's tables can then be used as before.

This technique for finding blue and red absorption was used by Bok (1956) to map absorptions for regions in Taurus and Ophiuchus.

Van Rhijn's star count tables were made many years ago, under less than perfect conditions, and the accuracy of his counts has long been in question. Consequently, absorptions should be found in this manner only when no magnitude sequences are available. The application of the van Rhijn tables to visual or red counts is even more dubious, since we should really use basic tables (not available) prepared for red or visual $\log N(m)$'s. The only safe procedure is really to proceed via a sequence of standard magnitudes in B, V, and R established for the purpose.

IV. ABSORPTIONS AND DISTANCES OF DARK NEBULAE

A. Wolf Diagrams

A fairly easy way to find the absorption of a dark nebula is through the use of a Wolf Diagram (Bok, 1937). Star counts yield values of A(m), the number of stars in the magnitude interval m - ½ to m + ½. A graph of $\log A(m)$ versus m (Wolf Diagram) is made for both the comparison region and the obscured field. At a distance less than that of the nebula the curves are parallel. At the point where the nebula begins to affect the counts in the obscured region, say at m_1 the curve for the obscured region has the lesser slope and separates from the curve for the comparison field, until some magnitude, m_2, beyond which the curves again run parallel. The difference in magnitudes, $m_2 - m_1$, is the absorption of the nebula.

It is generally not possible to use the Wolf curves derived from general star counts for a derivation of the approximate distance to a dark nebula. To find this distance, one should use one of the methods described in the sections that follow.

As spectral data for faint stars becomes more plentiful, one should not hesitate to represent the basic material through Wolf curves applicable to narrow intervals in spectral class, B8 to A3 for example. The Wolf curves continue to be the most effective way to illustrate the results that are obtained from counts in obscured fields and comparison regions.

An example of a pair of Wolf curves is shown in Figure 2.

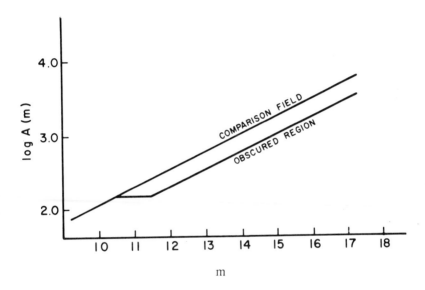

2. Example of a Wolf diagram. A total absorption of 1.0 mag. is shown here.

B. The (m, log π) Table

Distances to dark nebulae can be found in a straight forward way through the use of (m, log π) tables. Kapteyn, prior to 1918, first set up the (m, log π) table for teaching purposes. The potential of this type of presentation was later recognized by Bok who applied (m, log π) tables to finding absorptions and distances for dark nebulae. The details of this method can be found in the original paper by Bok (1932), with further ramifications in *The Distribution of Stars in Space*, pp. 27-30 and 46-47.

We consider space to be divided into shells concentric with the sun, with radii corresponding to the following values of the logarithms of the parallax:

$$\log \pi = -0.1, -0.3, -0.5 \ldots \tag{1}$$

The inner boundary of a shell k is given by

$$\log \pi_k = -\frac{2k-1}{10}, \tag{2}$$

and that of the outer boundary by

$$\log \pi_k = \frac{2k+1}{10}. \tag{3}$$

Without appreciable error we can take the average of these boundaries to give radii corresponding to

$$\log \pi_k = -\frac{2k}{10} \tag{4}$$

so that the relation between absolute and apparent magnitudes may, in the absence of interstellar absorption, be written as:

$$M = m + 5 - k. \tag{5}$$

An example of an $(m, \log \pi)$ table is shown in Figure 3. Each row of the table represents a certain shell; each column, a given apparent magnitude. The entries in the table, denoted by a_{mk}, are the products of ϕ (M), the luminosity function, and V, the volume of that part of the shell which covers one square degree of the sky. If the density in the space cone were uniform, and if general absorption is ignored, then the entries, a_{mk}, are the numbers of stars in shell k between magnitudes m - ½ to m + ½, per square degree. The total number of stars per square degree in this magnitude interval, A(m), would then be equal to the sum of the a_{mk}'s in column m.

The basic $(m, \log \pi)$ table is corrected for local density and general absorption, so that in the new table the a_{mk}'s at each m add up to the corrected A(m) values.

Let $\triangle_k$ represent the unknown density in shell k so that

$$A(m) = \sum_{k=-\infty}^{+\infty} \triangle_k a_{mk}. \tag{6}$$

log π	m = 8.0	m = 9.0	m = 10.0	m = 11.0	m = 12.0	m = 13.0	m = 14.0	m = 15.0	m = 16.0	m = 17.0	m = 18.0	m = 19.0	log V
	M=+9	M=+10	M=+11	M=+12	M=+13	M=+14	M=+15	M=+16	M=+17	M=+18	M=+19		
-1.000 M=+8													0.31-1
-1.200 M=+6	0.00	0.00	0.00	0.0	0.0	0	0	0	0	0	0	0	0.78-1
-1.400 M=+3	0.01	0.01	0.01	0.0	0.0	0	0	0	0	0	0	0	0.38
-1.600 M=+4	0.02	0.04	0.03	0.0	0.0	0	0	0	0	0	0	0	0.98
-1.800 M=+3	0.05	0.09	0.15	0.1	0.1	0	0	0	0	0	0	0	1.58
-2.000 M=+2	0.11	0.20	0.34	0.6	0.5	0	0	1	1	1	1	2	2.18
-2.200 M=+1	0.31	0.46	0.81	1.3	2.5	2	2	1	2	2	3	5	2.78
-2.400 M= 0	0.37	1.23	1.82	3.2	5.4	10	8	6	6	9	10	12	3.38
-2.600 M=-1	0.36	1.48	4.90	7.2	12.9	21	39	32	24	23	34	39	3.98
-2.800 M=-2	0.48	1.44	5.89	19.5	28.8	51	85	155	129	96	93	135	4.58
-3.000 M=-3	0.66	1.90	5.75	23.4	77.6	115	204	339	617	513	380	372	5.18
-3.200 M=-4	1.07	2.63	7.59	22.9	93.3	309	457	813	1 349	2 455	2 040	1 514	5.78
-3.400 M=-5	1.05	4.27	10.47	30.2	91.2	372	1 230	1 820	3 236	5 370	9 772	8 128	6.38
-3.600 M=-6	0.19	4.17	16.98	41.7	102.2	363	1 419	4 898	7 244	12 880	21 380	38 900	6.98
-3.800		0.76	16.60	67.6	166.0	479	1 445	5 888	19 500	28 840	51 290	85 110	7.58
-4.000			3.02	66.1	269.2	661	1 905	5 754	23 440	77 620	114 800	204200	8.18
-4.200				12.0	263.0	1 072	2 630	7 586	22 910	93 330	309000	457100	8.78
-4.400					47.9	1 047	4 266	10 470	30 200	91 200	371500	1230000	9.38
	-6	-5	-4	-3	-2	-1	0	+1	+2				

3. (m, log π) table for van Rhijn's general luminosity function for photographic magnitudes (After Bok, 1932).

By trail and error, the density values which best fit the observed A(m)'s are found, and the new entries in the table are given by $\triangle_k\, a_{mk}$ for each square. A simplification can be made by assuming $\triangle_k = 1.0$ for $k \leqslant 10$ and $\triangle_k = 0.0$ for $k \geqslant 21$. This is valid since stars within 100 parsecs of the sun contribute little to the total number of stars with m = 9.0 to 10.0 and fainter near the galactic plane. At large distances stars are not seen due to absorption and there is a decrease in star density, so that we may safely ignore the shells with $k \geqslant 21$. McCuskey (1956) was successful in using computers to calculate density functions.

To correct for general absorption in the comparison field (which must either be determined from spectro-photometric studies, or be assumed known), the entries for each shell k are shifted to fainter magnitudes by an amount corresponding to the absorption for that shell. The addition of the entries in each column should give the observed values of the A(m)'s for the comparison field.

The basic (m, log π) table that is obtained through the procedure just described is presumably a table applicable to the comparison field. We note that it may be necessary to correct the table further for effects of general interstellar absorption, which — if basic information is available at all — should be applied.

Turning to the field of the dark nebula, we have for it available values of log A(m) reduced to the same area of the sky that was used in the analysis for the comparison field. We now make the simplifying assumption that the only difference between the comparison field and the dark nebula is that the region of the dark nebula is affected by an absorbing sheet, absorbing ϵ magnitudes and located at a distance r from the sun. Our problem is to attempt to find the values of ϵ and r which will best represent the A(m) counts in the obscured region.

Suppose that the distance r corresponds to the inner boundary of some shell k_1. Then for $k \leqslant k_1$ the entries of the (m, $\log \pi$) table for the comparison region and the nebula should be the same, and for shells with $k > k_1$ the entries should be shifted towards fainter magnitudes by the amount the nebula absorbs, ϵ. One chooses several values of r and ϵ until the entries in the (m, $\log \pi$) table best represent the observed counts for the dark nebula. In this way we arrive at a distance and absorption of the dark nebula.

Distances are more accurately determined if stars in a narrow spectral interval are considered. In this case, the mean value of the absolute magnitude and dispersion around the mean are well known, so that the entries in the (m, $\log \pi$) table are accurately determined. Smaller magnitude and parallax intervals can be chosen—for example, intervals of 0.5 in magnitude and 0.1 in $\log \pi$—to obtain higher resolution in the analysis. For stars with a narrow range of absolute magnitudes the entries in the (m, $\log \pi$) table should cluster along diagonal lines, which facilitates the calculation of space densities and the making of corrections for general absorption.

Care must be exercised in choosing a spectral interval so that enough stars are in the comparison and obscured fields to permit statistical analysis.

The general (m, $\log \pi$) table for photographic magnitudes was calculated by Bok (1932) and is presented in Figure 3. A general (m, $\log \pi$) table for photored magnitudes was constructed by Franklin (1955).

C. Pannekoek's Method

The first statistically sound method for determining the absorption and distance of a dark nebula was developed by Pannekoek in 1921. The principles behind this method are fundamental for all dark nebulae investigations.

We denote the luminosity function for the stars in question by ϕ (M), and define D(r) as the number of stars per unit volume of space at distance r. The number of stars per unit volume with absolute magnitudes between M and M + dM at distance r is given by:

$$D(r) \, \phi \, (M) dM.$$

The number of stars per square degree between apparent magnitudes m and m + dm, falling inside a spherical shell with inner radius r and thickness dr, may be written as:

$$a(m,r)dmdr = \omega r^2 D(r) \phi (m+5-5\log r)dmdr \qquad (7)$$

where

$$\omega = \frac{4\pi}{41,253} \qquad (8)$$

As before, we denote by $A(m)$ the number of stars per square degree of the sky between apparent magnitudes m and $m + dm$. $A(m)$ and $a(m,r)$ are then related by the equation:

$$A(m)dm = dm \int_0^\infty a(m,r)dr \qquad (9)$$

and the fundamental equation for $A(m)$ in the absence of general interstellar absorption is:

$$A(m) = \omega \int_0^\infty r^2 D(r) \phi (m + 5 - 5\log r)dr. \qquad (10)$$

If we have a sheet absorbing ϵ magnitudes at a distance r_1, then the fundamental equation for the representation of the counts in the obscured region will be:

$$A'(m) = \omega \int_0^{r_1} r^2 D(r) \phi (m + 5 - 5\log r) \, dr$$

$$\qquad (11)$$

$$+ \omega \int_{r_1}^\infty r^2 D(r) \phi (m + 5 - 5\log r - \epsilon)dr.$$

We may write:

$$A'(m) = \gamma_1 A(m) + \gamma_2 A(m - \epsilon) \qquad (12)$$

where

$$\gamma_1 = \frac{\omega \int_0^{r_1} r^2 D(r) \phi (m + 5 - 5\log r) \, dr}{A(m)}$$

and

$$\gamma_2 = \frac{r \int_0^\infty r^2 D(r) \phi (m + 5 - 5\log r - \epsilon)dr}{A(m - \epsilon)} \qquad (13)$$

From star counts we get values of $A'(m)$ and $A(m)$ for the obscured and clear regions respectively. If we can assume a luminosity function ϕ (M), and have obtained a density function $D(r)$ for the comparison field, then we can proceed by trial and error to find the best values for r_1 and ϵ which represent the observations, very much in the manner of the $(m, \log \pi)$ method described above. Several values of r_1 and ϵ are tried in the computation of γ_1 and γ_2, and the derived numbers are then compared with the observed $A'(m)$ until the combination of best fit is found.

In a way the method developed by Bok is really the same as Pannekoek's method developed ten years earlier. The principal difference is that Bok's method is basically a numerical one, and need not be tied to analytical expressions for the density and luminosity functions. General absorption of light in space can readily be taken into account in the Bok approach; it was not considered in Pannekoek's early studies.

D. Malmquist's Method

In Pannekoek's work general space absorption was ignored. Malmquist (1939) has developed a method by which distances and absorptions can be found which are corrected for general space absorption.

Let $N(r)$ be the number of stars up to a distance r for the comparison field. Then,

$$N(r) \quad = \quad \omega \int_0^r r^2 D(r) dr. \tag{14}$$

$D(r)$ is again the density distribution function, and ω is a solid angle. Let $a(r)$ be the general space absorption in magnitudes, assumed to be known for the comparison field, at distance r. Define a fictitious r_0 such that

$$5 \log r_0 = 5 \log r + a(r). \tag{15}$$

In other words,

$$r_0 = r \times 10^{0.2 a(r)}. \tag{16}$$

From these relations we find

$$N_0(r_0) \quad = \quad \omega \int_0^{r_0} r_0^2 \, D_0(r_0) dr_0 \quad = \quad N_0(r \times 10^{0.2 a(r)}) \tag{17}$$

or

$$N(r) = N_0(r_0). \tag{18}$$

It is convenient to introduce a new variable

$$y = 5 \ell o g r, \tag{19}$$

so that

$$N(y) = N_0(y + a(y)). \tag{20}$$

In this fundamental equation $N_0(y)$ is the number of stars within a space cone considered up to distance $r = 10^{0.2y}$, computed without regard to any space absorption; $N(y)$, the real number of stars up to the same distance.

The computation of $N(r) = N(y)$ is generally less complicated than the determination of a density function $D(r)$. In most cases we derive $D(r)$ by dividing the computed number of stars between consecutive limits of distance by the corresponding elements of volume. That is, if $n(r_1)$ is the number of stars between distances 0 and r_1; $n(r_2)$, the number between distances r_1 and r_2 etc., and $v(r_1)$, $v(r_2)$...are the corresponding elements of volume, then the density at distances between r_{i-1} and r_i is given by:

$$D(r_i) = \frac{n(r_i)}{v(r_i)}. \tag{21}$$

On the other hand, the function $N(r)$ is obtained from

$$N(r_i) = \sum_{j}^{i} n(r_j). \tag{22}$$

From star counts we find $N_0(y)$. If we plot $\ell o g\, N_0(y)$ against y, we can obtain the $\ell o g\, N(y)$ curve by simple geometrical construction if the amount of space absorption for different distances is known. For example, the ordinate $\ell o g\, N_0(y + a(y_1))$, where $a(y_1)$ is the absorption at distance $r_1 = 10^{0.2\ y_1}$, is equal to he ordinate $\ell o g\, N(y)$ at the point $y = y_1$. In this way, the $\ell o g\, N(y)$ curve is constructed, an example of which is shown in Figure 4.

Analogously, we can determine the distance and amount of absorption of a dark nebula. From star counts in the obscured region we derive the function $\ell o g\, N_0(y)$, and from clear neighboring regions, $\ell o g\, N(y)$. In order to find the absorption at the distance $r = 10^{0.2y}$, the value of $\ell o g\, N(y)$ is found from the curve and the horizontal line from this point to the $\ell o g\, N_0(y)$ curve gives

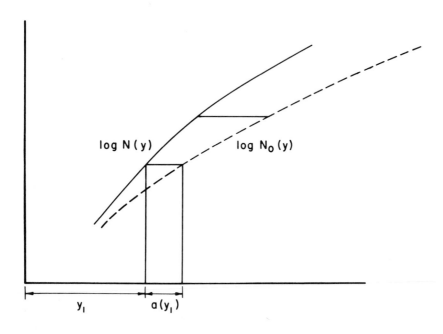

4. Malmquist curves of $\log N(y)$ and $\log N_0(y)$ versus y. At the distance $r_1 = 10^{0.2y_1}$ the absorption is $a(y_1)$.

directly the amount of the absorption at this distance. The assumptions are that the real distribution in space is the same in the obscured region as in the neighboring unobscured regions, and that the distribution of the absolute magnitudes is known. Malmquist's method can be applied most effectively when counts made with reference to small spectral subdivisions are available.

E. Becker's Color-Difference Method

Three-color UBV photometry may be used in a manner proposed by W. Becker (1938) and applied by Rodgers (1960) to find the absorption and distance of a dark nebula. The basis of this method is the color-difference diagram, a plot of (U-B) - (B-V) against (B-V). The intrinsic color-difference diagram for bright, and unreddened photoelectric standards is shown in Figure 5, with spectral types designated at different positions in the diagram. This diagram for unreddened stars is used as a reference standard.

For slightly reddened stars, we find approximately

$$E_{U-B} / E_{B-V} = 0.76,$$

so that the slope of the reddening line in the color-difference diagram is given by

$$E_{(U-B) - (B-V)} / E_{(B-V)} = -0.24.$$

This implies that if a star is reddened, its intrinsic colors $[(U-B - (B-V)]_0$ and $(B-V)_0$ may be found by extrapolating along the reddening slope (-0.24) to the

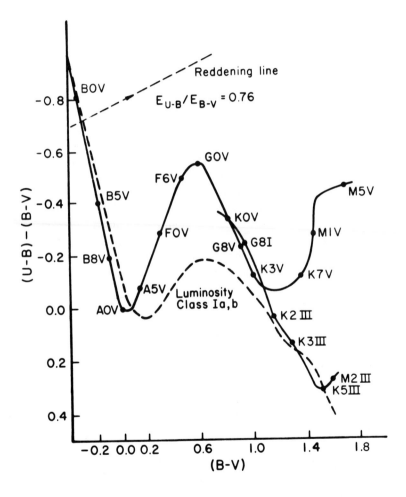

5. The intrinsic color-difference diagram for stars on the UBV system. (After Rodgers, 1960).

standard curve from the point given by its observed colors, (U-B) - (B-V) and (B-V).

The Becker-Rodgers' method has certain obvious weaknesses. The supposedly constant and known slope for the reddening line involved a rather restrictive assumption for the properties and dimensions of the particles that make up the dark nebula—we assume basically that the same distribution functions apply to particles in the dark nebula and those of the interstellar absorbing medium close to the galactic plane. We note also that the curve for the early-type stars in Figure 5 is for luminosity class V. Without luminosity classification, we naturally assume that all observed stars are of class V and we assign absolute magnitudes accordingly. Dispersion in absolute magnitudes may cause difficulties in analysis.

When the color-difference diagram is used for studying dark nebulae, it is assumed that the nebula acts as a thin absorbing sheet, and not as an extended dust cloud. In the former case, nearby stars in front of the nebula lie along the standard color-difference curve, whereas stars behind the nebula are displaced along reddening lines. If the dark nebula is at several hundred parsecs from the sun, then the foreground stars may become affected by general space reddening. The dark nebula will then stand out as a discontinuity in the diagram of A_v versus m - M, which is discussed in this section. If the cloud were extended, the observed color-difference curve would be distorted as well as displaced, and it would then be very difficult to analyze the situation from colors alone.

Rodgers applied Becker's method to several uniformly obscured regions in the Southern Coalsack. The methods of photoelectric calibration followed by photographic photometry were applied to find UBV colors for all the stars. (Note that observations of a comparison field are not required.) Relatively few stars brighter than V = 14.4 mag. were found in the regions of the dark nebula. Rodgers chose apparent magnitude intervals of V<9, 9 - 9.99, 10 - 10.99, etc., for separate analyses. For each interval, color-difference diagrams are plotted, and intrinsic colors are found for each star by following from observed positions along parallel reddening lines to the standard curve. These intrinsic colors yield absolute magnitudes, so that distance moduli are known. An example of an observed color-difference diagram is shown in Figure 6.

The color excess, E_{B-V}, is equal to the component of the reddening shift along the (B-V) axis. From this color excess, an absorption is found by assuming a standard extinction ratio of A_V/E_{B-V} = 3.0. Plots of A_V against m - M are then made for each region, and from these the absorption and distance of the obscuration are found. Figure 7 gives an example of the drawing of the final curve, absorption against distance. The conclusion is that the dark nebula is at a distance corresponding to m - M = 6.0 and that general absorption becomes appreciable beyond m - M = 8.0, with A_V= 2.6 for m - M = 12.

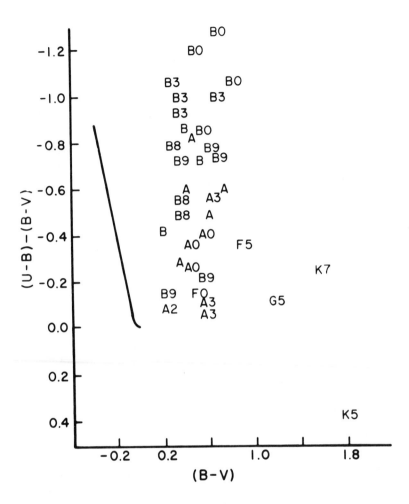

6. Observed color-difference diagram for a region in the Southern Coalsack (after Rodgers, 1960). The magnitude range is $10.00 < V < 10.99$. Spectral types (for illustration only) are from the H.D. Catalogue and Extension. Part of the intrinsic curve is shown, for B0V to A0V stars, as a solid line.

F. Observational Aspects

A word should be said about the kinds of basic observations which can be made with existing equipment. A first requirement is that a faint magnitude

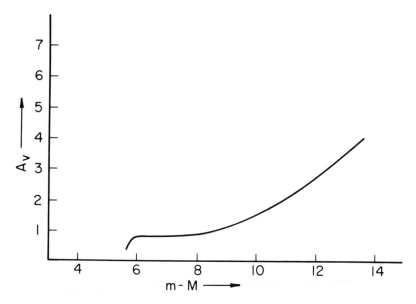

7. Example of one of Rodgers' (1960) curves for the Southern Coalsack. This plot of
absorption against distance shows a nebula at a distance of 174 parsecs, corresponding to
m - M = 6, with an absorption of 1.0 mag. Beyond this nebula, there is a clear region
extending to 800 pc. At greater distances, absorption again sets in.

sequence be established in or near the section of the sky under investigation. The
photoelectric magnitude limit of a telescope depends on its aperture size, seeing
conditions and, of course, on the experience of the observer. With the 36-inch
reflector at Cerro Tololo Interamerican Observatory, visual magnitudes from V =
9 to 16 can be obtained without offsetting when the seeing is of the order of 1
second of arc. With the 60-inch, under similar conditions, one is able to go
(without offsetting) as faint as 17th magnitude. With the Steward Observatory
90-inch telescope, direct photoelectric observations to V = 18 are possible on
good nights. Even fainter magnitudes can be reached by offsetting, which,
however, is a difficult technique for use in rich Milky Way fields. Claims have
been made that by offsetting, V = 20 can be reached with a 50-inch reflector
even in fairly crowded fields, but this is far from simple. No matter which
telescope is used, obtaining photoelectric magnitude sequences is a laborious
task and it would be very beneficial to have standard sequences established for
many positions along the band of the Milky Way. Commission 25 of the
International Astronomical Union has a Sub-Committee especially charged with
the listing of available standard sequences for faint stars.

Spectral counts are especially valuable for distance determinations of dark nebulae. McCuskey and Houk (1971) give a good example of spectral counting in the interval B8 - A3. They classify these stars from objective prism plates, and count down to V = 13 for regions near the galactic plane between galactic longitudes $\ell = 50°$ to $\ell = 150°$. The mean absolute magnitude and dispersion (per unit volume of space) adopted for the B8 - A3 are $M_o = +0.9$ and $\sigma_o = 0.7$, respectively. These have been taken from the tabulation by Blaauw (1963). For future reference, we list in Table 1 mean absolute magnitudes and dispersions per unit volume of space for several spectral intervals. Average corrections for interstellar absorption can be taken from the catalog by Neckel (1967), and the paper by FitzGerald (1968). These references give diagrams and charts showing E_{B-V} and/or A_V as functions of distance.

Table 1

Spectral Group	M_o	σ_o
B5	-1.0	0.5
B8 -A0	+0.2	0.5
A1 - A5	+1.6	0.4
A7 - F5	+2.8	0.5
F8V - G2V	+4.5	0.4
G5V	+5.0	0.3
F8III - K3III	+0.9	0.8
F8IV - K3IV	+3.2	0.8
G8III - K3III	+1.0	0.6
K5III - M5III	0.0	0.6

Mean absolute visual magnitudes and dispersions per unit volume of space. (After McCuskey, 1956)

Stars in the spectral interval B8 - A3 are both bright and plentiful enough to be used for spectral counts in large dark nebulae complexes. These stars are fairly easy to classify from objective prism Schmidt photographs.

For spectral-luminosity classification the limiting magnitude is about 12, but for straight spectral classification, stars to V = 13 can be reached with modern Schmidt telescopes. Color studies permit one to go at least 3 magnitudes fainter with a good Schmidt telescope.

G. Practical Applications

Different types of dark nebulae lend themselves to different analyses. The large nearby complexes cover enough area of the sky so that stars in narrow spectral ranges can be counted. Since we shall want to penetrate through the absorbing screen, we need to choose intrinsically bright stars, such as B8 - A3, which are present in large numbers. In a narrow spectral interval the mean absolute magnitude and the dispersion around the mean are quite well known, and an (m, $\log \pi$) analysis can easily be applied to find the distance to the absorbing cloud.

Absorptions are best found by counting stars to faint magnitudes in the colors desired, and by then applying a Wolf curve analysis. If no magnitude sequence is available for calibration, then, as a last resort, one can count all stars down to the magnitude limit of the plate and obtain a rough photographic absorption by use of van Rhijn's tables.

Large dark clouds can also be analyzed by the color-difference method of Becker and Rodgers to find distances and absorptions. UBV magnitudes are required for a plot of (U-B) - (B-V) against (B-V). By noticing at which spectral type a shift along reddening lines occurs, one can find the absorption and distance of an obscured region.

An alternate color method is discussed by Karlsson (1971), in his analysis of a large dark complex in Monoceros. Certain spectro-photometric quantities were derived from objective prism plates. The observed quantities used by Karlsson are:

m_{4400}	monochromatic magnitude at 4400 Å,
$H_{\gamma\delta}$	weighted mean of apparent depths of Hγ and Hδ,
K	apparent depth of K line,
C_l	color equivalent equal to m_{4030} - m_{4600}.

Calibration with stars of known MK types gives absolute magnitudes, M_{4400}, and intrinsic colors $(C_1)_o$ from the observed quantites $H_{\gamma\delta}$ and K, the procedure being equivalent to a one-dimensional classification. Then, using m_{4400} and C_1, distance moduli and color excesses are derived, and the interstellar extinction and space distribution of stars are studied as a function of distance.

Under certain special conditions, there are simple ways to find the distance to a dark nebula. If a star is observed as being associated with the dark nebula, then the distance to that star may be taken as the distance to the nebula. A spectral-luminosity classification plus color measurements for the star will yield an absolute magnitude and approximate photographic absorptions, hence a true distance modulus for the star. One must, of course, have good reason for supposing that the star is truly associated with the dark nebula. The best proof comes in the case where a small reflection nebula is found close to the star.

Large globules (from 2' to 20' in diameter) cannot be as accurately analyzed as the large dark complexes. Usually, very few stars are seen through them, so that star counts in a narrow spectral range are impossible. We can obtain a rough total photographic absorption with the aid of the (m,$\log \pi$) table, Figure 3, provided we see any stars shining through the dark nebula. A guess at red absorption may be made by noting the ratio of red to blue stars in the comparison field and by applying this ratio to the dark region.

Distance estimates are hazardous at best, but some postulation can be made by noticing the number of foreground stars in the obscured region. If there are as many blue stars as red stars seen in the globule, then these stars are most likely unreddened foreground stars.

With the aid of the (m ℓog π) Table, Figure 3, we can readily estimate the expected number of foreground stars shown by a globule of known angular diameter for a variety of assumed distances from the sun. To the limit of the Blue Palomar Sky Survey Prints, the following numbers of foreground stars are thus calculated:

Distance of Globule (Parsecs)	Expected Number of Foreground Stars
100	1.5
160	5
250	15
400	40

These numbers are calculated for a globule with an angular diameter of 20′; hence they apply only to the larger globules listed in Table 3. If a globule is totally opaque, then roughly equal numbers of foreground stars can be expected on the blue and red Palomar Sky Survey plates. If the globule has a total absorption in the range 2.5 to 5 magnitudes on the Blue Print, then the number of stars, foreground and background stars combined, seen projected against the globule, should be considerably greater on the Red Print than on the Blue Print. It becomes thus a relatively straightforward process to distinguish between totally opaque and rather transparent globules and, furthermore, we can make a rough estimate of distance for most of the larger opaque globules in Table 3. Because of statistical fluctuations in numbers, the above considerations apply nicely to those globules of Table 3 that have diameters of 20′ or greater, and they are still of some value for diameters in the range 10′ to 20′; they do not apply to the smaller globules of Table 3.

If the red stars seen in the globule far outnumber the blue stars, the these stars are probably background stars for which the light has been reddened by dust in the globule, and an estimate of the number of foreground stars cannot be made. However, a good total absorption value can be found if many stars lie beyond the nebula.

The third category of dark nebulae includes tiny globules seen against emission nebulae. These are usually less than 2 minutes of arc in size and no detailed absorption studies are possible. Apparently these globules are usually totally opaque to the limiting magnitude of the plate. The distance to these globules is taken to be the distance to the emission feature.

V. CATALOGS AND ATLASES OF DARK NEBULAE

Several photographic atlases of the sky are available for searches of dark nebulae. E.E. Barnard and M.R. Calvert (1927) have published a series of 50 positive photographs in blue light of various regions of the Northern Milky Way. Each print covers an area of approximately 7° by 7° and has a scale of about 2.7'/mm. Barnard has drawn contours for the dark nebulae on his photographs, which are published in a separate volume. He lists coordinates, apparent dimensions and individual characteristics of each dark nebula. His listing includes a total of 349 dark nebulae.

In the early 1900's J. Franklin-Adams undertook a photographic survey of the whole sky. The Royal Astronomical Society has published the 206 blue photographs, each of which covers an area of 15° by 15°, with a scale of 3'/mm. To date this is the only photographic survey of the entire sky and, although some of the prints are not of the best quality, the uniformity of the atlas is valuable for searches of both northern and southern dark clouds. Lundmark (1926) graphically displays the sizes and distribution of the 1550 dark nebulae which he and P.J. Melotte discovered using the original Franklin-Adams plates.

The Ross-Calvert Atlas (1934), which also is in blue light, covers a large section of the Northern Milky Way. The atlas contains forty positive prints 21° on a side with a scale of 3'8/mm. Schoenberg (1964) prepared a catalog of 1456 dark clouds based on these prints. Khavtassi (1955) used the Ross-Calvert Atlas and others to catalog 797 dark nebulae over the whole sky between latitudes -20° ⩽ b ⩽ + 20°. He lists old galactic coordinates, equatorial coordinates, the value of the visible surface area in square degrees, the position angle and individual comments for each nebula. In addition, Khavtassi (1960) has drawn in three different shades, colored contour maps of these nebulae. The different shades correspond to apparent obscuration.

The National Geographic Society-Palomar Observatory Sky Survey prints are very well suited for the study of dark nebulae. The prints are 6° on a side and have a relatively open scale of 1'11/mm. The two colors, red and blue, yield a contrast that the single prints from the other atlases cannot accomplish. The resolution is very good, and stars of 20th magnitude can readily be seen on the blue prints. This limit is about 3 magnitudes fainter than that reached by the atlases previously mentioned. Lynds (1962) compiled a list of dark nebulae from studies of the Palomar prints. Positions, areas and relative visually estimated opacities on a scale from 1 to 6 are given for each nebula.

The lack of a southern counterpart to the Palomar Sky Survey is severely felt in all studies of dark nebulae for the Southern Milky Way. Fortunately the European Southern Observatory is now placing into operation at La Silla in Chile a 40/60-inch Schmidt Telescope with a focal length identical to that of the Palomar Schmidt. This telescope is scheduled to undertake as one of its first assignments the preparation of the extension of the Palomar Schmidt Survey to southern declinations.

The minimum size of the dark nebulae listed in the catalog of Lynds and others described earlier in this paper is 2 square minutes of arc, which excludes the tiny globules. However, Sim (1968) has published a detailed catalog of globules observed in and near 66 OB clusters and associations which were investigated by Reddish (1967) for dust-embedded stars. She finds 63 certain and 63 probable globules using the Palomar Sky Survey prints. The positions, shapes, orientations and angular sizes of the globules are tabulated.

Several photographic surveys of the southern hemisphere have been published. Rodgers, Campbell, Whiteoak, Bailey and Hunt (1960) present a mosaic in Hα light of the southern sky on a small scale. Haffner and Nowak (1969) have edited the Würzburg Atlas which covers the Southern Milky Way from -15° to -70° declination. The prints reach as faint as B = 16, and have a scale of about 2.8'/mm. The original plates were not intended for general publication, hence the quality of the prints is not, in many cases, as fine as that of the Franklin-Adams charts, which have nearly the same scale.

Wray and Westerlund (1971) have prepared an atlas based on a series of 159 103aD plates with GG 14 filters taken with the Uppsala Schmidt telescope at Mount Stromlo Observatory. The area covered extends from $\ell = 237°$ through 360° to 7° within the band $-3° \leq b \leq +3°$ (new galactic coordinates). The original plates have been enlarged onto a dimensionally stable Gravare film, to a scale matching the Palomar Observatory Sky Survey plates. The Palomar Survey coordinate grids may be used directly with these prints.

In addition to the more extensive catalogs of dark nebulae, selected lists have been published. Bok (1937) gives a brief survey of major dark features seen along the Milky Way. Becker (1942) lists 18 large dark complexes of area 1 square degree or greater for which approximate distances and absorptions are available from star counts. Lynds (1968) has updated Becker's list and added to it.

Bok, Cordwell and Cromwell (1971) have cataloged a few representative dark nebulae of various kinds and have given approximate absorptions, distances and masses for some of these.

VI. REPRESENTATIVE DARK NEBULAE

In the following three sections we present lists of representative dark nebulae of all sizes. None of the lists are exhaustive, but each list is meant to draw attention to some dark nebulae, or complexes of nebulae, for which further study would be of interest, both in the optical and radio range.

A number of distinct dark complexes is listed in Table 2, together with their known properties.

Several large globules, mostly Barnard objects, are cataloged in Table 3. In general, comparable globules situated at far southern declinations are not listed because of unavoidable lack of information. The current Atlases of the Southern Hemisphere do not offer the resolution necessary for the discovery of the small nebulae of the Barnard variety. For instance, the object Barnard 72 (The Snake) and its associated globules show only faintly on the Franklin-Adams Chart (no. 61), whereas they are quite clear in Barnard's Atlas (plate 20) and particularly on the Palomar prints (-24° 17:20). These objects are not evident in the prints of the Wurzburg Atlas (plate 23).

Barnard 335, a particularly opaque and rather small globule, does not appear on the Franklin-Adams Chart (no. 136), whereas it is clearly present on the print from the Barnard Atlas (plate 41) and on the Palomar prints (+6° 19:36). It is evident that the discovery and cataloging of southern globules must be delayed until the Palomar Sky Survey is extended to the Southern Hemisphere; the Schmidt survey at the European Southern Observatory, about to be undertaken, should provide the necessary basic photographic prints.

Tiny globules are listed in Table 4. These are abundant in many emission nebulae; only a few examples are given here.

Optical and radio astronomers have studied the large complexes to some extent, but the smaller nebulae and globules remain virtually untouched. In all cases, much work remains to be done in the realm of dark nebulae studies.

A. Introduction to Table 2: The Large Complexes

A few representative large, dark nebulae have been cataloged in Table 2. Most of these objects have been previously listed in tables by Becker (1942) and Lynds (1968). In compiling our list, we referred to several Atlases. For the northern hemisphere extensive use was made of the Ross-Calvert, Barnard and Palomar Sky Survey Atlases. In the southern hemisphere reference was made to the Rodgers et al. Atlas (1960) and the Franklin-Adams charts. In some cases the Wray-Westerlund (1971) and Würzburg (Haffner and Nowak, 1969) Atlases were used.

TABLE 2. THE LARGE COMPLEXES

Object	Coordinates (1975)	New Galactic Coordinates	Print and Position in Centimeters	Approx. Angular Size	Approx. Photographic Absorption	Approx. Distance	Comments
Dark Nebulae in Taurus (1)	$\alpha = 4^h39^m$, $\delta = +25.5$	$\ell = 174°$, $b = -13°$	P+24°4:20, x = 3.0, y = 24.5	2° x 1°	6.7 mag.	100 pc.	1
(2)	$\alpha = 4^h17^m$, $\delta = +28.3$	$\ell = 169°$, $b = -16°$	P+30°4:20, x = 30.0, y = 6.5	56'x45'	4.7	100	1
Dark Nebula in Orion	$\alpha = 5^h45^m$, $\delta = -1.1$	$\ell = 207°$, $b = -15°$	P -°5:36, x = 13.0, y = 11.0	1°7	2.4	300-400	2
Dark Nebula in Vela	$\alpha = 9^h19^m$, $\delta = -48.3$	$\ell = 264°$, $b = +7°$	F-A #33, x = 13.0, y = 8.4	2°	>2	500-750	3
Southern Coalsack (1)	$\alpha = 12^h57.6^m$, $\delta = -61°14'$	$\ell = 304°$, $b = -3°$	F-A #19, x = 12.2, y = 11.5	45'	≥2.2	150	4
(2)	$\alpha = 12^h30.3^m$, $\delta = -63°37'$	$\ell = 301°$, $b = -6°$	F-A #19, x = 16.7, y = 7.8	20'	≥2.4	175	4
Dark Nebula in Norma	$\alpha = 15^h26^m$, $\delta = -57.0$	$\ell = 324°$, $b = 0°$	F-A #21, x = 21.0, y = 17.2	8°x3°	1.7	>500	5
ρ Ophiuchi Complex	$\alpha = 16^h27.0^m$, $\delta = -24°34'$	$\ell = 354°$, $b = +16°$	P-24°16:28, x = 27.0, y = 15.5	1°5	8.0	200	6
Barnard 78 (Near θ Ophiuchi)	$\alpha = 17^h32.4^m$, $\delta = -25°35'$	$\ell = 1°$, $b = +4°$	P-24°17:20, x = 9.8, y = 9.8	3°	6.5	200	7

Table 2, continued

Object	Coordinates (1975)	New Galactic Coordinates	Print and Position in Centimeters	Approx. Angular Size	Approx. Photographic Absorption	Approx. Distance	Comments
Dark Nebula in Scutum	$\alpha = 18^h46^m$ $\delta = -3°1$	$l = 39°$ $b = +5°$	P -6°18:48 x = 26.6 y = 31.9	1°5	4.2	180–265	8
Dark areas in Alq. Rift (1)	$\alpha = 19^h27^m$ $\delta = +18°0$	$l = 55°$ $b = +1°$	R-C #13 x = 11.3 y = 21.0	3°7	2.3	500	9
(2)	$\alpha = 18^h29^m$ $\delta = -2°6$	$l = 28°$ $b = +3°$	P 0°18:24 x = 17.0 y = 3.0	1°5	5	150–250	9
Barnard 144 "Fish on the Platter"	$\alpha = 19^h59^m.4$ $\delta = +34°39'$	$l = 71°$ $b = +3°$	P+36°20:04 x = 27.0 y = 9.5	55'	2.2	1550	10
Dark Cloud West of North American Nebula	$\alpha = 20^h55^m.4$ $\delta = +43°41'$	$l = 84°$ $b = -1°$	P+42°21:00 x = 26.0 y = 24.0	55'	4	600	11

Table 2 contains the following information:

Column 1: The name of the object.

Column 2: The 1975 equatorial coordinates of the central core of the dark nebula.

Column 3: The new galactic coordinates of the object.

Column 4: The Atlas and print number on which the object may be found (Palomar, P; Franklin-Adams, F-A; Ross-Calvert, R-C), and the horizontal and vertical distance in centimeters (x and y) from the inside lower left corner of the print.

Column 5: The approximate angular size of the dark nebula. For very large objects the angular size of the darkest core is given.

Column 6:, The approximate photographic absorption of the darkest part of the nebula. These values are taken either from the literature, or from star counts on the Palomar or Franklin-Adams prints, using van Rhijn's (1929) tables for purposes of reductions. Such star counts were made when the existing literature provided clearly inadequate information, or when no previous work had been done.

Column 7: The approximate distance to the dark nebula, usually taken from the literature, sometimes an educated guess. In most cases the distance values need to be carefully checked.

Column 8: Comments about the individual nebula. The comments follow the table.

Comments for Table 2

(1) Absorption: Bok (1956). Distance: McCuskey (1941).
   These two centers were chosen as representative of the many knots of absorption in the dark lanes of Taurus.

(2) Absorption: Star counts. Distance: Asklöf (1930).
   Large dark area of uniform absorption lying directly east of the easternmost belt star of Orion.

(3) Absorption and distance: Greenstein (1936).

(4) Absorption and distance: Rodgers (1960).
   The small, dark centers listed appear more opaque than the larger areas studied by Rodgers.

(5) Absorption: Star counts. Distance: Educated guess of lower limit.

(6) Absorption and distance: Bok (1956).
   Further information in article by Bok, Cordwell and Cromwell (1971).

(7) Absorption: Star counts. Distance: Assumed to be that of the ρ Ophiuchi complex. Red Palomar print shows core clearly.

(8) Absorption: Star counts. Distance: Becker (1942).
   Many more stars seen in red than blue.

(9) Center (1): Absorption and distance: Baker (1941).

Center (2): Absorption: Star counts. Distance: Weaver (1949).

Part of the Great Rift runs southwest from Vulpecula through Aquila, +23° 19h 30m to -12° 18h 24m. Center (1) is representative of the general rift; center (2) shows the strongest absorption in this area.

(10) Absorption and distance: Franklin (1955).

(11) Absorption and distance: Miller (1936).

Nebula is cross-shaped on red Palomar print, and occupies a substantial portion of the "Atlantic Ocean" and the "Gulf of Mexico" with reference to the North American nebula.

B. Introduction to Table 3: Large Globules

The objects listed in Table 3 are small, isolated dark nebulae ranging in size from $2'$ to $20'$. Most were found on the prints of Barnard's Atlas, and then examined on the Palomar prints. Lynds' (1962) catalog was used when Barnard's list was lacking. For objects far to the south we referred to the Franklin-Adams charts, as well as selected Schmidt plates taken at Cerro Tololo, Chile.

Table 3 contains the following information:

Column 1: The name of the object.

Column 2: The 1975 equatorial coordinates of the object.

Column 3: The Atlas and print number on which the object may be found (Palomar, P; Franklin-Adams, F-A), and the horizontal and vertical distance in centimeters (x and y) from the inside lower left corner of the print.

Column 4: The approximate shape of the globule.

Column 5: The approximate angular size of the globule.

Column 6: Comments about each globule. The comments follow the table.

Comments for Table 3

(1) Fairly isolated globule in Taurus. Not as definite in red as blue.

(2) Photographic absorption = 3 mag. (Bok, Cordwell, Cromwell, 1971).

(3) Clumpy. Opaque in red and blue.

(4) Photographic absorption = 5 mag. (Bok et al. 1971) Chain of 3 globules lie to the south of Barnard 227.

(5) Opaque globule at head of cometary structure, which is an area rich in molecular compounds. Region has been studied optically by Karlsson (1971).

(6) Two definite, round globules seen clearly on Schmidt plates taken at Cerro Tololo Interamerican Observatory.

(7) Large opaque core in red and blue. Studied by Sancisi (1971) for neutral hydrogen.

(8) Has round, opaque core with diameter 3'.
(9) More transparent in red than blue. Seems to have several foreground stars.
(10) Not totally isolated. Opaque in blue, nearly so in red.
(11) Not totally isolated. Opaque in blue and red.
(12) Near region of much nebulosity. Transparent in red and blue.
(13) Contain three round centers bunched together.
(14) Sharply defined. Opaque in red and blue.
(15) Contains two round centers.
(16) Lies in front of the Great Star Cloud in Sagittarius. Nearly opaque.
(17) Has streamers. Transparent in red and blue. (Probably distant.)
(18) Stands out sharply against star cloud in Sagittarius.
(19) One of many tiny, opaque globules on this plate.
(20) Well-defined but filamentary. Opaque in red and blue.
(21) Opaque head 2' in diameter, Diffuse tail 12' long.
(22) Dark core of larger, indefinite nebula.
(23) In front of dark complex. More transparent in red than in blue.
(24) Connected to Barnard 110. Core opaque in red and blue.
(25) Clumpy. Linked to large nebula in Scutum. Has two cores opaque in red and blue.
(26) Linked to nebula in Scutum. Core opaque in red and blue.
(27) End of chain of small nebulae. Opaque in blue and red. Projected against Scutum star cloud.
(28) Lies in front of Scutum star cloud. Isolated. Opaque in red and blue.
(29) Isolated. Transparent in red.
(30) Tail of long, sweeping nebula. Definite. Transparent in red.
(31) Isolated, definite. Opaque in red and blue.
(32) Distant. Lies north of the "Fish on the Platter."
(33) Opaque and distant. Situated in region of turbulence. Several tiny, round globules to northeast seen in red.
(34) Photographic absorption = 3 mag. (Bok *et al*, 1971). Moderately isolated from nearby nebulae.
(35) Somewhat clumpy. More transparent in red than blue.
(36) Near other nebulosity. More transparent in red than blue.
(37) Nearly opaque in red and blue.
(38) Situated in an area containing many small, irregular globules. Opaque in red and blue.
(39) In same region as Barnard 161, 163. Moderately opaque in red and blue.
(40) Double-nucleated globule, more transparent in red than blue. Connected to some nebulosity.
(41) Near other nebulosity. Opaque in red and blue.

TABLE 3

Object	Coordinates (1975)	Print and Position in Centimeters	Shape	Approx. Angular Size	Comment
Barnard 5	$\alpha = 3^h46^m.3$ $\delta = +32°49'$	P+30°3:28 x = 5.0 y = 30.0	Elliptical	22' x 9'	1
Barnard 34	$\alpha = 5^h41^m.8$ $\delta = +32°39'$	P+30°5:38 x = 21.8 y = 31.0	Round	20'	2
Lynds 1622	$\alpha = 5^h53^m.2$ $\delta = +1°48'$	P 0°5:36 x = 2.4 y = 26.7	Irregular	15'	3
Barnard 227	$\alpha = 6^h5^m.9$ $\delta = +19°29'$	P+18°6:00 x = 18.6 y = 25.5	Elliptical	10'	4
Globule in NGC 2264	$\alpha = 6^h39^m.9$ $\delta = +9°27'$	P+12°6:24 x = 5.0 y = 4.4	Square	1'	5
Twin Globules near η Carinae	$\alpha = 10^h35^m.2$ $\delta = -60°59'$	F-A #18 x = 20.4 y = 11.5	Round	2'	6
Lynds 134	$\alpha = 15^h52^m.3$ $\delta = -4° 30'$	P- 6°15:36 x = 4.1 y = 27.1	Elliptical	22' x 12'	7
Barnard 46	$\alpha = 16^h55^m.8$ $\delta = -22°47'$	P- 24°16:54 x = 22.5 y = 25.1	Irregular	12'	8
Barnard 49	$\alpha = 17^h01^m.0$ $\delta = 33°14'$	P- 30° 16:54 x = 16.1 y = 0.9	Elliptical	6'	9

Table 3, continued

Object	Coordinates (1975)	Print and Position in Centimeters	Shape	Approx. Angular Size	Comments
Barnard 57	$\alpha = 17^h\ 7^m.0$ $\delta = -22°51'$	P- 24° 16:54 x = 8.7 y = 24.4	Elliptical	5'	10
Barnard 63	$\alpha = 17^h 12^m.3$ $\delta = -21°53'$	P- 24° 16:54 x = 2.0 y = 29.4	Irregular	11'	11
Barnard 255	$\alpha = 17^h 19^m.2$ $\delta = -23°27'$	P- 24° 17:20 x = 25.6 y = 20.9	Round	6'	12
Barnard 67a	$\alpha = 17^h 21^m.0$ $\delta = -21°52'$	P- 24° 17:20 x = 23.7 y = 29.4	Irregular	16'	13
Barnard 68	$\alpha = 17^h 21^m.1$ $\delta = -23°48'$	P- 24° 17:20 x = 23.1 y = 18.9	Kidney	4'	14
Barnard 72 (Snake)	$\alpha = 17^h 22^m.1$ $\delta = -23°39'$	P- 24° 17:20 x = 21.7 y = 19.7	S-Shaped	4'	15
Barnard 86	$\alpha = 18^h\ 1^m.5$ $\delta = -27°52'$	P- 30° 17:46 x = 6.6 y = 29.1	Irregular	4'	16
Barnard 87	$\alpha = 18^h\ 2^m.4$ $\delta = -32°30'$	P- 30° 17:46 x = 5.8 y = 4.0	Round	12'	17
Barnard 90	$\alpha = 18^h\ 8^m.6$ $\delta = -28°17'$	P- 30° 18:12 x = 28.8 y = 26.4	Elliptical	2'	18

Table 3, continued

Object	Coordinates (1975)	Point and Position in Centimeters	Shape	Approx. Angular Size	Comments
Unlisted	$\alpha = 18^h11.^m1$ $\delta = -15°50'$	P- 18°18:00 x = 12.7 y = 29.1	Irregular	2'	19
Barnard 92	$\alpha = 18^h14.^m2$ $\delta = -18°16'$	P- 18°18:00 x = 9.1 y = 16.1	Elliptical	12' x 6'	20
Barnard 93	$\alpha = 18^h15.^m8$ $\delta = -18°3'$	P- 18°18:00 x = 7.2 y = 17.1	Cometary	12' x 2'	21
Barnard 95	$\alpha = 18^h24.^m2$ $\delta = -11°46'$	P- 12°18:24 x = 25.8 y = 18.1	Round	8'	22
Barnard 100	$\alpha = 18^h31.^m5$ $\delta = -9°13'$	P- 12°18:24 x = 16.1 y = 31.9	Irregular	12'	23
Barnard 107	$\alpha = 18^h48.^m2$ $\delta = -5°2'$	P- 6°18:48 x = 26.0 y = 22.0	Irregular	5'	24
Barnard 110	$\alpha = 18^h48.^m8$ $\delta = -4°51'$	P- 6°18:48 x = 25.2 y = 22.9	Irregular	9'	25
Barnard 113	$\alpha = 18^h50.^m1$ $\delta = -4°21'$	P- 6° 18:48 x = 23.5 y = 25.8	Irregular	11'	26
Barnard 118	$\alpha = 18^h52.^m6$ $\delta = -7°29'$	P- 6°18:48 x = 20.1 y = 8.8	Round	1'	27

Table 3, continued

Object	Coordinates (1975)	Point and Position in Centimeters	Shape	Approx. Angular Size	Comments
Barnard 133	$\alpha = 19^h04.8$ $\delta = -6°57'$	P- 6°18:48 x = 3.9 y = 11.8	Irregular Ellipse	3' x 10'	28
Barnard 134	$\alpha = 19^h5.6^m$ $\delta = -6°17'$	P- 6°18:48 x = 2.8 y = 15.2	Round	3'	29
Barnard 139	$\alpha = 19^h16.8^m$ $\delta = -1°30'$	P 0°19:12 x = 19.0 y = 7.9	Elongated	2' x 10'	30
Barnard 335	$\alpha = 19^h35.7^m$ $\delta = +7°31'$	P+ 6°19:36 x = 25.5 y = 24.0	Elliptical	4'	31
Barnard 145	$\alpha = 20^h1.9^m$ $\delta = +37°37'$	P+36°20:04 x = 24.0 y = 24.3	Elongated	6' x 35'	32
Barnard 343	$\alpha = 20^h12.5^m$ $\delta = +40°12'$	P+42°20:00 x = 8.5 y = 6.0	Irregular	8'	33
Barnard 361	$\alpha = 21^h11.9^m$ $\delta = +47°18'$	P+48°20:58 x = 8.0 y = 11.1	Round	17'	34
Barnard 362	$\alpha = 21^h23.4^m$ $\delta = +50°3'$	P+48°21:32 x = 28.5 y = 25.8	Elliptical	10'	35
Barnard 157	$\alpha = 21^h33.0^m$ $\delta = +54°34'$	P+54°21.32 x = 19.4 y = 17.7	Round	4'	36

Table 3, continued

Object	Coordinates (1975)	Print and Position in Centimeters	Shape	Approx. Angular Size	Comments
Barnard 161	$\alpha = 21^h39^m_.6$ $\delta = +57°42'$	P+54°21:32 x = 14.5 y = 34.4	Cometary	3'	37
Barnard 163	$\alpha = 21^h41^m_.3$ $\delta = +56°38'$	P+54°21:32 x = 13.1 y = 28.6	Irregular	4'	38
Barnard 367	$\alpha = 21^h43^m_.5$ $\delta = +57°4'$	P+54°21:32 x = 11.7 y = 31.3	Irregular	3'	39
Barnard 164	$\alpha = 21^h45^m_.8$ $\delta = +50°59'$	P+48°21:32 x = 9.3 y = 30.5	Kidney	12' x 6'	40
Lynds 1225	$\alpha = 23^h10^m_.9$ $\delta = +61°33'$	P+60°22:44 x = 2.8 y = 23.0	Round	6'	41

C. Introduction to Table 4: Small Globules

Most emission nebulae display small globules, either singly or in clumps and chains. These globules vary from tiny, round specks to windswept filaments. Table 4 lists a few emission nebulae for which globules are easily seen against the emission background. 1975 equatorial coordinates are given, as well as a brief description of the type of globules found. A final column gives the distance to the emission nebula. These distances are from Sharpless (1965) and Miller (1971).

Bok, Cordwell and Cromwell (1971) discuss some of these small globules in more detail. Sim (1968) has done a thorough job of cataloging tiny globules found near OB clusters and associations.

TABLE 4

Object	Coordinates (1975)	Description of Globules	Distance (Kpc)
NGC 2244 (Rosette Nebula)	$\alpha = 6^h30^m.8$ $\delta = +4°53'$	Round and isolated. Also in clumps and chains.	1.4
NGC 3576	$\alpha = 11^h10^m.8$ $\delta = -61°11'$	Tiny, single specks. Also chains.	3.7
NGC 3603	$\alpha = 11^h23^m.0$ $\delta = -61°7'$	Clumpy.	7.6
IC 2944	$\alpha = 11^h36^m.2$ $\delta = -62°38'$	Isolated, irregular,	2.2
NGC 6514 (Trifid)	$\alpha = 18^h00^m.6$ $\delta = -23°2'$	Windswept.	1.4
Messier 8 (NGC 6523)	$\alpha = 18^h01^m.8$ $\delta = -24°23'$	Windswept.	1.3

VII. EPILOGUE

Dark nebulae studies are becoming increasingly more important as more sophisticated theories of star formation are advanced. If accurate observational data can be made available, greater insight into pre-protostar conditions can be achieved and theories of star formation can be developed on the basis of well-established initial conditions. With high resolution telescopes we can make good progress toward finding absorptions and distances, especially through the proper use of assorted star count techniques. Progress is being made in the study of interstellar grains, although the question as to whether grains in dark clouds are the same as grains in the diffuse interstellar matter is still unanswered. Estimates of total masses and densities for dark nebulae depend in part on assumptions of grain characteristics, so that work on interstellar grains appears vital for those who study dark nebulae. The total mass and density of a dark cloud are heavily dependent on the amount of gas present. The ratio of gas to dust by mass has usually been assumed to be 100 for the interstellar medium. We note, however, that no neutral atomic hydrogen concentrations have been detected in dense dark nebulae. Because of the low temperatures involved, it has been postulated that the gas in these nebulae may well be in the form of molecular hydrogen, or possibly in the form of very cold HI. It may be that high resolution radio telescope arrays will detect the 21-centimeter line of HI in dense, dark nebulae, but as of now little is known about the amount of gas in such nebulae or what form that gas has taken.

The extremely low temperatures of dense dark nebulae, which are at approximately $10°K$ or less, provide excellent conditions for molecule formation. OH has already been found in the densest region of the ρ Ophiuchi complex (Heiles, 1969), and most certainly other molecules will be discovered which will give an indication of the chemical constituents and internal conditions in the interiors of dark clouds. An obvious and real need exists for extensive, high-caliber, optical and radio studies of dark nebulae.

BIBLIOGRAPHY

Asklöf, S. 1930, *Upp. Obs. Medd.*, No. 51.

Barnard, E.E. 1927, *Atlas of Selected Regions of the Milky Way*, eds. E.B. Frost and M.R. Calvert (Washington: Carnegie Institute of Washington).

Baker, R.H. 1941, *Ap. J.* **94**, 493.

Becker, W. 1938, *Zs. f. Astrophys.* **15**, 225.

Becker, W. 1942, *Sterne und Sternsysteme* (Dresden and Leipzig: ,T. Steinkopf) pp. 178-179.

Blaauw, A. 1963, *Basic Astronomical Data*, ed. K. Aa. Strand (Chicago and London: University of Chicago Press) p. 401.

Bok, B.J. 1932, *Harvard Obs. Circular*, No. 371.

Bok, B.J. 1937, *The Distribution of Stars in Space* (Chicago: University of Chicago Press).

Bok, B.J. and Reilly, E.F. 1947, *Ap. J.* **105**, 225.

Bok, B.J. 1956, *A. J.* **61**, 309.

Bok, B.J, Cordwell, C.S., and Cromwell, R.H. 1971, *Dark Nebulae, Globules and Protostars*, ed. B.T. Lynds (Tucson: University of Arizona Press) p. 33.

FitzGerald, M.P. 1968, *A. J.* **73**, 983.

Franklin, F.A. 1955, *A. J.* **60**, 351.

Franklin-Adams, J. 1936, *M.N.* **97**, No. 1, 89.

Franklin-Adams, J. 1936, *The Franklin-Adams Chart* (3d ed.; Royal Astronomical Society).

Greenstein, J. 1936, *Ann. Harvard Obs.* **105**, 359.

Haffner, H. and Nowak, Th., ed., 1969, *Atlas of Southern Milky Way* (Hamburg-Bergedorf: European Southern Observatory).

Heiles, Carl 1969, *Astron. Soc. of the Pac.*, Leaflet No. 482.

Karlsson, B. 1971, *Astron. and Astrophys. Supp.*, in press.

Khavtassi, J. Sh. 1955, *Bull. Abastumani Obs.*, No. 18, 29.

Khavtassi, J. Sh. 1960, *Atlas of Galactic Dark Nebulae* (Abastumani Astrophysical Observatory).

Kraus, J.D. 1966, *Radio Astronomy* (New York, McGraw-Hill) p. 456.

Lindsay, E.M. and Bok, B.J. 1936, *Ann. Harvard Obs.* **105**, 225.

Lundmark, K. 1926, *Upp. Obs. Medd.*, No. 12.

Lynds, B.T. 1962, *Ap. J. Supp.* **7**, No. 64, 1.

Lynds, B.T. 1968, *Nebulae and Interstellar Matter*, eds. B.M. Middlehurst and L.H. Aller (Chicago and London: University of Chicago Press) p. 119.

Malmquist, K.G. 1939, *Stockholm Obs. Ann.*, Band 13, No. 4.

McCuskey, S.W. 1941, *Ap. J.* **94**, 468.

McCuskey, S.W. 1956, *Ap. J.* **123**, 458.

McCuskey, S.W. and Houk, N.M. 1971, *A. J.*, in press.

Miller, E.W. 1971, *Ph.D. Thesis* (Tucson: University of Arizona).

Miller, F.D. 1936, *Ann. Harvard Obs.* **105**, 297.

Neckel, T. 1967, *Landessternwarte Heidelberg-Königstuhl Veröffentl.* **19**, 1.

Pratt, N.M. 1971, *The Royal Obs. Edinburgh Pubs.* **8**, 109.

Reddish, V.C. 1967, *M.N.* **135**, 251.

Rodgers, A.W. 1960, *M.N.* **120**, 163.

Rodgers, A.W.; Campbell, C.T.; Whiteoak, J.B.; Bailey, H.H. and Hunt, V.O. 1960, *An Atlas of H-Alpha Emission in the Southern Milky Way* (Mount Stromlo Obs.).

Ross, F.W. and Calvert, M.R. 1934, *Atlas of the Northern Milky Way,* (Chicago: University of Chicago Press).

Sancisi, R. 1971, *Astronomy and Astrophys.* **12**, No. 3, 323.

Schoenberg, E. 1964, *Veröffentl. Sternwerte Munchen* **5**, No. 21.

Schreur, J. 1970, *A. J.* **75**, 38.

Sharpless, S. 1965, *Galactic Structure,* eds. A. Blaauw and M. Schmidt (Chicago and London: University of Chicago Press) p. 141.

Sim, M.E. 1968, *The Royal Obs. Edinburgh Publ.* **6**, No. 8.

Van Rhijn, P.J. 1929, *Gronigen Publ.* **43**.

Walker, G.S. 1971, *The Royal Obs. Edinburgh Publ.* **8**, 103.

Weaver, H.F. 1949, *Ap. J.* **110**, 190.

Wray, J.D., and Westerlund, B.E. 1971, *Atlas of the Southern Milky Way,* (Dearborn Observatory, Northwestern University).

Zuckerman: Dark nebulae do not all show the same molecular lines. However, NGC2264 shows a large number of lines including those of HCN, HNC, CS, and "x-ogen" which should permit us to determine hyperfine intensity ratios, velocity dispersions, etc. Furthermore, a preliminary map of the molecular emission shows it to be more extended than the visual size of NGC2264 listed by Bok.

Bok: NGC2264 is the large nebula associated with the famous dark nebula which Max Wolf used to develope one of his original Wolf diagrams.

Crutcher: Kurt Riegel and I find OH radio emission to be definitely more extended than the visual size of NGC2264.

Solomon: Why do you refer to photographic absorptions of 10 mag, when the observations indicate minimum absorptions of approximately 5 mag? Could the actual absorption be much larger – say 100 mag?

Bok: Correct. The absorptions listed in Table 2 are lower limits.

Solomon: Absorptions as large as 100 mag are suggested by the H_2CO observations, which show a sharp decrease in intensity at the edge of dark clouds and thereby suggest a large change in photographic absorption at the cloud edges. On the other hand, CO observations show molecular regions to extend far beyond the boundaries of the dark clouds.

Stecher: What fraction of the interstellar gas is in the form of globules?

Bok: I know of no more than 200 globules within 150 pc of the sun. It is difficult to calculate the fractional mass of the ISM contained in globules principally because of uncertainties in the H_2 abundance.

Gordon: Is there such a thing as "invisible dust"? I am thinking of a tenuous distribution of very small particles which would have a large (Rayleigh) scattering cross-section in the UV, but small—and hence "invisible"— in the visual region of the spectrum. The existence of such dust could explain why apparently unprotected molecules are not photo-dissociated in the interstellar medium.

Bok: I know of no evidence for the wide-spread existence of such dust.

Field: Mike Werner estimates that up to half of the ISM may be in dust clouds of visual absorption greater than one. If the cloud temperatures are 5-10° K, self-gravitation is too large for the clouds to be stabilized by thermal pressures. Yet they must be stable because of the difficulty of replacing them if they were collapsing on a free-fall time scale of about 10^6 years. My paper mentions a number of mechanisms which would offset self-gravitation but I would now like to add rotation as another possible mechanism, as suggested by Mike Werner. It would be interesting to analyze the radio molecular observations in terms of a spinning disk model.

The Trifid Nebula, NGC 6514, in the constellation Sagittarius. A red light photograph by the 200-inch. (Hale Observatories photograph)

Chemical and Physical Properties of
Interstellar Dust*

J. Mayo Greenberg

Department of Astronomy and Space Science
State University of New York at Albany
and
Dudley Observatory, Albany, New York

*Work supported in part by grants from the Research Foundation of the State University of New York.

I. INTRODUCTION

The interstellar dust appears to play a critical role in the formation of the interstellar molecules. As is shown elsewhere in this volume the rate of formation of molecules directly out of the gas fails by orders of magnitude to compete with the rate of destruction in maintaining the observed abundances. Exactly what the mechanism of molecule production via dust may be is not clearly defined. But it seems clear that the chemical and physical character of the dust are critical in distinguishing between the relative importance of the various conceivable processes.

The distribution of the molecules under consideration is sufficiently far removed for stars that they can only have reached the regions in times much larger than their typical decay times as determined by the ultraviolet (or other disruptive) radiation.

A molecule with a radiative life limited to 100 years in a radiation field like that suggested by Habing would travel, at a speed of 10 km/sec, a distance of merely 10^{-3} pc before being destroyed. The existence of reasonably complex molecules well beyond this distance from the center of dark clouds seems to preclude the likelihood of their formation in the atmosphere of protostars.

Molecules may be formed on or in grain surfaces. The surface phenomenon has been a subject of both theoretical and experimental studies, (see Breuer, this volume). The possibility of molecules being an intrinsic part of the grain and appearing in space as the result of some grain disruption process or event is the focal point of this investigation. Both laboratory as well as theoretical investigation bearing on this process are considered.

The basic astronomical observations and their interpretation leading to the optical properties of interstellar grains and subsequently their chemical nature will be reviewed with the aim of leading to possible criteria for the alternative modes of molecule production.

II. DISTRIBUTION AND AMOUNT OF DUST

The question of what constitutes the interstellar dust depends on where and how the dust is formed. There is abundant evidence for the coincidence of dust and new stars. One possibility is that the dust as we see it appears directly as a consequence of star formation. Another possibility is that the dust is already in the gas and is merely substantially denser in the condensation regions of prestellar formation. Still another consideration is that the particles produced during the early stages of star formation are largely the condensation nuclei on which interstellar condensible atoms subsequently accrete to produce the observable (optical) manifestation of the interstellar grains. With this model we would not expect the optical properties of circumstellar dust to be the same as that of the interstellar component.

Several other theories of dust formation have been proposed. The prestellar theory of Herbig (1970) considers dust as condensing out of the gaseous material in the solar nebulae, leading to a chemical composition similar to that of the matrix material of Type 1 carbonaceous chondrites: Si, Fe, Mg, Al, *etc.* The theoretical argument of Dorschner (1967) based upon a continual grinding up of solid materials in an already developed planetary system leads to similarly constituted dust particles, that is, dust constructed largely of stony materials. If dust condenses in the atmospheres of cool stars, one would expect it to be largely silicate material.

Alternatively, the coincidence of dust and young stars may mean that dust is already present in the pre-stellar stage. The early contraction of gas and dust occurs at sufficiently low temperatures that O, C, and N can form non-volatile particles much like the dirty-ice grains of Oort and van de Hulst (1946). This condensation process may result from shock compression at the inner edges of the spiral density waves. Roberts (1969) suggests the waves accelerate the gas while it is being compressed relative to the interarm regions. The two phase model suggests the average density in the "Normal" HI region must be greater than the average of the two phases taken together so that the clouds may consist of even higher density with hydrogen number densities $N_H \approx 200$ being typical. In the vicinity of the sun the speed of the gas relative to the density pattern being about 1kpc in 10^8 years, the dust may achieve sizes of the order of 0.1μ in about 1/20 the width of the sprial arm (assumed to be $\sim$ 1 kpc). The appearance of a sharp onset of dust at the inner edges of the arms of spiral galaxies (Lynds 1970) appears to be consistent with this model. As the material moves on through the density wave, the gas and dust condense giving birth to hot bright stars which both eject and evaporate dust from their neighborhood as well as causing further secondary compressions which initiate subsequent regions of dust accretion and star birth. This sequence occurs a number of times leading to bands of stars and dust until the outer edge of the pattern is reached at which point the overall density of the gas is down to sufficiently low levels that no further generation of dust out of gas occurs before the material reaches the next density maximum. Does the dust become less dense overall as the gas decompresses or does it undergo degradation during the 10^8 years or so it takes to pass through this region or both? When we look at external spiral galaxies we do not see dust concentration between the arms — except perhaps in spurs. However, is the lack of visibility of the dust due not to its absence but because there are too few bright stars to give the necessary contrast?

Certainly the gas density between the arms is insufficient to produce grain accretion. However, if the interarm region is mostly of the low density, high temperature, 'HI' type we must consider the possibility that at the relatively high temperature ($T \sim 1000°$ - $10,000°K$) atomic bombardment may lead to a significant amount of sputtering. If P is the probability that the colliding H atom causes a molecule of average molecular weight M_{mol} to be sputtered off, the rate of decrease of radius of the grain is

$$a_{sp} = -P \frac{N_H \ V_H \ M_{mol}}{4s}$$

where s is the grain material density.

At a gas temperature of $5000°K$, using $N_H = 0.2$ and assuming $M_{mol} = 26$ m_H and $s = 1$, we find that $a_{sp} = -P(5 \times 10^{-6})$ cm per 10^6 yrs. For $P = 10^{-2}$, all grains smaller than 0.05μ would be gone in the 10^8 yrs passage time between arms.

We note that if this sputtering process is correct it would affect all grain types to at least some degree and would appear observationally through the production of smaller than average dust gains in the region prior to reaching the inner edge of spiral arms.

It is generally assumed that within the arms there is a correlation between gas and dust density given by gas/dust = 100. This working estimate is used whenever no direct information is available. However, it is by no means well-established to be universally correct, and it must always be remembered that this value is based on rather local distributions of dust — an observational necessity — and is furthermore not observationally determinable for the very high density dust clouds which are most important in the molecule problem.

III. OBSERVATIONAL CRITERIA FOR GRAIN MATERIALS

In this section we propose to give reliability estimates for various interstellar dust compositions, based on the observational evidence. The most desirable and direct evidence would be spectroscopic; *i.e.* direct identification by means of specific radiation absorption or emission uniquely characteristic of the atom, molecule, or solid as the case may be. However, we shall see that the nature of the dust seems to be such as to prevent clear-cut interpretation of this sort.

A. Extinction

The broad continuum of observational properties of the dust are the extinction (or reddening), the polarization, and the scattering of star light. The presently accepted curve for the extinction is as shown in Figure 1. The shape of the curve in the range $1 \leqslant \lambda^{-1} \leqslant 3\mu^{-1}$ has been well established for many years. The ease of matching of this part by models of interstellar particles of a wide

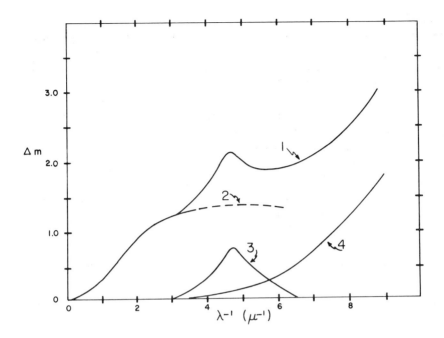

1. Representative extinction curve and schematic separation into several contributions. Curves are labeled as follows: (1) Typical OAO extinction curve, (2) Dashed extension showing contribution by classical sized particles (as discussed in text), (3) Contribution of absorption in the .22μ band, (4) Contribution by very small particles.

diversity of optical properties has been one of the frustrating features of trying to determine the fundamental chemical ingredients. With appropriate size distributions of any given particle type whether dielectric, as are dirty ice or silicates, or metallic as are graphite or iron, we find it fairly easy to reproduce the shape of the extinction in its broadest features. There are some detailed extinction features which may give clues and these will be included in the discussion of spectral characteristics.

A really perturbing problem of interpretation appears when we consider the far ultraviolet portion of the curve. The continued rise beyond $\lambda^{-1} = 5\mu^{-1}$ is not easily accounted for in terms of solid particles with physically realizable indices of refraction. This has been a persistently misunderstood situation although the theoretical basis is well-known.

In the very first attempts to account for the continued rise of the extinction beyond $\lambda^{-1} = 5\mu^{-1}$, it was seen that the particle sizes which dominate in producing the extinction at longer wavelengths were such as to produce a saturation or even drop in the extinction at the wavelengths in question. The reason for this is that the change in curvature of this extinction curve from concave to convex at about $\lambda^{-1} = 2.3\mu^{-1}$ is a dominant signature of an average particle size radius of about 0.05 to 0.2μ (see the next section) for the grain models under consideration. This means that at $\lambda^{-1} = 5\mu^{-1}$ the particles have values of $2\pi a/\lambda$ from 1.5 to 6. If the imaginary part of the complex index of the material of the particle is as it is for all real materials, the total or extinction cross section for these particles will have saturated and there would be no monotonously continued rise in the interstellar extinction. This was true for the dirty ice, the graphite, and the dirty-ice-coated graphite, because of the ice mantle, (Greenberg and Shah 1969). The most recent attempt by Gilra (1971) to match the far U-V extinction by a mixture of graphite, silicates, and silicon carbide, neglects the fact that the silicates (see Figure 2) have a rapid increase in the absorptivity beyond $\lambda^{-1} = 5\mu^{-1}$ and that actual particles of this material will

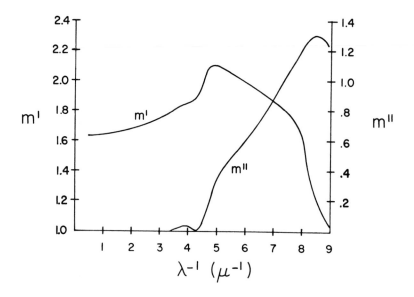

2. Complex indices of refraction for silicates from Huffman and Stapp (1970).

not produce the continued extinction rise which he calculates for the fictitious silicate material possessing constant imaginary index of only $m'' = 0.07$ throughout the visible and ultraviolet (see Figure 3). As has already been stated, this point has a very fundamental origin based on the absorptive properties of actual dielectric materials and their scattering properties. A criterion for the combination of size and wavelength dependence of the material required to

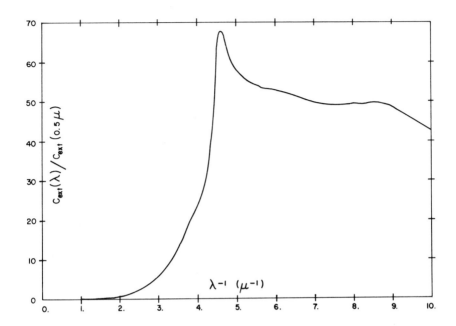

3. Wave length dependence of extinction by silicate spheres of size a = 0.05μ with complex index of refraction as given in Figure 2.

produce the continued extinction rise (with curvature upward) as far as λ^{-1} = $9\mu^{-1}$ is that if the onset of rapid increase in m'' occurs at about $\lambda^{-1} = 9\mu^{-1}$ (λ = 1100Å) the typical size, a, of the particle must be such that $2\pi a/9 < 1$. This implies a < 0.02 μ. It should be noted that the reason it is difficult to find optical windows for UV work is that even very tightly bound molecules seldom have bond breaking energies greater than 5 ev (λ = 2480Å) and most systems ionize at E < 10eV (λ = 1240Å). The above criterion is probably a too high upper bound estimate of particle size not only because we have used a very short wavelength for the absorptivity increase but also because most of the extinction curves still show no onset of a convexity at $\lambda^{-1} = 9\mu^{-1}$. Combining a more likely limiting value of x < ½ with an absorptivity rise at $\lambda^{-1} < 8$, we find the required particles to have radii less than 100Å! Particles as small as these are no longer purely classical scatterers and, to say the least, the Mie theory is no longer appropriate in this region. In any case, there is yet no simple solution to the problem of the monotonic extinction rise in the far UV. In view of the fact that it has not yet been observed except for very hot bright stars, it is perhaps possible that it is special to this situation and is not representative of normal interstellar regions. The high radiation density may be evaporating dust to form molecules which produce this absorption or may be modifying the dust material in some as yet undetermined fashion.

B. Scattering

The colors and polarization of reflection nebulae and the diffuse galactic light provide information on the scattering properties of the grains. The theoretical interpretation of the Merope nebulosity in terms of various grain models and nebular geometry appears to preclude both graphite and silicate grains when comparison is made with observational data. (Greenberg & Hanner 1970, Hanner 1971). For particles whose index of refraction in the visible portion of the spectra is like that of ice, the difference between the observed and theoretical results is apparently small enough to be correctible by small model changes.

The studies of diffuse galactic light provide information on the angular distribution and albedo of the dust particles. We shall not discuss this problem with respect to the wavelength region longward of $\lambda^{-1} = 3\mu^{-1}$. However, it may be extremely important to note that recent OAO observations indicate that the 2200Å hump is due to a highly absorbing component and that for shorter wavelengths the albedo appears to be very close to unity (Witt and Lillie 1971). In view of the fact that "normal" solid materials would have albedos considerably less than unity, we suggest the possibility that the far ultraviolet portion of the extinction curve is produced by a component of the interstellar material whose optical properties are not obtainable from Mie theory for normal materials. In the context of the current investigation we should not restrict ourselves to the properties of materials as we are accustomed to measure them in normal laboratory conditions. The very low temperatures of the interstellar particles and the diluted radiation fields may produce conditions "frozen" into the grains which very significantly modify their optical properties.

C. Discrete Features

Let us now examine some of the more discrete spectral characteristics of the extinction curve. As far back as 14 years ago, Whitford (1958) noted that he could represent the extinction curve by two straight lines intersecting at about $\lambda^{-1} = 2.2\mu^{-1}$. This feature has since been noted by a number of other observers (Nandy 1964).

We shall show how this apparent discontinuity in the slope is almost entirely due to the change in curvature of the extinction curve and how, as a matter of fact, it gives the size information referred to in the previous section (Hayes, Greenberg, Mavko, and Rex 1971).

We start with the low index of refraction approximation for extinction by a spherical particle of radius a given by (van de Hulst 1957)

$$C_{ext} = \pi a^2 \ [2 - \frac{4}{\rho} \ \sin \rho + \frac{4}{\rho^2} \ (1 - \cos \rho)] \tag{1}$$

where $\rho = 4\pi a \lambda^{-1} (m - 1)$, m = real. For purposes of carrying the calculations through analytically we consider a Gaussian size distribution of particles given by

$$n(a) = A \ e^{-a^2} a^2 \tag{2}$$

where the typical size is given by $\langle a^2 \rangle^{1/2} = \alpha^{-1} \ 2^{-1/2}$. The total extinction is given by

$$\frac{\Delta m}{1.086} \ = \ \int_0^\infty n(a) \ C_{ext} \ (a) \ da \tag{3}$$

which finally becomes (Greenberg 1968)

$$\frac{3a\Delta m}{2.172 \ \pi^{3/2} A} \ = 1/4 = 1/2 e^{-\zeta^2/4} + \zeta^{-2} \ (1 - e^{-\zeta^2/4}) \tag{4}$$

where we define

$$\zeta \ = \ \frac{4\pi \ (m-1)}{\lambda \ a}$$

The change in curvature in the extinction occurs for the value of ζ for which

$$\frac{d^2}{d\zeta^2} \ \Delta m \ = 0$$

Performing the indicated differentiation and numerically solving the resulting equation gives $\zeta = 2.52$. Letting the kink in the curve be at $\lambda^{-1} = 2.2\mu^{-1}$ gives the typical size as

$$\overline{a} \ = \ \frac{0.055}{m - 1} \ \mu \tag{5}$$

The index of refraction of dirty ice being about m = 4/3 gives a =
.165μ. For silicates m = 1 2/3 giving a = .082μ. In Figure 4, we have plotted the

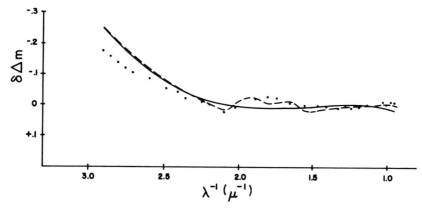

4. Comparison of observations (Whiteoak 1966) with theoretical calculations of extinction.
The dashed curve is for infinite dielectric cylinders with a distribution of sizes and m =
1.33-0.05 i (except for the region of the broad structure). The solid curve is obtained
from equation (4) with a value of ζ chosen to give a best fit.

values of the appropriate analytical expression and compared the results with
exact calculation for aligned infinite cylinders and with observational results.

Although this gives the curvature change there is still no slope discontinuity
because analytical expressions of the sort given in Eq. (4) are continuous in all
derivatives. However, the observational results shown in Figure 4 as a difference
between a λ^{-1} curve and the observed extinction also does not show two straight
lines but rather something very similar to the theoretical calculations (here for
constant index infinite cylinders) with a small modification in the particle
absorptivity, longward of λ^{-1} = 2.2μ^{-1}. It is by running an average straight line
through the structure that Nandy arrives at a slope discontinuity point at λ^{-1} =
2.2μ^{-1}. The actual change of curvature would occur (with constant index of
refraction) at a wavelength of λ^{-1} = 1.8μ^{-1}. The dirty ice "a" for this wavelength
is a = 0.2μ. The broad shallow feature centered on λ^{-1} = 1.8μ^{-1} was noted by
Whiteoak (1966). It is proposed that the combination of this with the curvature
shift point produces the apparent slope discontinuity. Finally we note, in
passing, that the change in the slope of the index of refraction of graphite
(Wickramasinghe and Guillaume 1965) occurs far to the blue of not only the
spurious structure but even farther from the real structure of the extinction
curve and therefore can not be the cause of the structure.

The next broad feature is that at λ^{-1} ≈ 4.7μ^{-1} (λ ≈ .22μ). Judging by the fact that this has been attributed with equal reliability to silicates (Huffman and Stapp 1971) and graphite (Gilra 1971) plus the fact that the dirty ice and similar materials have an absorption edge at approximately the same region we do not believe that there is yet sufficient reliability to this particular feature in uniquely determining the chemical composition of the dust. Furthermore, as will be considered next in greater detail, the fact that in the far UV even small particles become comparable in size to the wavelength makes it rather difficult to get a clearly defined correlation between extinction cross section and material absorptivity.

The unidentified diffuse bands have such a high degree of correlation with the dust extinction that it is generally assumed that they are produced by the dust itself. Consequently, it would appear that their identification should prove important in establishing the chemical nature of the dust. The identifications suggested to date have varied from imbedded Ca atoms for 4430 alone (Greenberg 1968) to simple compounds like iron oxides (Huffman and Stapp 1971) to very complicated molecules (porphyrines, Johnson 1970). There are objections raised to each of these and no agreement exists. However, the question of whether the bands are indeed produced within the dust has been reinvestigated recently with a view to establishing a strong criterion for this mechanism. The λ 4430 band shows up in extinction as a very asymmetric shape (Bruck, Nandy, and Seddon 1969). Theoretical calculations of the polarization produced by absorbing material in a dust grain show that the same type of asymmetry should occur there also (Greenberg and Stoeckly 1970, Greenberg and Stoeckly 1971). As a matter of fact the asymmetry in the polarization is a better defined observational quantity than the asymmetry in extinction because it is not perturbed by detailed spectral characteristics of the star's radiation unless the star exhibits intrinsic polarization – which is the exceptional case and is generally well identifiable. It seems that the polarization asymmetry exists (Walker 1970, Nandy and Seddon 1970) but further and more precise observations are needed to make a decisive statement. The reason for the expected asymmetry of the absorption and polarization is that the addition of a solid solution material with a symmetric absorption dispersed in a dielectric particle whose size is comparable with the wavelength produces an additional extinction which is more like a dispersion curve, *i.e.* it produces a decrease in extinction at wavelengths shorter than the band center and an increased extinction at longer wavelengths (van de Hulst 1949, Greenberg 1968). Importantly, it is to be noted that the shape of the extinction and polarization within the band is a function not only of the particle size but also of the material refractive index. It turns out that a metallic particle like graphite would not give rise to asymmetric band shapes and the character of the asymmetry for dielectric particles like silicates is distinctly different from that for ices. Thus the existence of asymmetries not only helps to establish that the diffuse bands are produced intrinsically by grains but also the nature of the asymmetry will assist in defining acceptable values of the index of refraction of the grain materials.

The only spectral region which is theoretically suitable for producing grain material identification lies in the infrared. The 3μ band of ice was the first spectral feature of a possible material constituent of grains to be investigated. The band was not seen and the very low allowable upper limit to the amount of ice in grains seemed to make the dirty ice model (the interstellar accretion model) unacceptable. With regard to the negative result obtained for μ Cep (Danielson, Woolf, Gausted 1968), a possible source of misinterpretation may be found in the use of a reference continuum extinction curve of theoretical origin rather than an observational one specifically for μ Cep. In figure 5 are shown three continuum extinction curves one of which, due to H. L. Johnson, applies to μ Cep. The others are based on a theoretical and an average observational extinction curve. In Figure 6 we show Johnson's curve added to those of Figure 4 in the reference paper (Danielson et al.). It appears that this curve fits the observational points very well in that it very nearly corresponds to absorption bands. However, in the neighborhood of the ice band at about 3.0μ a somewhat anomalous effect is seen in that the observed points exceed the continuum on the short wavelength side. This dispersion phenomena if real and taken in combination with the anomalous shape of the Johnson extinction curve in the far infrared leads one to consider that some of the extinction may be due to additional particles much larger (effectively in "a" or in m or both) than the ones producing the theoretical continuum extinction as used. The shape of the ice extinction band produced by these large (or high index) particles would be distorted very similarly to the form indicated by the difference between the curves labeled H.L.J. and D.W.G. Of course, another possible explanation for the negative result for the ice band is that μ Cep is reddened by a circumstellar dust cloud whose particles are at too high a temperature to contain ice.

The negative 3μ ice result for such highly reddened stars as CIT 11 and VI Cygnus No. 12 may possibly be explained away by effects of ultraviolet radiation in modifying the chemical composition of the dirty ice mixtures. If the "ice" (H_2O, NH_3, CH_4) has been photolyzed to be either in the form of free radicals, or replaced by different and possibly more complex molecules of similar composition, then the OH-stretching band is not significantly shifted from its gas phase (H_2O vapor) position and therefore is to be found in the same spectral region as that of atmospheric water vapor (at about 2.8μ) and is consequently totally obscured in ground-based observations. Circumstellar or near stellar dust is likely to be completely photolyzed.

The only possible spectral identification of a particular interstellar dust component has been given by 10μ emission around stars with an infrared excess

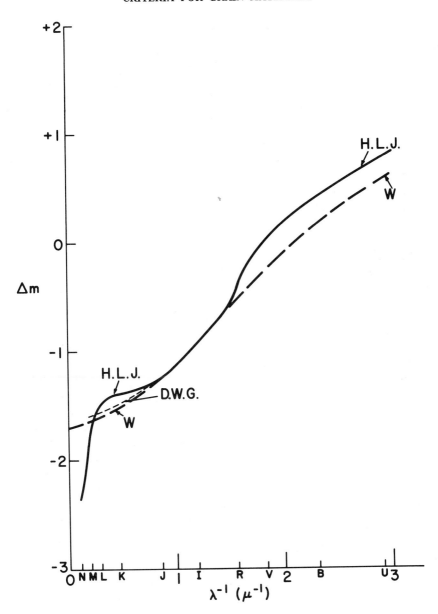

5. Extinction curves used for μ Cep. infrared data. The curve labeled W is a Whitford curve. The curve labeled D.W.G. is a theoretical one used by Danielson, Woolf, and Gaustad. The curve labeled H.L.J. is the one determined observationally by H.L. Johnson.

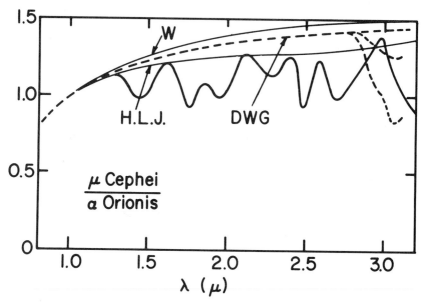

6. Comparison between the observed ratio of emissivities of μ Cep to α Ori and those obtained from the extinction curves as given in Figure 5.

(Knacke *et al.* 1969). Silicate materials such as olivene have an absorption band (or bands) in this region and if dust grains are sufficiently close to the star they are heated to high enough temperatures so that their emissivity at 10μ becomes appreciable. While there could still be a substantial question whether this silicate material is the major contributor to the optical manifestations of the interstellar grains far from stars, it seems likely that they contribute at least in the form of condensation nuclei in the interstellar medium (Greenberg, 1970). As we have seen in the previous section, the distribution of color and polarization of the Merope nebula do not appear to be consistent with a silicate grain model.

Before passing on to other optical features of the dust such as polarization, it is worth noting that all of the theoretical dust modeling has used scattering theories for smooth particles. It is not clear why the interstellar grains should be smooth and, in fact, there are numerous reasons to expect that the surface of the dust could be sufficiently rough that the smooth particle prediction may be misleading at least in the ultraviolet where the wavelength of the radiation is short enough to begin to "see" surface perturbations. There is no simple theoretical means of predicting such effects. However, microwave analog studies show that it is a point of concern both with regard to the detailed structure of the extinction curve as produced by specific absorptivity effects as well as the question of the general shape of the extinction curve in the far UV (Greenberg *et al.* 1971). It is not to be excluded that small perturbations on larger particles act partially in the direction of adding small particles to the distribution of particle sizes although this is hardly more than a conjecture at this time. Considerably more experimental evidence is needed.

D. Polarization

The amount and wavelength dependence of polarization has been studied for several of the grain models. Broadly speaking, the observations are reasonably well matched although there may be yet some problem in accounting for the fact that the observed decrease of polarization in the infrared is more rapid than produced by the theoretical grain models (as in Greenberg 1968). In view of the fact that the variability of the degree of grain alignment with size has not been included in the theoretical calculations, the problem may not be insurmountable but rather may lead to an observational criterion for helping to determine the alignment mechanism which produces the appropriate size dependence of alignment.

The magnetic alignment mechanism still appears to be the most likely one for orienting grains sufficiently to produce the maximum observed ratio of interstellar polarization to extinction. A required magnetic field of 0.8×10^{-5} Gauss was obtained by Greenberg (1969). The more detailed numerical work of Purcell and Spitzer (1971) employing the Monte Carlo method implies the need for a magnetic field approximately 1.8 times higher (B = 1.4×10^{-5} Gauss).

This factor of 1.8 (relative to that of Greenberg's result for B) arises as a consequence of the spin axis of the grain being incompletely aligned orthogonally (for elongated particles) with respect to the angular momentum of the grain. If one ignores the radiation emitted and absorbed by the grain and considers only the collisional and magnetic term, this is correct. However, if one considers, say an elongated particle whose emission (or absorption) cross section is greater for radiation directed normal to the grain axis than for that directed along the grain axis, we see that, as a result of the interaction to be discussed later (Harwit 1970), there is an additional radiation "torque" tending to make the grain spin about its short axis. The exact incorporation of this effect in the theory has not been carried out.

Although the observational evidence for galactic magnetic fields does not exclude the value of B = 1.5×10^{-5} gauss existing over sufficiently extended regions to supply the grain orientation needed, this field strength is at the limits of being physically acceptable for dynamical and energy reasons. One way to avoid this problem is to invoke superparamagnetic properties for the grains (Jones & Spitzer 1967). However, because of the uncertainty of this property prevailing for a substantial fraction of the grains, other mechanisms are still being investigated with a view (Salpeter and Wickramasinghe 1969, Harwit 1970) to determine their significance either alone or in combination with the ordinary paramagnetic relaxation mechanism which is certainly substantial. The question still remaining is whether it is dominant.

It is important to note that both the starlight radiation mechanism of Harwit and the streaming mechanisms (proposed by Gold 1952) produce a type of grain orientation distribution which is quite different from the magnetically based one. In the latter case an elongated particle tends to spin with its short axis (and angular momentum) along the orienting field (magnetic field) direction. For the streaming case, an elongated particle would tend to spin with its angular momentum perpendicular to the direction of the orienting (velocity) field, *i.e.*, a planar rather than an axial type of alignment. With the assumption of a planar anisotropy of the external (galactic) radiation field the Harwit mechanism also produces a planar alignment of the spin axes. For these cases, the parameter ξ as defined in Greenberg (1968, 1969) is greater than unity and the Rayleigh reduction factor formula becomes

$$R(\xi) = -\frac{3}{2(\xi^2 - 1)} - 1/2 - 3/2 \frac{\xi}{(\xi^2 - 1)^{3/2}} \left[\ln \; \xi^2 \sqrt{\xi^2 - 1} \right] \quad (6)$$

The quanity ξ is approximately given by (Greenberg 1968, 1969)

$$\xi^2 = \frac{T_o}{T}$$

where T_o is an "effective" orientation spinning temperature associated with the plane (or axis) of orientation and T is the rotation temperature along the axis (or plane) of disorientation (random effects only).

If we have several mechanisms simultaneously acting to orient a grain we may express the effective orientation temperature as an average of the sort (Greenberg 1970)

$$T_o = \frac{T_m/\tau_m + T_v/\tau_v + T_g/\tau_g + \ldots\ldots}{\dfrac{1}{\tau_m} + \dfrac{1}{\tau_v} + \dfrac{1}{\tau_g} + \ldots\ldots} \quad (7)$$

where as an example we have shown the case of a combination of magnetic (m) and streaming (v) orientation with respective equilibrium temperatures T_m and

T_v and respective relaxation times τ_m and τ_v. The quantity τ_g is the relaxation time for the disorientation by collision in a gas at temperature T_g. A detailed calculation by Derringh (1971) for the case of orientation by high speed gas particles streaming past a sphere gives finally $\xi^2 = (T_d + T_v)/T_d \approx T_v/T_d$, where T_d is the internal temperature of the dust grain and T_v is the streaming "temperature" defined by $V = (8 k T_v/\pi m)^{1/2}$ where m is the mass of the gas molecule.

This result differs from that in Greenberg (1970) in the substitution of the dust temperature T_d for the gas temperature T. However, the basis for the latter was that the collision process is rough elastic while for the former it is assumed to be by sticking and evaporation. In any case, if $T_v \gg T_d$ or T, we find $\xi^2 \gg 1$ and the Rayleigh reduction factor become R = -½ (which is equivalent to R = ½ for axial orientation but for an orthogonal direction of the polarization). When two clouds collide, a relative velocity of 10 km/sec implies a $T_v \approx 10,000°$ which is clearly much larger than $T_d \approx 10°$ and thus should produce substantial orientation of the grains being swept by one cloud through the other. Of course, the grains are slowed down within a short distance and the mechanism is effective only within a thin boundary whose thickness is of the order of t = $a/N_H \times 10^{24}$ cm where a = the particle radius and N_H is the number density of the hydrogen atoms. For a typical grain size of 10^{-5} cm and $N_H = 10$ cm^{-3} we find the region of effective orientation to be within a thickness of the order of 1/30 pc which would include a rather small fraction of the total number of grains.

The mechanism proposed by Harwit in which the non-isotropic character of the galactic radiation field is used to supply the alignment may be similarly considered.

We note first of all that the fundamental basis of the mechanism is that every photon absorbed or emitted by the grains carries with it an angular momentum of ħ regardless of the frequency of the radiation.

Consider the plane of the dominant external radiation field to be the x-y plane. Then the angular momenta imported to the grain by this radiation would be only in this plane. We define T_r as the temperature corresponding to the equilibrium angular momentum distribution which the grain would achieve if it were in a uniform bath of the radiation and it achieved thermodynamic equilibrium between internal energy and spinning energy. The rate at which the grain achieves this spin is defined by a relaxation times. τ_r. In the actual case, the grain is at another temperature $T_d \ll T_r$.

The radiation of the grain is random in direction (we consider a spherical grain), therefore the spinning corresponding to T_d is isotropic. Thus in the x-y plane, we arrive at an effective temperature.

$$T_{x,y} = \frac{T_r/\tau_r + T_d/\tau_d + T_g/\tau_g}{\dfrac{1}{\tau_r} + \dfrac{1}{\tau_d} + \dfrac{1}{\tau_g}} \tag{8}$$

Along the axis we have then

$$T_z = \frac{T_d/\tau_d + T_g/\tau_g}{\dfrac{1}{\tau_d} + \dfrac{1}{\tau_g}} \tag{9}$$

The orientation parameter ξ^2 is given by $\xi^2 = T_{x,y}/T_z$. We see that it is only if T_r/τ_r is much smaller than either T_g/τ_g or T_d/τ_d that we can achieve values of ξ^2 much greater than unity which, according to equation (6) and the results of Table III in Greenberg (1969) would appear to be needed to give the required ratio of polarization to extinction. For each photon absorbed by a spherical grain the change in the square of the angular momentum along the photon proprogation direction is $(\delta J_k)^2 = 2I\omega_k \hbar + \hbar^2$. Therefore, the average rate of change of $J_k{}^2$ is given by

$$\langle\!\langle (\Delta J_k)^2 \rangle\!\rangle = 1/3\, N_{ab}\hbar^2 \tag{10}$$

where N_{ab}, the total number of photons per second absorbed by the grain, is given by

$$N_{ab} = \int_0^\infty 4\pi a^2\, Q_{ab}\,(\nu)\, n_\nu d_\nu \tag{11}$$

Where Q_{ab} is the absorption efficiency and n_ν is the number of photons per unit area impinging on the grain from the radiation field.

If we consider a black body radiation field at temperature T we may write

$$n_\nu = 2\pi\, \nu^2 c^{-2}\, /\, (e^{h\nu/kt} - 1)$$

The normal grain temperatures are quite low so that the radiation emitted by the grain is at wave lengths much greater than the size a. Therefore we may use the Rayleigh limit for the cross section. For such a situation if we wish to establish an upper limit to the radiation effect it is convenient to use for Q_{ab} the maximum value theoretically attainable when $a/\lambda \ll 1$; namely (Greenberg and de Jong 1970)

$$(Q_{ab})_{max} = 8\pi a/\lambda$$

We find then that

$$N_{ab} = 3a^3 \ cT_d^4 \ x \ 10^3 \tag{14}$$

On the other hand, for radiation absorbed from the normal high temperature (but diluted) radiation field we approximate Q_{ab} (ν) over the spectral region by Q_{ab} where Q_{ab} may be the the range 10^{-2} to 1 as seen from Mie theory calculations of small particle grain temperatures (see for example, Greenberg 1971). For a radiation field at temperature T_r and dilution W_r we obtain

$$N_{ab} = 19 \ x \ 10^{11} a^2 Q_{ab} T_r^3 W_r \tag{15}$$

The relaxation time for the radiation effect is defined as the time it takes for the random transfer of angular momentum to the grain to equal the angular momentum in thermodynamic equilibrium with the radiation field. We have then from equation (10),

$$1/3 \ N_{ab} \ \hbar^2 \tau = J_k^2 = IkT \tag{16}$$

Substituting the appropriate values of N_{ab} in equations (14) and (15), we obtain respectively for the relaxation times corresponding to the grain temperature and the external radiation field

$$\tau_d = 1.04s \ a^2 T_d^3 \ x \ 10^{25} \ sec \tag{17}$$

$$\tau_r = (0.49s/Q_{ab}) \ a^3 \ T_r^{-2} \ W_r^{-1} \ x \ 10^{27} sec \tag{18}$$

where s is the dust density and the subscripts "d" and "r" apply to emission of radiation from dust and absorption of radiation from the external field.

In actuality, the dust temperature T_d does not really define a black body radiation field but rather one diluted by a factor W_d. This may be obtained by knowing the actual radiation energy density in the grain. A reasonable approximation is given by letting (Greenberg 1971)

$$W_d T_d^4 = W_r T_r^4 \tag{19}$$

which gives $W_d = 10^{-2}$ for $W_r = 10^{-14}$, $T_r = 10,000°$, $T_d = 10°$.
We therefore should replace equation (17) by

$$\tau_d = 1.04 \text{ s } a^2 T_d{}^{-3} W_d{}^{-1} \times 10^{25} \text{ sec.} \qquad (20)$$

We now wish to apply the above results to equation (8). As an example, we shall let $a = 10^{-5}$ cm, $s = 1$, $T_d = 10°$, $W_d = 10^{-2}$, $Q_{ab} = 0.1$ (See Greenberg 1971), $T_r = 10,000°$, $W_r = 10^{-14}$, $T = 100$. For these conditions the relaxation times are

$$\tau_d = 1.04 \times 10^{14} \text{ sec} \qquad (21)$$
$$\tau_r = 4.9 \times 10^{18} \text{ sec}$$
$$\tau_g = 5 \times 10^{12} \text{ sec} \ll \tau_{d, } \tau_r$$

and we get

$$\frac{T_{xy}}{T_z} = \left(\frac{T_r(\tau_g/\tau_r) + T_d(\tau_g/\tau_d) + T_g}{T_d(\tau_g/\tau_d) + T_g} \right) \cdot \left(\frac{(\tau_g/\tau_d) + 1}{(\tau_g/\tau_r) + (\tau_g/\tau_d) + 1} \right) \approx 1 \qquad (22)$$

which as already noted, implies very little orientation.

IV. THE ULTRAVIOLET RADIATION FIELD

Before discussing the effects of radiation on the interstellar grains and the subsequent production of complex molecules it will be useful to decide on some representative frequency distribution as well as spatial distribution of the radiation, particularly in the ultraviolet. The hitherto "standard" representation of the interstellar radiation field was that of a black body distribution at 10,000° diluted by a factor of $W = 10^{-14}$. For many purposes this distribution is adequate (Greenberg 1971). However for specific processes involving the far ultraviolet it is undoubtedly too crude. The most recent attempt to arrive at a more realistic ultraviolet distribution is that of Habing (1968) who considered the region between 912 Å and 2400 Å. His recommended values for the energy densities are (in ergs cm^{-3} Å-1) 40 x 10^{-18} at 1000 Å, 50 x 10^{-18} at 1400 Å, and 30 x 10^{-18} at 2200 Å. In the following discussion we will, for simplicity, further approximate this distribution by using a constant value of 40 x 10^{-18} erg cm^{-3} Å-1 throughout the given ultraviolet spectral region. It is perhaps worth noting that the total energy density in this approximate field between 900 and 2500 Å is 4 x 10^{-2} ev cm^{-3} as compared with 7.5 x 10^{-2} ev cm^{-3} for the classical Eddington form, $W = 10^{-14}$, $T = 10,000°$ black body field; in which the total energy density would be 0.75 ev cm^{-3}. The frequency distribution is however quite different in that the latter falls off very rapidly with increasing frequency whereas the former is rather flat.

Assuming the above representation for relatively clear regions in space what should be the case for interiors of dark clouds? It is usual to think of the dark clouds as being immersed in a homogeneous radiation field with the ultraviolet penetration to the interior being limited by the absorptivity of the dust in the clouds. As a first approximation we calculate the ultraviolet flux at a wave length λ as being $e^{-\tau(\lambda)}$ where $\tau(\lambda)$ is the optical depth $(\Delta m(\lambda)/1.086)$ between the outside and the point in question. We see immediately that the attenuation is determined both by the shape of the extinction curve for the dust and the geometrical shape of the cloud. A spherical cloud is the easiest to treat and we will use it for numerical purposes with the provision for nonsphericity always in mind. As mentioned earlier the extinction curve in the far ultraviolet (beyond λ ⁻ 2000Å) may possibly be due to a component of interstellar matter which scatters but does not absorb the light. In view also of the fact that the observation of this spectral region is limited to bright nearby stars and may not be representative of the extinction in dark clouds we shall consider both of the extinction curves given in Figure 1 in estimating ultraviolet fluxes in dark clouds. Using a rough linear approximation of the extinction curve out to the far UV (Greenberg 1971) leads to $\Delta m(\lambda) = 0.5 \times \Delta m_v \lambda^{-1}$ with λ in microns. On the other hand if we assume that both the 2200 Å bump and the far UV do not contribute to the dark cloud extinction we arrive at a roughly constant value of $\Delta m(\lambda) = 2\Delta m_v$ for $\lambda < 2000$ Å.

V. ULTRAVIOLET ABSORPTION BY SOLIDS AND MOLECULES

Because we are interested in the ultraviolet conversion of interstellar accreted materials we consider only the absorptivities of solid water, methane, and ammonia. In order to do the problem fully we should properly include also consideration of the absorptivities of the new materials which are generated as a result of photolysis. However, for lack of complete information on this we will neglect such subsequent modifications in the following calculations.

The molar absorption coefficients of solid and vapor water, methane, and ammonia are shown in Figure 7. One obtains the imaginary part of the refractive index from the equation

$$m'' = \frac{\lambda}{4\pi} \, \alpha D \qquad (23)$$

where α is the molar absorption coefficient of the solid and D is the number of moles/liter for the material. The value of D for solid water, is $D_w = 55.5$. The values of m'' for ice and for methane and ammonia may be seen in Greenberg (1968) and in Field, Partridge & Sobel (1968). The wavelengths at the

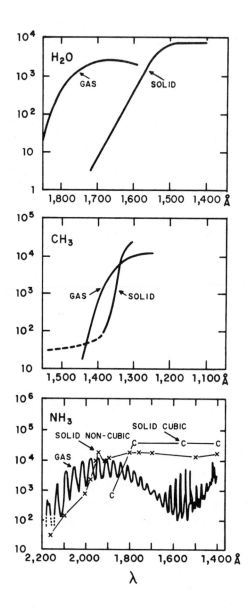

7. Molar absorption coefficients for water, methane and ammonia. From Dressler & Schnepp, 1960, J. Chem. Phys. 33, 270.

absorption edges correspond approximately to the bond-breaking energies for the molecules.

The molar absorption coefficient for the vapor may be used to find the photon absorption cross section of the molecule in the gaseous or solid state from the fact that

$$\sigma n = aD = \frac{4\pi}{\lambda} \, m"$$ (24)

where n is the number of molecules per cm^3.

We note that the onset of absorption occurs at longer wavelengths for the gas than for the solid.

From Equation 24 we find representative, approximate, average, gas molecule cross-sections

$$\bar{\sigma}_{H_2O} = 2 \times 10^{-18} \, cm^2 \, (\lambda \leqslant 1800 \, \text{Å})$$

$$\bar{\sigma}_{CH_4} = 4 \times 10^{-18} \, cm^2 \, (\lambda \leqslant 1350 \, \text{Å}), \bar{\sigma}_{NH} = 6 \times 10^{-18} \, cm^2 \, (\lambda \leqslant 2000 \, \text{Å})$$

VI. RADIATION EFFECTS ON DUST

The significance of radiation effects on the dust is determined by the time it takes for all the molecules in a dust grain to have absorbed a photon compared with characteristic dust lifetimes which are of the order of 10^7 years. The number of photons impinging per second on a dust grain of radius a is

$$\frac{dN_{h\nu}}{dt} = \frac{\pi a^2}{h} \int_{912}^{\lambda_t} u_\lambda d_\lambda$$ (25)

which, for the distribution we are using, gives

$$\frac{dN_{h\nu}}{dt} = 30 \, \pi a^2 \, (\lambda_t^2 - 912^2)$$ (26)

where λ_t is in angstroms.

Thus for $\lambda_t \leqslant 2000\text{Å}$ the number of photons impinging on a 0.1μ size grain is

$$\frac{dN_{h\nu}}{dt} \leqslant 33 \times 10^{-3} \ s^{-1} \qquad (27)$$

If we assume that the absorption efficiency of the grains is of the order of one, the total time for every molecule to have been subjected to a bond-breaking photon is $(dN_{h\nu}/dt/n_{mol})^{-1}$ where n_{mol} is the number of molecules in the grain. For an average molecular size $\simeq 2\text{Å}$, we find the saturation time to be of the order of 130 years. This is indeed very short compared with 10^7 years.

As a consequence of the ultraviolet irradiation of the interstellar grains a number of phenomena take place (Bass and Broida, 1960), among which are the formation and trapping of free radicals, and the formation of complex organic molecules. This latter problem has been investigated by many workers largely inspired by the initial work on the prebiogenic synthesis of organic molecules in the earth's atmosphere (see R. M. Lemon, 1970 for a bibliography). The very low temperature solid state conditions prevailing in interstellar grains have not been considered in this connection for obvious reasons. However, low temperature solids have sufficiently low diffusion rates that they are most ideally suited for the trapping of free radicals.

The variety of possible processes to be considered are: (1) Creation of free radicals by either bond-breaking or ionization (we will limit ourselves here to bond-breaking), (2) immediate recombination of the original molecule, (3) migration of the free radical, (4) recombination of different radicals, (5) combination of radicals with molecules. The latter process is like polymerization.

Schematically the system is represented by (as modified from Morawetz 1960):

Radical production $\quad m \rightarrow r \quad$ rate: p

Growth of large unsaturated molecules: $M + r \rightarrow r \quad$ rate: $k_g [\![M]\!][\![r]\!]$

Termination to saturated molecules: $r + r \rightarrow S \quad$ rate: $k_t [\![r]\!]^2$

Where $[\![M]\!]$ and $[\![r]\!]$ are the numbers of reacting molecules and free radicals. In steady state

$$\frac{d[r]}{dt} = p - k_t [r]^2 \quad = 0 \qquad (28)$$

$$[r] = (p/k_t)^{1/2}$$

The rate of consumption of "reactive" molecules is

$$\frac{d[M]}{dt} = (k_g/k_t^{1/2}) \; [M] p^{1/2} \tag{29}$$

and the average number $\overline{s}$, of "reactive" molecules in the large molecules, S, is given by

$$\overline{s} = - \frac{d[M]/dt}{d[S]/dt} = fk_g(k_t p)^{-1/2} \; [M] \tag{30}$$

where f is a fraction varying between ½ and 1.

We have already shown that $p \approx 2.5 \times 10^{-10}$. In order to achieve a reasonable concentration of free radicals we see that the recombination rate must also be of this order. This rate will depend on the molecular composition already achieved because it will be significantly less as the molecule and radical sizes increase to produce reduced diffusion rates. In some cases ESR measurements have indicated that, subsequent to irradiation, free radical concentration remained unchanged for a month (Bijl and Rose-Innes 1955) implying quite small values of k_t.

The two extreme possibilities for a grain are that either it maintains a high concentration of free radicals or it forms predominantly saturated complex molecules. It should be noted that from a purely theoretical point of view the absolute maximum of free radical concentration cannot be greater than about 10% (Jackson and Montroll 1958). Experimentally the maximum values achieved are less than 1%. However, there does not appear to be sufficient data or theory at the present to exclude the possibility of substantial free radical concentration in the grains sufficient to give rise to exotic optical properties (Greenberg *et al* 1971). In any case the formation of large organic molecules is certain to take place during irradiation. It is interesting to note that if a grain containing 1% free radical concentration is triggered to ignition of a chain chemical reaction, the total energy released is sufficient to evaporate the grain.

With this process of grain disruption by evaporation we may anticipate the presence of a wide range of complex organic molecules, many appearing originally with extremely high molecular weights. Subsequent degradation of the initially very large molecules would occur as a result of photon disruption. On this basis the molecules observed would be the resulting products of large molecule breakup rather than the growth from smaller molecules either in space or on grain surfaces. The equilibrium density of molecules is then given by equating the creation rate for grain disruption with the destruction rate by radiation. We have then equality between

$$\left(\frac{d\,N_{mol}}{dt}\right)_{creation} = n_{mol}N_d/\tau_d$$

and   (31)

$$\left(\frac{d\,N_{mol}}{dt}\right)_{destruction} = N_{mol}/\tau_{mol}$$

where N_{mol} = number of molecules per dust grain, n_d = number of dust grains per cm^3, τ_d = disruption lifetime for dust and τ_{mol} = radiative lifetime of a typical molecule, and N_{mol} = number of molecules per cm^3 in space.

Using a grain size of 0.1 micron and a ratio of dust to hydrogen density of 10^{-2} we get

$$\tau_d = 4 \times 10^{-3}\, \tau_{mol}N_H/N_{mol} \qquad (32)$$

In dense clouds, where molecules are found, τ_{mol} may be of the order of 10^4 years or greater and $N_H/N_{mol} \approx 10^{-8}$. For these values τ_d is of the order of 4×10^9 years which is extremely large compared with the formation and growth lifetimes of dust, and would not deplete the dust density significantly.

Photochemical experiements have been made on simulated interstellar materials. In the initial stages of our studies we have irradiated solid mixtures of water, methane (or ethane), and ammonia with non-ionizing radiation between 1400 and 2000Å wavelength and have looked at the product distribution using a mass spectrometer. The experiments are carried out by preparing various mixtures of gaseous water, methane or ethane and ammonia in "cosmic abundance" proportions of oxygen, carbon, and nitrogen, and then depositing the mixture on a cold finger in a cryostat (See Figure 8). After depositing approximately 20 mg of the mixture onto the cold finger, which results in a "dirty ice" pattern of about 25 mm in diameter, the radiation source (Cary hydrogen arc lamp fitted with a sapphire window) is turned on. The pressure in the system is maintained below 10^{-6} torr throughout the irradiation process. We have performed experiments at 28, 42, and 77°K with irradiation times varying between 4 and 6 hours. The experiments at 77°K required the use of ethane instead of methane since methane has an appreciable vapor pressure at this temperature. After irradiation is complete the cold finger is allowed to warm up to room temperature and the sample cryogenically pumped into an evacuated tube and subsequently injected into a mass spectrometer (AEI ms 902 Cl) for analysis. The mass spectra of the samples clearly indicate the presence of high

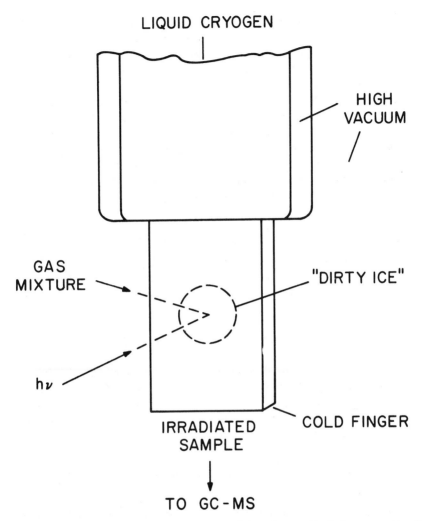

LIQUID CRYOGEN

HIGH VACUUM

GAS MIXTURE

"DIRTY ICE"

$h\nu$

IRRADIATED SAMPLE

COLD FINGER

TO GC-MS

8. Schematic diagram of apparatus used in studying photosynthesis in very cold solid mixtures of water, methane (or ethane) and ammonia.

molecular weight (mass > 106 and possibly some with mass > 200) organic material. In order to insure against organic contamination we have employed a mercury diffusion pump with liquid nitrogen cold trapping between it and the mechanical pump and the system itself. Even so, we routinely run blanks before each experiment to check for possible contamination. In general, two blanks are taken. One consists of depositing a sample onto the cold finger and then removing it without any radiation. The second blank obtained by irradiation of the cold finger without any sample and subsequent collection of any material. Both of these blank samples are run on the mass spectrometer immediately prior to the analysis of our irradiated sample and the results compared with the mass spectrum of the irradiated sample.

The results obtained thus far indicate the presence of numerous organic materials, the total amount of products being of the order of 10 micrograms (implying a photon efficiency of the order of 0.1% .)We have not been able to make positive identification of these species. The degree of complexity of the final composition of the material does not appear to depend sensitively on the temperature and the implication is that complex products will be produced at temperatures of $10°K$ or less with efficiencies which make the process astronomically important.

At present we are conducting our experiments using nitrogen-15 ammonia, deuterated methane, and deuterated water in further attempts to clarify our results. In addition, we have interfaced a gas chromatograph (Hewlett Packard 5750) with the mass spectrometer to enable us to separate our products prior to being mass analyzed. This should prove to be very helpful in our attempts to identify the irradiation products.

VII. RADIATION EFFECTS ON MOLECULES

In the last section we derived an expression for the equilibrium density of complex molecules which depends on the radiative lifetime of the molecules. We will give below some estimates of lifetimes for the water, methane, and ammonia molecules based on the absorption cross sections given at the end of section V.

The probability for molecule disruption is

$$P_D = c \int_{912}^{\lambda_t} n(\lambda) \sigma_D (\lambda) d(\lambda) \qquad (33)$$

when $n(\lambda)$ is the photon number density and λ_t is the maximum (threshold) wavelength for photodisintegration. In the terms of the energy density and, using an effective constant cross section $\langle \sigma_D \rangle$ we get

$$P_D = h^{-1} \langle \sigma_D \rangle \int_{912}^{\lambda_t} u_\lambda \lambda d\lambda$$

Substituting our approximation to the Habing radiation field $u = 40 \times 10^{-18}$ ergs cm^{-3} $Å^{-1}$ give for the radiative lifetime:

$$\tau_D = P_D^{-1} = \frac{2h}{\langle\sigma_D\rangle\ u(\lambda_t^2 - 912^2)} \qquad (34)$$

which, for $\langle\sigma_D\rangle$ in units of 10^{-16} cm², and λ_t in microns, is

$$\tau_D = 0.11\ \langle\sigma_D\rangle^{-1}\ (\lambda_t^2 - 0.0083)^{-1}\ \text{yrs.} \qquad (35)$$

For water, methane, and ammonia we get respectively $\tau_{H_2O} = 230$ yr, $\tau_{CH_4} = 180$ yr, $\tau_{NH_3} = 84$ yr.

These values are probably not much better than a factor of two but they give a working basis for the general discussion of creation and destruction and also of the effect of dust shielding on the radiation field and consequently on lifetimes. The effective cross section of formaldehyde and carbon monoxide are, respectively, smaller and larger than the values used for water, methane, and ammonia. Furthermore CO is very stable and its threshold for photodissociation is at much shorter wavelengths ($\lambda_t = 1100$Å) while H_2CO dissociates for wavelengths longward of 2000Å.

The degree of increase in photodissociation lifetimes in dark clouds depends on the degree of ultraviolet attenuation which, in turn, depends on the shape of the cloud as well as the wavelength dependence of the absorption. We use the term absorption here because a pure scattering process which gives rise to extinction would not be effective in attenuation of the radiation penetrating within the cloud. If we assume that the interstellar absorption is as more or less classically given in the dashed curve in Figure 1, then if the molecules are in a region where the visible extinction is Δm_v, its lifetime would be increased by a factor of the order of $e^{2\Delta m_v}$. Thus for 2 magnitudes of visual extinction the radiative lifetime is increased by about 50 times, and for $\Delta m_v = 3.5$ the increase is about 1000 times. With the assumption that the continually rising extinction represents absorption as well, these factors would be increased by amounts depending on the threshold wavelength for dissociation. Very approximately, Equation (35) would be modified to give

$$\tau_D = 0.11 \langle\sigma_D\rangle^{-1} \left[e^{-0.5\lambda_t^{-1}} - e^{-0.046}(0.0083) \right]^{-1} \qquad (36)$$

VIII. DUST AROUND STARS

It is well established that many stars are intimately associated with distribution of dust. The existence of infrared excesses and intrinsic polarizations are generally interpreted in terms of circumstellar dust grains. In addition to such observations there are intensity variations of stars which indicate variations in shells of dust. Theoretical models of the condensation of dust particles in stellar atmospheres and their subsequent injection into interstellar space have been adduced to explain the total or at least partial composition of the particles which give rise to the observed interstellar extinction. Finally the residue of condensation of dust in the prestellar nebula would lead to the possibility of significant portions of this material remaining in close proximity to the young stars.

In interpreting the observational properties of dust grains which have an obvious physical relationship with a star, one is clearly led to consider the fact that the standard scattering and extinction properties do not take into account the situation in which the telescope "sees" the grains along with the star. This is due to the finiteness of the conical field of view of the telescope which implies that the dust within the field scatters the radiation back into the reception cone as well as out of it. This property applies even when the dust is not intimately associated with the star so long as it is contained within the telescope reception cone.

Because the observations in the medium and far infrared and in the ultraviolet from balloons or satellites are performed with relatively small telescopes the reception cones to be considered are considerably larger than for most ground-based observations. We note that at a stellar distance R, a telescope of diameter D includes within its field those dust grains within a radius R_c about the star given by $R_c = R\lambda/D$.

This means, for example, that at $\lambda = 2000\text{Å}$ and R = 200 pc, the field of view of a 6in. telescope extends out to about 10^4 $R_\odot$ or about 50 AU. This same value scales to a telescope of diameter 120 in. for a wavelength of 4 microns. Thus, a ballon-borne telescope of 12 in. aperture observing at 8 microns would subtend about 1000 AU—a quite extended region. Using a temperature of the dust which radiates efficiently at 10μ to be $T_d \approx 300°$ we see that such a region extends to about $R_d = RT/600$. The values of R_d for an M or B supergiant are 50 AU and 140 AU, respectively. The latter value is out of the range 10^{13} - 10^{14} cm obtained by Gillett and Stein (1971) for the B_e stars they investigated.

The procedure for calculating the extinction of starlight by dust particles must be modified when there is a substantial number of the dust particles which are within a radius about the star subtended by the reception cone of the telescope. The circumstellar particles act like a reflection nebula and a contribution of the received light is in the form of scattered radiation. Consequently the extinction efficiency of these dust particles approaches their absorption efficiency. The wavelength dependences of these two processes (extinction and absorption) are generally quite different. Several grain types and dust distributions have been considered by Greenberg and Wang (1972). A very important effect occurs in spectral regions where the particles are nonabsorbing because there the extinction cross section is entirely due to scattering. Thus particles whose imaginary index of refraction is small in the visual ("clean" ice or silicates) may make a negligible contribution to infrared absorption or radiation. Polarization by truly circumstellar dust distributions give small polarizations. Stars with very high intrinsic polarizations may be shown in some circumstances to appear to be considerably overluminous by as much or more than 1.5 magnitudes. Absorption bands are amplified relative to extinction by circumstellar dust.

REFERENCES

Bass, A. M., and Broida, H. P. 1960 Editors, *Formation and Trapping of Free Radicals*, (New York. Academic Press).

Bijl, D. and Rose-Innes, A. C. 1955, *Nature* **175**, 82.

Breuer, H. D. 1972, (this volume).

Bruck, M. T.; Nandy, K. and Seddon H. 1969, *Physica* **41**, 128.

Danielson, R. E.; Woolf, N. J. and Gaustad, J. E. 1965, *Ap. J.* **141**, 116.

Derringh, E. 1971, (private communication).

Dorschner, J. 1967 *Astr. Nachr.* **290**, 171.

Field, G. B.; Partridge, R. B. and Sobel, H. 1967, in *Proc. IAU Colloquium in Interstellar Grains*, ed. J. M. Greenberg and T. P. Roark, NASA, **AP-140**, 207

FitzGerald, M. 1968, *Astr. J.* **73**, 983.

Gillett, F. D. and Stein, N. A. 1971, *Ap. J.* **164**, 77.

Gilra, D. P., 1971 *Nature* **229**, 237.

Gold, T. 1952 *M. N. R. A. S.* **112**, 215.

Greenberg, J. M. 1968, in *Nebulae and Interstellar Matter*, ed. B. M. Middlehurst and L. H. Aller, (Chicago: University of Chicago Press), 221.

————1969, *Physica* **41**, 67.

————1970, in *Space Research X,* Amsterdam, North Holland Publishing Co., 225.

————1970, *Proc. IAU Symposium No. 39, Interstellar Gass Dynamics* ed. H. J. Habing, (Dordrecht: Reidel), 306.

————1971, *Astr. & Ap.* **12**, 240.

————and de Jong, T. 1969, *Nature* **224**, 251.

————and Hanner, M. S. 1970, *Ap. J.* **161**, 947.

————and Shah, G. A. 1969, *Physica* **41**, 92.

————and Stoeckly, R. 1970, *IAU Symposium No. 36, Ultraviolet Stellar Spectra and Related Ground-Based Observations,* ed. L. Houziaux and H. E. Butler, (Dordrecht: Reidel), 306.

————and Stoeckly, R. 1971, *Nature Phys. Sci.* **230**, 15.

————and Wang, R. T. 1972, *Mem. Soc. Roy. des Sci. de Liege,* 6e serie, **3**, 193.

————Wang, R. T., and Bangs, L. 1971, *Nature Phys. Sci.* **230**, 110.

————Yencha, A. J.; Corbett, J. W. and Frisch, H. L. 1972, *Mem. Soc. Roy. des Sci. de Liege,* 6e serie, **3**,

Habing, H. J. 1968, *B. A. N.* **19**, 421.

Hanner, M. S. 1971, *Ap. J.* **164**, 425.

Harwit, M. 1970, *Nature* **226**, 61.

Hayes. D. S.; Greenberg, J. M.; Mavko, G. E. & Rex, K. H. 1971, *BAAS* **3**, No. 3, Part 1, 389.

Herbig, G. H. 1970, *Mem. Soc. Roy. Sci., Lieg* **19**, 16.

Huffman, D. R. and Stapp, J. L. 1971, *Nature Phys. Sci.* **229**, 45.

van de Hulst, H. C. 1949, *Recherches Astronomiques de l' Observatoire d' Utrecht,* XI, Part 2.

————1957, *Light Scattering by Small Particles,* (New York: J. Wiley & Sons).

Jackson, J. L. and Montroll, E. W. 1958, *J. Chem. Phys.* **28**, 1101.

Johnson, F. M. 1970, *Bull. Am. Astr. Soc.,* **2**, 323.

Jones, R. V. and Spitzer, Jr., L. 1967, *Ap. J.* **147**, 943.

Knacke, R. F.; Gaustad, J. E.; Gillett, F. C. and Stein, W. A. 1969, *Ap. J.* **155**, L 189.

Lemon, R. M. 1970, *Chem. Rev.* **70**, 95.

Lynds, B. T. 1970, *IAU Symposium No. 38, The Spiral Structure of Our Galaxy,* ed. W. Becker and G. Contopoulos, (Dordrecht: Reidel)

Morawetz, H. 1960, in *Formation and Trapping of Free Radicals,* ed. A. M. Bass and H. P. Broida, (New York: Academic Press), 363.

Nandy, K. 1964, *Pub. Roy. Obs. Edinburgh* **4**, 57.

————and Seddon, H. 1970, *Nature* **227**, 264.

Oort, J. H. and van de Hulst, H. C. 1946, *B. A. N.* **10**, No. 376, 187.

Purcell, E. M. and Spitzer, Jr., L. 1971, *Ap. J.* **167**, 3.

Roberts, W. W. 1969, *Ap. J.* **158**, 123.

Salpeter, E. E. and Wickramasinghe, N. C. 1969, *Nature* **222**, 442.

Walker, G. A. H. 1970, (private communication).

Whiteoak, J. B. 1966, *Ap. J.* **144**, 305.

Whitford, A. E. 1958, *Astr. J.* **63**, 201.

Wickramasinghe, N. C. and Guillaume, C. 1965, *Nature* **207**, 366.

Witt, A. N. and Lillie, C. F. 1971, Paper presented at the OAO Symposium, Amherst, Mass.

Molecules in Stars

George Wallerstein

Astronomy Department
University of Washington
Seattle, Washington

I. INTRODUCTION

Molecular lines and bands are found in stars of types F, G, K, M, R, N, and S; i.e. in all except the O, B, and A stars. Hence, a review of molecules in stars should cover more than half of stellar astronomy; in fact one might say it covers the more complicated half, since the cooler stars are surely more difficult to analyze than their hotter brothers. We will therefore restrict the discussion to two subjects within the field of molecular stellar astronomy. The first is the chemical analysis of cool stellar atmospheres. The second will be the role of molecules in the extended envelopes of very cool stars. Even in these fields there is insufficient time to be complete so I will cite some particular examples rather than try to refer to everything that has been done in the field.

II. THE COMPOSITION OF COOL STARS

Molecular features were first seen in stellar spectra about a century ago when Angelo Secchi, S.J., divided stellar spectra into four types. His Type II included the G and K stars which show the bands of CH and CN. Spectra of Type III are now known as the M stars characterized then as now by the bands of TiO, while his Type IV are now the carbon stars with their very strong C_2, CH, and CN bands.

Only within the past 10 years have quantitative analysis of stars with molecular bands been possible. Not only have stellar atmosphere methods become available during the last decade but data on dissociation potentials as well as vibrational and electronic levels have accumulated only recently. In addition, extensive tables of individual rotational lines for astrophysically important molecules are becoming available. Without such tables it is not possible to identify the features seen in stellar spectra.

In analyzing cool stars, especially those with extensive molecular features, one-layer atmospheres are inadequate. The large dissociation potentials of some molecules, especially CO, cause the molecular concentrations to depend sensitively upon temperature and hence upon optical depth. Most model atmospheres consist of some 20 or 30 layers between optical depth 10^{-3} and 10. For many purposes models with only 4 or 5 layers would do the trick and could be constructed at much lower cost. My reason for suggesting this is the huge variety of chemical compositions that we find in cool stars since the chemical composition itself has major effects on the atmospheric structure. This comes about through the opacity which depends upon composition, the blanketing by lines and bands, and the dependence of the specific heat upon dissociation and ionization of various species. Thus we must be able to construct models whose starting parameters are not only the usual effective temperature and gravity but also critical abundance ratios such as He/H, C/H, N/H, O/H, and metals/H. These are not the only parameters, unfortunately. The micro-turbulent velocity effects the line blanketing when the lines are saturated, as many of the atomic and molecular lines usually are. Finally even the ratio of C^{12}/C^{13} effects the line blanketing in carbon stars because a high C^{13} abundance results in bands of $C^{12}C^{13}$, $C^{13}C^{13}$, and $C^{13}N$, which add to the blanketing.

All of the parameters listed above are known to vary over several orders of magnitude as shown in Table 1, which lists abundances of various interesting stars. HD 201626 is a metal-poor, carbon rich star. RY Sgr is hydrogen poor and carbon rich. RU Cam, a former variable of 22 day period, has a high abundance of carbon and C^{13}. HD 30353 is too hot to show molecular features but is very different from the other stars because of its low hydrogen content and very high nitrogen abundance. If it were sufficiently cool it might have quite a variety of nitrogen compounds in its atmosphere. Could we even recognize such a star for what it is if the effective temperature were $3000°K$ rather than $10,000°K$? The last star is also hot but it is amusing to contemplate the strength of the CN bands in a cool star with such a low hydrogen content and such a high abundance of carbon and nitrogen. The interpretation of these abundances is discussed elsewhere (Wallerstein 1968).

Returning now to model atmosphere analysis of stars with molecular spectra, we would like to mention a number of recent examples of this type of study.

The sun has been treated by Schadee (1964), who has also applied his method to G and K giants with various ratios of carbon to oxygen (Schadee 1968). A similar study of four K giants has been published by Greene (1969) who found significant differences in the nitrogen abundance amongst otherwise "normal" stars. A good example of using first order abundances obtained from a curve of growth is the grid of models for R CrB by Myerseough (1968) based upon the abundances of Searle (1961). It would be desirable now to run Searle's equivalent widths through the new atmospheric model to derive more accurate abundances.

III. OH & H_2O IN RED GIANTS

After the discovery of OH emission in interstellar gas, stellar astronomy was jolted by the discovery of very strong OH in the infra-red star NML Cygnus. (Wilson and Barrett, 1968). Subsequently, at least a dozen stars have been found to emit in the 18 cm lines of OH (Wilson and Barrett, 1971).

The following year H_2O emission at 1.35 cm was found in a number of the same stars (Knowles, et. al., 1969). At least eight stars are now known to show H_2O emission (Schwartz and Barrett, 1971).

All of the stars showing OH/H_2O are of late type, about M4 or later, and

MOLECULES IN STARS

TABLE 1

ABUNDANCES IN PECULIAR STARS

Star	Type	H	He	C^{12}	C^{13}	N	O	Fe
Sun	GO V	12.0	11.2	8.6	6.6	8.0	9.0	7.5
HD 201626	CH	12.0	—	8.3	<6.6	—	<7.7	6.0
RY Sgr	RCrB	8.3	11.6	9.7	<8.0	7.6	8.0	7.1
RU Cam	Carbon	12.0	—	9.0	8.2	—	<9.0	7.5
HD 30353	A pec	7.6	11.6	6.2		9.2	7.5	(7.5)*
BD10°2179	B pec	8.4	11.2	10.7		8.9	<8.2	(7.5)*

*The entry refers to other metals whose ratio to iron is assumed to be normal.

All entries are logarithms of the number of atoms.

show an excess in infra-red wavelengths, i.e. somewhere in the 2-20 micron region, above the radiation as would be predicted by the photospheric effective temperature. Stars never show the 1720 line of OH. We can divide them into two groups: Those with the 1612 MHz line the strongest (the most common situation) and those in which the main lines, 1665 and 1667 MHz dominate. Each group can be further subdivided into those stars which show H_2O emission and those which do not, but the classification of a star as not showing H_2O may be temporary, since the H_2O intensity is variable. There is some tendency for H_2O to be strongest in stars with main line OH emission and the velocities of 1665,7 are usually (but not invariably) similar to the H_2O velocities. One star, the semi-regular variable RX Boo shows H_2O emission but not OH. Stars with OH and/or H_2O emission are listed in Table 2.

TABLE 2

Stars Showing OH and/or H_2O Emission

Group 1 (1612 Dominant)

NML Cyg	(1)		WX Ser	
VYCMa	(1)	(2)	R Aql	(1)
VX Sgr	(1)	(2)	RR Aql	
NML Tau			UX Cyg	

(1) H_2O detected

(2) Main lines also present with variable intensity and polarization

Group 2 (Main Lines Dominant)

U Ori	(1)	R Cas
W Hya	(1)	R Hor
S CrB	(1)	

(1) H_2O detected

Group 3 (H_2O present but no OH)

RX Boo

In order to understand the phenomena at hand we must first try to establish a kinematic model, i.e., figure out where the emitting material is located and how it is moving. Only then can we proceed toward a physical model to predict the line intensities and polarization. We will discuss only kinematic models in this paper, since even that first step is very uncertain at this time.

The following figures will illustrate some of the systems for which both optical and microwave radial velocities are available. All velocities are in km/sec with respect to the sun.

The first star is RX Boo, which is a semi-regular variable with H_2O emission. Optical emission lines are present now and then with hydrogen and SiI the strongest. A simple spherical model can explain the data with the absorption lines from excited levels representing the velocity of the photosphere. The emission lines and strong absorption lines from the ground state appear to come from material leaving the star at about 5 km/sec hence we assign them to a

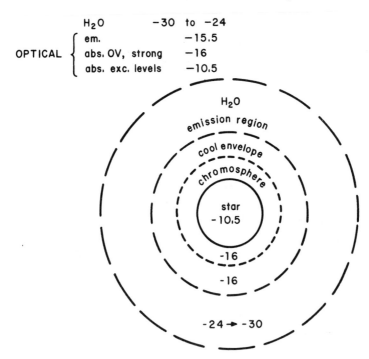

1. Radial Velocities (in km/sec with respect to the sun) and a schematic spherical model of RX Boo.

U ORIONIS (LPV, 360 days)

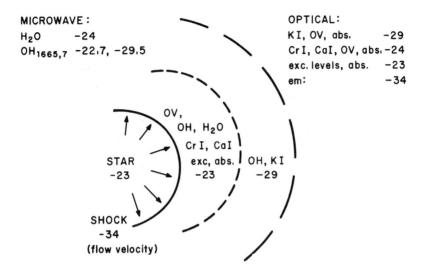

MICROWAVE :
H$_2$O -24
OH$_{1665,7}$ -22.7, -29.5

OPTICAL:
K I, OV, abs. -29
Cr I, Ca I, OV, abs.-24
exc. levels, abs. -23
em: -34

OV,
OH, H$_2$O
Cr I, Ca I
exc, abs.
-23

STAR
-23

SHOCK
-34
(flow velocity)

OH, K I
-29

2. Radial Velocities and a Schematic Spherical Model for U Orionis.

chromosphere and envelope. The H2O emission has a more negative velocity and may come from material further out and expanding more rapidly.

Figure 2 shows similar data and a proposed model for the 360 day long period variable U Ori. The emission lines at -34 km/sec must come from a sub-photospheric shock as demonstrated by Merrill many years ago. The observed velocity is the flow velocity behind the shock, not the velocity of the shock itself. In this star most of the zero-volt lines show the same velocity as the absorption lines from excited levels and presumably they all originate in the photosphere or not far from the photosphere. The H$_2$O emission and one component of OH has the same velocity. Only the zero-volt lines of potassium and the second component of OH shows an expansion with respect to the photosphere.

R Aql seems to be a more complicated system. My own velocities were obtained during rising light and I have used measurements by others during declining light. In R Aql the flow behind the shock, seen during declining light, shows an expansion of 17 km/sec. One component of OH agrees with the photospheric velocity so we tentatively suggest that some of the OH emission comes from the photospheric region itself. During rising light there are few emission lines and their velocity is somewhat different from the velocity during declining light while more observations, especially near minimum, are needed; I would like to suggest that these lines come from a chromosphere outside the

R Aql (LPV, 300 days)

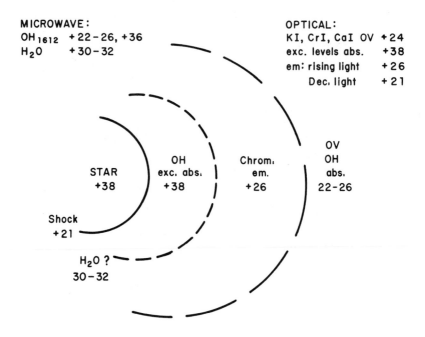

MICROWAVE:
OH$_{1612}$ +22-26, +36
H$_2$O   +30-32

OPTICAL:
KI, CrI, CaI OV +24
exc. levels abs.   +38
em: rising light   +26
Dec. light   +21

STAR
+38

OH
exc. abs.
+38

Chrom.
em.
+26

OV
OH
abs.
22-26

Shock
+21

H$_2$O ?
30-32

3. Radial Velocities and a Schematic Spherical Model for R Aql.

photosphere. The zero-volt lines of KI, CaI, and CrI show a velocity a little more negative than the chromospheric emission lines and very similar to the more negative component of OH. It seems reasonable to assign these lines to formation in a cool, extended, outer region beyond the emission region. The velocity of H$_2$O originates somewhere in the region between the photosphere and the outer envelope, at least that would preserve the continuity of velocities.

Now we turn to one of the more complicated systems, VX Sgr. Figure 4 shows the velocity data for VX Sgr plotted as a profile of the antenna temperature in the 1612 OH line. VX Sgr also show 1665 and 1667 as well as H$_2$O whose velocities are also shown. Below we show the stellar velocities. In October 1970 (i.e., during rising light) the absorption lines of the star agreed with the positive component of OH, and the zero-volt lines of CaI and KI agreed with the negative component of OH. This can be explained once again as a photosphere near +12 km/sec and an expanding shell at -20 km/sec.

A spherical model for VX Sgr is shown in Figure 5. The intermediate velocity

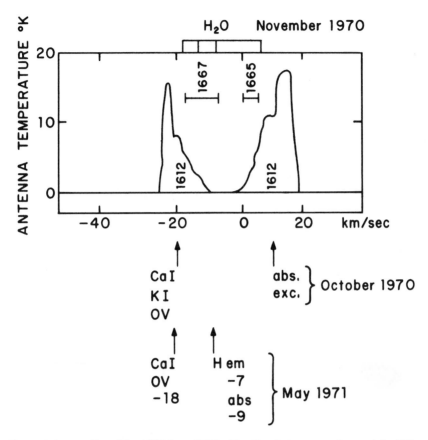

4. The emission profile of the 1612 line of OH with other features superimposed for VX Sgr.

of the 1665 and 1667 lines of OH along with H_2O is difficult to understand but is shown as emanating from a region between the two OH, 1612, regions. One possible theory for the pumping of H_2O is also depicted in Figure 5.

In May 1971 the spectrum began to look like a long period variable with emission in Hγ and Hδ. The period is very long, about 800 days, and the star is both faint and awkward to observe so it will take quite some time to understand it. Possibly it is a transition object between a long period variable and the peculiar object VY CMa.

VY CMa is the most complex system of the group under discussion. It is a cool star immersed in a reflection nebula. The system shows a very high degree of polarization in the optical region due almost entirely to the high polarization of the nebula which reaches 70% in one region (Serkowski 1971, Herbig 1971).

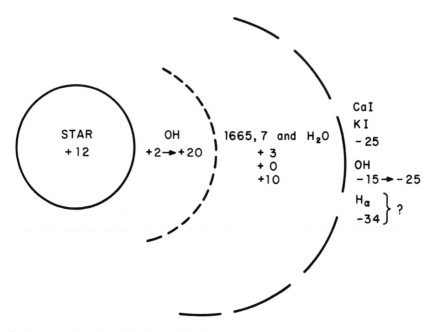

VX Sgr
MODEL (OCTOBER 1970)

5. A proposed Spherical Model for VX Sgr based upon optical observations in October 1970.

The velocities are shown in Figure 6 where the profiles of 1612 line of OH are shown and the velocities of the other features are also indicated. This star has everything. In addition to OH emission at 1612 it shows highly variable lines at 1665 and 1667 mHz and two components of H_2O whose velocity do not agree with the velocities of 1665,7 of OH. The optical spectrum shows stellar absorption lines near + 69 km/sec, i.e., very near one of the 1612 OH maxima. The optical emission spectrum is unique consisting of NaI, CaI, KI and RbI. The molecules TiO, VO, ScO and YO are also in emission. The NaI and KI lines show absorption cores that agree in velocity with the short wavelength component of 1612 OH. Hα emission is sometimes present with a velocity near that of the short wavelength 1612 OH from 1957 until December 1969 when it began to shift to the violet. By December 1970 it had shifted some 35 km/sec to the violet. Also last winter the star was fainter by about a magnitude, so some new activity may be beginning.

Two models have been proposed for this system. Hyland et. al. (1969) suggested a spherically symmetric system while Herbig (1970) has proposed a

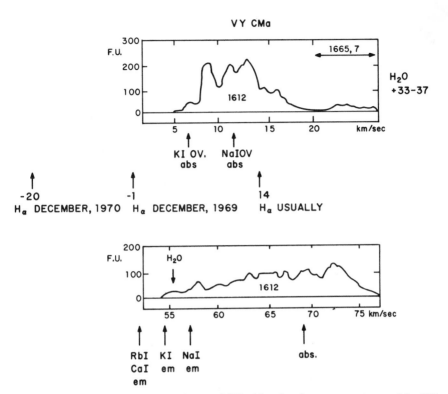

6. The emission profile of the 1612 line of OH with other features superimposed for VY CMa.

model consisting of a rotating, expanding, disk of dust and gas around a newly formed star.

The spherically symmetric model is shown in Figure 7. The structure is similar to the model for VX Sgr and need not be described in detail except to note the difficulties. These are the H_2O velocities which do not agree with any other velocities in the system and $H\alpha$ whose velocity places it in the region farthest from the star where the excitation is extremely low and hence inadequate to excite $H\alpha$ unless a temperature inversion is hypothesized. The outward acceleration of $H\alpha$ during the past two years would also indicate that the hydrogen emission is more likely to be formed closer to the star.

Herbig's disk model is shown in Figure 8. He considers the star's velocity to be half-way between the velocity of the two 1612 OH components. The OH profiles are accounted for by rotation and expansion of the nebula. The low excitation emission spectrum comes from the far side which is visible because the disk is tilted with respect to the line of sight. The H_2O emission then comes

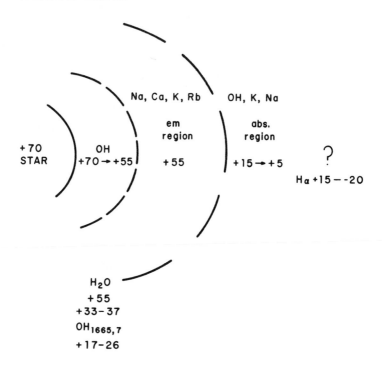

7. A Spherical Model for VY CMa as suggested by Hyland *et. al.* (1969).

from near the star and hence has a velocity between the velocities of the OH components. The velocity of Hα is still a problem.

A full analysis of these cool envelopes will allow us to understand and tie together several important astrophysical phenomena ranging from mass loss in red giants through the pumping of H_2O and OH and perhaps including the origin of interstellar dust. A real understanding of the nature and origin of interstellar dust may tie stellar and galactic astronomy together in ways that could not have been imagined a decade ago and whose character are only beginning to take shape now.

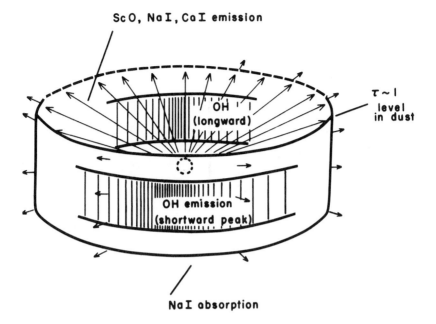

8. A rotating, expanding, disk model for VY CMa as proposed by Herbig (1970)

REFERENCES

Greene, T. F., *Ap. J.*, **157**, 737.

Herbig, G. H., 1970, *Mem. de la Soc. R. des Sci. de Liege*, **19**, 13.

Herbig, G. H., 1971 (preprint).

Hyland, A. R.; Becklin, E. E.; Neugebauer, G. and Wallerstein G. 1969, *Ap. J.*, **158**, 619.

Knowles, S. H.; Mayer, C. H.; Chueng, A. C.; Rank, D. M. and Townes, C. H. 1969, *Science*, **163**, 1055.

Myerscough, V. P. *Ap. J.*, **153**, 421.

Schadee, A. 1964, *B. A. N.*, **17**, 311.

Schadee, A. 1968, *Ap. J.*, **151**, 239.

Schwartz, P. H. and Barrett, A. H. 1971. *Proceedings of the Conference on Late-Type Stars*. Contribution No. 554 of the Kitt Peak National Observatory. P. 95.

Searle, L. *Ap. J.*, **133**, 531.

Serkowski, K. 1969. *Ap. J. Ltrs.*, **158**. L107.

Wallerstein, G. 1968. *Science*, **162**, 625.

Wilson, W. J. and Barrett, A. H. 1968, *Science*, **161**, 23.

Wilson, W. J. and Barrett, A. H. 1971, *Proceedings of the Conference on Late-Type Stars*. Contribution No. 554 of the Kitt Peak National Observatory. p. 77.

Sagan: What are the most complex polyatomic molecules found to date in cool stars? Has acetylene been looked for?

Wallerstein: Acetylene has not been identified, but other polyatomic molecules have been identified. Some of these are H_2O in M-type stars, C_3 and SiC_2 in carbon stars. Recently, Querci and Querci (*Astron. Astrophys.* 9, 1, 1970) have reported HCN and CH_2 in the carbon star UU Aur.

Woolf: The most complex molecules yet seen in stars are the silicates which show up by the double emission peaks at 10 and 20 microns. These molecules are bound in grains which might themselves be considered as molecules containing 10^7 or 10^8 atoms.

Donn: Herbig has emphasized that objects such as VY CMa are possible sources of interstellar grains. Many objects show circumstellar dust and also mass ejection which would eject material into the interstellar medium. I have always found it difficult to understand grain formation at low densities. The high densities of stellar atmospheres or regions of star formation appear more favorable.

Greenberg: VY CMa has too large an intrinsic polarization to fit the Herbig model. I have done an approximate radiative transfer calculation for a dish distribution of dust and find that the 20 per cent or so polarization in the UV is impossible to achieve. It can be achieved by putting clouds of dust to the side of the star, which would then give scattered light greater than the star light.

Sullivan: You mentioned the large shift in the radial velocity of Hα in VY CMa during 1970. It is interesting to note that the 1.35-cm H_2O emission in this star remained constant in intensity during 1969-70, but sometime during the first half of 1971, it underwent a decrease in intensity by a factor of 2 or 3. The weak H_2O feature at 55 km/s, which you mentioned, disappeared in early 1970.

Wallerstein: The decrease in H_2O emission agrees with the correlation found by Schwartz that H_2O emission is greatest at maximum light in long period variables. VY CMa was approximately a magnitude fainter in the spring of 1971 than a year earlier.

Robinson: Johnston, Robinson, Caswell, and Batchelor have recently failed at Parkes to find 22 GHz H_2O emission from R Hor, which is an OH/IR object which emits only at 1665 and 1667 MHz.

The 140-foot diameter telescope of the National Radio Astronomy Observatory, at Green Bank, West Virginia. This telescope is used at wavelengths longer than 1.35 cm for observations of line and continuum radio emission from cosmic sources. The reflector surface focusses the incoming signals to the receiver mounted at the apex of the quadrupod support legs.

Session 2

Observations of Molecules in the
Interstellar Medium

The first successful radio astronomer, Karl G. Jansky, appears with his steerable radio telescope which detected extra-terrestrial radio noise in 1933. This telescope, operating at 20.53 MHz, was designed to study the characteristics and origin of atmospheric signals interrupting early long distance communication. Jansky, who worked for the Bell Telephone Laboratories, discovered a strong source of radio noise coming from a position very near to the galactic center.

The Distribution of Interstellar Molecules

B. E. Turner

National Radio Astronomy Observatory*
Green Bank, West Virginia

*Operated by Associated Universities, Inc., under contract with the National Science Foundation.

I. INTRODUCTION

Some 20 interstellar molecules have now been detected at radio wavelengths and another 5 at optical wavelengths. The picture which has customarily been adopted is that the radio molecules reside in dense clouds (total densities > 1000 cm^{-3} for most radio molecules), and that the clouds containing the different species have roughly the same positions and velocities. It is recognized that the spatial distribution is very non-uniform on a broad scale, and also throughout individual clouds.

Actually, very little specific information exists on whether or not the different molecular species coexist, and whether the presence of molecules is correlated with the presence of extinction (grains), of enhanced gas density, or of continuum sources. The lack of adequate spatial resolution for most observations prevents a study of the fine-scale relationship for the various molecules. And the lack of adequate survey data, in a statistical sense, has prevented a definitive study of the types of astrophysical environments with which interstellar molecules are associated. For example, molecules have recently been found outside of the optically observed dark dust clouds with apparent densities about as large as occur inside the clouds; thus the conclusion that some molecules (OH and H_2CO) occur only in regions of large obscuration has been based on insufficient data. Similarly, it has been customary to associate many types of OH clouds, particularly those having anomalous emission, with continuum sources although all surveys of OH sources in the general galactic plane have been strongly selective toward the directions of strong continuum sources.

In this report we present briefly the results of surveys of OH, H_2CO, H_2O, and NH_3 in the galactic plane. The details can be found in the references.

II. THE DISTRIBUTION OF OH IN THE GALAXY

A survey of the Galaxy in the region $337° \leqslant \ell \leqslant 75° \leqslant b \leqslant +2°$ has revealed OH absorption or emission in 180 new sources comprising 424 individual clouds, in a total of 264 directions searched to a limit of 0.7 f.u. (Turner 1972). The survey is somewhat biased toward directions of continuum sources, as are the recent surveys in the 6 cm line of formaldehyde (Zuckerman *et al.* 1970; Wilson 1971; Turner and Heiles 1972). However, the number of OH and H_2CO clouds now known is large enough to warrant a statistical study of the association of OH clouds with H_2CO clouds and with continuum sources.

A. Large-scale Distribution of OH

The large-scale distribution of OH in the Galaxy can be characterized as follows. The apparent opacity $\langle \tau_{OH} \rangle$ does not appear to fall off with increasing

|b| in the interval $-2° \leqslant b \leqslant +2°$. Within the limited degree to which it is possible to convert b to linear distance z above the galactic plane, it appears that $\langle \tau_{OH} \rangle$ does decrease with increasing z and at a rate somewhat faster than that of atomic hydrogen. OH clouds with $|b| \geqslant 0°5$ are somewhat more closely correlated in velocity with HII regions than are those with $|b| < 0°5$. Several lines of evidence indicate that OH probably occurs whenever the total gas or dust density exceeds a critical value. At low values of |b| this condition is met in relatively many regions throughout spiral arms which are not necessarily associated with HII regions; however at higher |b| it appears that the condition is more likely to be met only in the vicinity of HII regions. OH would therefore appear to be increasingly less widespread as the distance from the plane increases.

As a function of galactic longitude, $\langle \tau_{OH} \rangle$ has a well-known peak extending out to roughly $\pm 5°$ on either side of the Galactic Center. $\langle \tau_{OH} \rangle$ assumes intermediate values in the intervals $337° \leqslant \ell \leqslant 355°$ and $5° \leqslant \ell \leqslant 35°$ but there is another sharp drop in $\langle \tau_{OH} \rangle$ at $\ell \approx 35°$ such that for $\ell > 35°$ the typical values of $\langle \tau_{OH} \rangle$ are much less than half those in the interval $5° \leqslant \ell \leqslant 35°$. The discontinuity at $\ell \approx 35°$ may correspond to the other edge of the Norma-Scutum arm.

The galactic distribution of anomalous emission sources is somewhat different than that of the absorption sources in that it appears generally to be random in both ℓ and b. The one possible exception is a group of 6 emission sources near ℓ = 19°, which may possibly be a manifestation of the 3-kpc arm which is seen tangentially in this direction.

B. Is OH Associated With Continuum Sources?

In the absence of either line or continuum observations of high spatial resolution, one has to answer this question statistically. For continuum sources whose angular size is greater than the observing beam size, one can map the variations in $\langle \tau_{OH} \rangle$ over the source, and it is found that $\langle \tau_{OH} \rangle$ does not correlate in general with the distribution of the continuum brightness T_c. Most of the continuum sources are smaller in size than the observing beam. For these, one can only ask whether $\langle \tau_{OH} \rangle$ is correlated with T_c in a statistical sense, when all of the 424 OH clouds are considered. The answer is that it is not. This conclusion remains valid even when one considers only those OH clouds whose velocity is within 5 km/s of the HII region velocity.

An alternative approach to the question whether OH is associated with continuum sources is to look for a correlation of $\langle \tau_{OH} \rangle$ with the difference in velocity $|V(OH)-v(109a)|$ is directly associated with the continuum source. Again one finds no correlation. Such a correlation might be expected if the strongest OH arises in the densest clouds and if the densest clouds are those most likely to form HII regions. However, the expected correlation in this case could be destroyed by two possibilities. First, the outer portions of such clouds, which form neutral shells around the HII regions and in which the OH would have to exist, might well be less dense than other OH clouds in an earlier evolutionary phase, which have not yet formed stars or HII regions. Second, some OH might be formed in shock fronts which are moving rapidly with respect to the nebulae, but at different relative velocities from nebula to nebula; this would tend to randomize the difference in OH and nebular velocities. This possibility is considered unlikely, however, because for OH seen in absorption there should, in such a case, be a larger number of negative than positive values of $v(OH)$-$v(109a)$ and this is not observed.

A third approach is to examine the number of OH clouds as a function of $|v(OH)$-$v(109a)|$. Here there is a definite correlation: more OH clouds have velocities close to HII region velocities than do not. This correlation should not be taken to indicate an actual physical association for two reasons. First, a significant fraction of the sources in this survey lie in the interval $0° \leqslant \ell \leqslant 25°$. In this region, the velocities expected for either OH clouds or HII regions on the basis of galactic rotation vary extremely slowly with distance from the Sun, for a range of distances about the tangential point that covers up to one-third the total line of sight. OH clouds or HII regions lying in this vicinity would be expected to have small differences in velocity, even though they might well not be physically associated at all. Second, any part of the correlation not caused by galactic rotation probably indicates only that OH and HII regions exist in the same general regions of spiral arms where the gas and dust densities are highest. Therefore the velocities of OH and HII regions should in general be similar, but not identical. It should be pointed out that no greater fraction of the OH clouds have velocities similar to the HII region velocities when the sample of sources is restricted to the strongest HII regions. Thus conditions for forming OH appear no more favorable near the strong HII regions than near weak ones. This conclusion is consistent with the idea that there is no close physical association of OH clouds and HII regions beyond the fact that both tend to occur in the regions of largest density in spiral arms.

C. Relation of OH with Extinction

A fundamental problem is encountered in studying the relationship of interstellar molecules with extinction, namely, that in most regions where the density of molecules is large enough to detect via radio frequency transitions, the obscuration is also large enough to render the region completely opaque at optical wavelengths so that no direct measures of the extinction can be derived. There are only a few regions that do not fall into this category, and these give conflicting results. Several HII regions of sufficiently large angular size to permit mapping of the OH absorption are also characterized by only modest extinction at optical wavelengths. For several of these (M8, M16, M17, M20, IC1805) the extinction has been determined with high spatial resolution by combining the Hα distribution with the radio continuum distribution (Ishida and Kawajiri 1968). In all of these cases the distribution of $\langle \tau_{OH} \rangle$ is not correlated with that of the extinction, at least on a size scale comparable with or larger than the observing beamwidth (18 arc min). On the other hand, OH and H_2CO have recently been observed and mapped throughout two large areas (each of size $\sim 2^{\circ}$ by 3°) in the Taurus region of the sky (Heiles and Turner 1972); in these areas the extinction is modest enough to permit a direct determination of its magnitude (Bok 1956). The distribution of both $\langle \tau_{OH} \rangle$ and $\langle \tau_{H_2CO} \rangle$ is found to correlate very highly with the distribution of extinction.

III. THE DISTRIBUTION OF FORMALDEHYDE IN THE GALAXY

Surveys similar to those in OH have been made of the galactic plane in the 6 cm line of H_2CO (Zuckerman et al. 1970; Wilson 1971; Turner and Heiles 1972). As yet, these surveys ary only about 60% as complete as the OH surveys, and have been restricted almost entirely to the directions in which OH has been found. Within these limitations, it is found that H_2CO is equally as widespread throughout the galactic plane as is OH. Attempts to detect H_2CO absorption from gas in our galaxy against high latitude, extragalactic sources have failed. Thus H_2CO seems highly concentrated toward the plane (possibly more so than OH) and also there can be no large quantity of H_2CO gas close to the sun. The main conclusions drawn above for OH can also be applied to H_2CO; namely, that a)$\langle \tau_{H_2CO} \rangle$ is not correlated with the brightness, T_c, of continuum sources; b) $\langle \tau_{H_2CO} \rangle$ is not correlated with $|v(H_2CO)\text{-}v(109a)|$; c) the number of H_2CO clouds as a function of $|v(H_2CO)\text{-}v(109a)|$ increases as $|v(H_2CO)\text{-}v(109a)|$ decreases, although not as sharply as in the OH case; d) $\langle \tau_{H_2CO} \rangle$ appears unrelated to the magnitude of extinction in the HII regions M8, M16, and M17; however $\langle \tau_{H_2CO} \rangle$ is correlated with the extinction in the large Taurus regions mentioned above.

Although neither OH nor H_2CO appear physically associated with continuum sources, we might well expect a more intimate association of OH with H_2CO throughout the galactic plane if, for example, a fairly critical range of physical conditions is needed for the formation and protection of interstellar molecules.

A. Relationship of OH and H_2CO

Three approaches indicate, however, that OH and H_2CO themselves are not closely associated over much of the Galaxy. First, a direct plot of $<\tau_{OH}>$ vs. $<\tau_{H_2CO}>$ shows no correlation, even when one includes only those OH and H_2CO clouds whose velocities differ by less than 2 km/s. Secondly, one can attempt to separate the thermal and turbulent contributions to the overall linewidths by combining data respectively for OH and H, H_2CO and H, and OH, and H_2CO. The results for these three cases compare very poorly, indicating that the basic assumption required in this procedure – that the H, OH, and H_2CO coexist and are homogeneously mixed in the cloud – is apparently invalid. In at least some cases in which physically unreasonable kinetic temperatures are obtained by this procedure, the problem lies with the lack of coexistence of OH and H_2CO. Thirdly, of a total of 99 clouds observed in both OH and H_2CO, 6 have velocity differences in excess of 30 km/s and thus cannot be physically related. In addition, 29 OH clouds are seen in directions where there is no observed H_2CO, and 9 H_2CO clouds have no OH counterparts.

An interesting relation is the number of clouds N observed in both OH and H_2CO as a function of the difference in velocity, $\Delta v \equiv v(H_2CO)\text{-}v(OH)$. There is a strong peak in N at $\Delta v \approx O$ but the decrease in N with increasing $|\Delta v|$ is much faster for $\Delta v < O$ than for $\Delta v > O$. In fact at $\Delta v = 10$, N still is about one-third of its maximum value. This behavior applies equally well to those clouds seen in the directions of the strongest continuum sources as to the entire data sample. The most probable explanation for this asymmetry in the velocity distribution is that more H_2CO is seen at large distances than OH, as would tend to be the case if the H_2CO excitation temperature were less than the $3^\circ K$ background while that of OH is not. In the northern hemisphere, this corresponds to observing more positive velocities in H_2CO than in OH. Based as it is on a statistical argument, this is the first indication that the anomalous "refrigeration" of the 6 cm line of H_2CO (observed directly in the dark dust clouds), may be a universal phenomenon.

B. The General Excitation of H_2CO

Other indications also suggest that galactic H_2CO may be commonly refrigerated. Against some continuum sources, the observed temperature in the absorption line exceeds the continuum source brightness temperature; the excitation temperature of the 6 cm transition must in these cases be less than the $3^\circ K$ cosmic background. In other continuum sources whose angular size is $< 6'$, the antenna temperature in the absorption lines is as large when observed with an $11'$ beam as with a $6'$ beam; this can only occur if a significant contribution to the antenna temperature in the line comes from absorption outside the region of the continuum source, that is, against the $3^\circ K$ background. Finally, direct mapping of the H_2CO absorption around several continuum sources has shown that the absorption persists well off the source.

One must keep in mind the probable anomalous absorption of H_2CO in most if not all of the sources with which we are comparing the OH properties. For example, such effects could mask an actual correlation of $\langle \tau_{H_2CO} \rangle$ with $\langle \tau_{OH} \rangle$. If the anomalous excitation of H_2CO is caused by collisions, so that $\langle \tau_{H_2CO} \rangle$ is proportional to the square of the total density, then the correlation might not be destroyed; however a dependence of the H_2CO excitation on the spectrum of the local radiation fields could be expected to mask any actual correlation in the abundances of the two species. In addition, if the H_2CO is observable against the $3°K$ background while the OH is not, then gas clouds behind the continuum sources may contribute to the H_2CO absorption spectra but not to the OH spectra.

As mentioned above, the H_2CO clouds with the highest values of $\langle \tau_{H_2CO} \rangle$ are found to have velocities close to that of the HII regions. The most likely interpretation of this is merely that H_2CO clouds form in the same regions as HII regions because both require high gas densities. An alternative interpretation is that the H_2CO molecules are excited anomalously by radiation from the HII regions. To test this hypothesis, one must be able to separate the effects of excitation temperature T_s and H_2CO density on the observed $\langle \tau_{H_2CO} \rangle$. One method is to assume that the hyperfine components of the 6 cm transition have LTE ratios and that the anomalous excitation affects only the overall K-doublet populations. Then observed deviations in the hyperfine strengths from LTE strengths are attributed to the optical depth which may then be derived separately from T_s. This method has been applied to the source Ori B, and it is tentatively found that $T_s \approx 1.5°K$ on source and as one moves away from the continuum source T_s rises toward $3°K$ at a faster rate than the H_2CO density falls off. This might argue in favor of radiative pumping for H_2CO, since if the H_2CO is collisionally pumped one might expect $|T_s - 3|$ to be proportional to the total density. Such a conclusion depends of course on the H_2CO-to-total density ratio being independent of the distance from the source.

The very dark and dense dust cloud known as Cloud 2 (Heiles 1969) lies within one of the large Taurus regions of modest extinction throughout which H_2CO and OH have recently been mapped (Heiles and Turner 1972). The measured extinction in the $\sim 2°$ by $3°$ region outside Cloud 2 is $\lesssim 5$ magnitudes, while it is indirectly estimated at 50 to 100 magnitudes inside Cloud 2. Nevertheless, the observed strengths of OH and H_2CO are only slightly weaker outside Cloud 2 than inside it. Outside Cloud 2, $\langle \tau_{H_2CO} \rangle$ and $\langle \tau_{OH} \rangle$ both correlate well with the magnitude of the extinction. This correlation evidently must weaken or cease as the extinction approaches values as high as those inside Cloud 2. 21 cm maps of atomic hydrogen outside Cloud 2 show that the line brightness T_L does not correlate at all with extinction. If T_L is principally a measure of the H abundance (rather than of excitation temperature), and if this in turn is a measure of the total gas density outside Cloud 2, then we conclude that $<\tau_{OH}>$ and $<\tau_{H_2CO}>$ do not correlate with total gas density, although they correlate with extinction. One might then argue again in favor of radiative rather than collisional pumping; the role of the extinction would be to provide catalysts for formation of the molecules as well as protection against photodestruction.

IV. OTHER MOLECULES

With the exceptions of OH, H_2CO, and probably H_2O, our knowledge of the distribution of all other known interstellar molecules is as yet inadequate to allow conclusions on where they are found, how well their properties correlate with those of OH and H_2CO, and so on. Of the 18 identified radio molecules, the 9 most complex ones have all been observed in only a single source – Sgr B2. From the properties chiefly of the NH_3 observed in Sgr B2, it is thought to be the most dense molecular cloud known (total density $\gtrsim 10^7$ cm^{-3}) and thus, as one might expect, the complex molecules would seem, observationally, to require very high densities to form in detectable amounts. Other than OH, H_2CO, and H_2O, only the identified molecules NH_3, CO, HCN, and CS have been observed in more than two regions in the galaxy. Even these four molecules have been sought and found only in the directions of the strongest OH and H_2CO lines. These in turn correspond to the directions of the brightest HII regions. In several of these sources, however, the velocities of CO, CS, and HCN agree very well but are noticeably different from the OH and H_2CO velocities. This is the best evidence so far that either physical and chemical conditions vary considerably within individual sources, or that several clouds with distinctly different properties may occupy the same region of space. A similar conclusion is reached for the Sgr B2 region, where the various molecular lines indicate different degrees of turbulence, temperature, or both, as well as having somewhat different velocities in some cases. And in the two cases for which high spatial resolution ($\sim$1') maps exist, namely for CO (Penzias *et al.* 1971) and for CH_3OH (Turner *et al.* 1972), no similarly in the spatial distribution is found.

A. Ammonia

A partial map of the (1, 1) transition of NH_3 has been made in the Galactic Center region with a 6' beam (Morris *et al.* 1972). This survey used as a guide the map of the same region by Scoville *et al.* (1972) in the 6 cm H_2CO line, also with a resolution of 6'. In the few clouds studied, neither the intensities nor the velocities of the NH_3 appear to be related to those of H_2CO. Similarly, the NH_3 parameters seem uncorrelated with those of CO and CH_3OH, the only other molecules studied with high spatial resolution in this region.

In addition to the Galactic Center region, NH_3 has been observed in the direction of 5 other continuum sources all of which are strong OH emission sources, and in a dust cloud (Cloud 4). The spatial distribution of NH_3 is unknown in these regions. In some sources the NH_3 velocity agrees with that of OH and H_2CO, but not with the HII region velocity (e.g. DR21OH) while in other sources the NH_3 velocity is that of the HII region and differs with the other molecular velocities (e.g. W51). There is perfect agreement of all molecular velocities in Cloud 4.

B. Water Vapor

In terms of the 6_{16}-5_{23} transition of H_2O, which exhibits strong maser action in all observed cases, a definite and simple pattern for its distribution has emerged. H_2O emission is associated with only two types of objects: a) with type I anomalous OH emission sources (which usually are observed in the directions of HII regions); and b) with late-type "infrared" stars, specifically M supergiants and Mira Ceti variables, which usually, but not always, also exhibit type II(b) anomalous OH emission. Nineteen H_2O sources are at present known in the first category, and about a dozen in the second category. Because considerable energy by interstellar standards is needed to excite the 6_{16}-5_{23} transition of H_2O ($\sim$450 cm^{-1} above ground), it is not surprising that H_2O emission always coincides with anomalous OH emission. It is probable that the association of H_2O and OH emission is due to more than the presence of high energy. For at least the H_2O sources associated with type I OH emission, it appears that the OH is formed by the collisional dissociation of H_2O into specially excited states which lead to the anomalous OH emission (Gwinn et al. 1972). For the H_2O/IR star sources, the excitation of both H_2O and OH emission is more likely caused by infrared pumping, and it is not clear whether the two molecules have any chemical relationship.

It is unlikely that H_2O will be observed in absorption in the interstellar medium, because it has no lower-energy transitions at wavelengths where there is any appreciable background continuum radiation from radio sources. This is unfortunate, for we are constrained to observe H_2O only in very special, high-energy regions. Thus we have no way of discovering whether H_2O molecules are distributed as widely throughout the Galaxy as are OH, H_2CO, and CO.

V. CONCLUSIONS

Despite the large-scale observations of OH and H_2CO and the rapidly growing body of information on several other molecules, our understanding of the distribution and of the physical environment of these molecules is very limited. Most radio molecules appear to be observed only in the most dense regions of the interstellar medium, and, with the possible exception of those that exhibit anomalous excitation, probably in the coolest regions. It is not clear that the large nearby dark dust clouds in which OH, H_2CO, and CO have been studied, are characteristic of the majority of molecular sources seen throughout the Galactic plane. For example, all molecules observed within any given dark dust clouds have very similar velocities while the velocities of these same molecules differ noticeably among themselves in the majority of other regions throughout the Galactic plane. Certainly the small sample of foreground dust clouds that lie close enough to the sun to be observed optically are less dense than the regions which exhibit the most complex molecules, by a factor as high as 10^4 when compared with Sgr B2.

One of the most important goals in establishing a clear understanding of the distribution of molecules is to test theories of formation and destruction of interstellar molecules. Observationally, it seems well-established that the more complex the molecule, the more dense (and hence the fewer) the clouds in which it is found. This conclusion is consistent with the idea that the principle destruction mechanism for especially the more complex molecules is photodissociation by the interstellar UV. If it is adopted that the more complex interstellar molecules require surfaces (grains) for their formation, then one also concludes that the higher the gas density, the larger the grain density. Although the concept of a more or less constant gas-to-dust ratio is an old one, it has been established only in the low-density realm that pertains to optical observations. The new molecular observations at radio wavelengths might suggest, within the above assumptions, that the concept of constant gas-to-dust ratio extends to much higher densities.

On this picture, it is not surprising that radio and optical molecules are never observed in the same regions. "Optical" molecules (CH^+, CH, CN, CO) are simpler in structure than radio molecules and can apparently be formed in the observed quantities by gaseous reactions alone (Solomon and Klemperer 1972). Because they must be observed against bright stars they are seen only in regions of very low obscuration. Radio molecules have never been observed in these regions, presumably because the density of grains and gas is too low to form these molecules in detectable quantities. Radio molecules have been seen only in regions so highly obscured that optical molecules could not be observed even if they were present. To test the hypothesis that optical molecules can form by gaseous reactions whereas the more complicated radio molecules require surfaces to form, one would like to show observationally that optical molecules correlate fundamentally with gas density and not necessarily with extinction, while radio molecules correlate with extinction alone. Unfortunately, the regions that can be sampled for optical molecules are so restricted that no reliable observational information exists on whether optical molecule abundances correlate with either gas density or extinction. In the regions of radio molecules, both gas densities and extinction are known generally to be very large; however, neither quantity can be estimated very accurately in the majority of cases, hence it is not clear how well the radio molecule abundances correlate quantitatively with gas density or extinction. Only in the Taurus region, where the extinction is modest and where OH and H_2CO are widespread, may these questions eventually be answered. In these regions it will be very important to decide whether the 21 cm line is a measure of the total gas density, as well as to try to detect optical molecules against bright stars.

REFERENCES

Bok, B. J. 1956, *A. J.,* **61** 309.

Gwinn, W. D.; Turner, B. E. and Goss, W. M. 1972, *Ap. J.,* (submitted).

Heiles, C. E. 1969, *Ap. J.* **160**, 51.

Heiles, C. E. and Turner, B. E. 1972, in preparation.

Ishida, K. and Kawajiri, N. 1968, *Publ. Astr. Sci. Jap.,* **20**, 95.

Morris, M.; Zuckerman, B.; Turner, B. E. and Palmer, P. 1972, in preparation.

Penzias, A. A.; Jefferts, K. B. and Wilson R. W. 1971, *Ap. J.,* **165**, 229.

Scoville, N. Z.; Solomon, P. M. and Thaddeus, P. 1972, *Ap. J.* **172**, 335.

Solomon, P. M. and Klemperer, W. 1972, *Ap. J.,* (submitted).

Turner, B. E. 1972, *Ap. J.* **171**, 503.

Turner, B. E.; Gordon, M. A. and Wrixon, G. T. 1972, *Ap. J.* (submitted).

Turner, B. E. and Heiles, C. E. 1972, in preparation.

Wilson, T. L. 1971, paper presented at 133rd Meeting of the AAS, Tampa, Florida.

Zuckerman, B.; Buhl, D.; Palmer, P. and Snyder, L. E. 1970, *Ap. J.* **160** 485.

The Kinetic Temperature of Dense Interstellar Clouds determined from CO Observations

P. M. Solomon

School of Physics and Astronomy
University of Minnesota
Minneapolis, Minnesota

I. INTRODUCTION

Carbon monoxide has been observed extensively in emission at λ = 2.6 mm from the 0-1 rotational transition in a wide variety of objects including HII regions (Penzias *et al.*, 1971), IRC + 10216, (Solomon *et al.*, 1971) and throughout the galactic center region (Solomon *et al.*, 1972). CO is the most abundant molecule detected by radio astronomical techniques and it is the only molecular constituent of the interstellar medium except H_2 which contains a large fraction of the available atoms.

The long lifetime of the J = 1 level in CO, τ_{10} = 1.7 x 10^7 sec, which results from the small permanent dipole moment μ = 0.1 debye, means that this level will be populated at a much smaller excitation rate than rotational levels from other molecules such as CN, HCN and CS. If the excitation mechanism is collisions with hydrogen atoms or molecules, the required density sufficient to thermalize the J = 1 level by collisions ($\eta_{H_2} > 10^3$) is about equal to the expected density in dark interstellar clouds. Carbon monoxide is therefore an excellent candidate for detection in dark interstellar clouds in which OH (Heiles, 1968 and 1969) and H_2CO (Palmer *et al.*, 1969) have been observed.

It is the purpose of this paper to briefly describe recent observations of $^{12}C^{16}O$ and $^{13}C^{16}O$ emission from dark interstellar clouds, a search for $H_2^{13}C^{16}O$ which was made to determine the $^{13}C/^{12}C$ ratio, and to present a short discussion of the relation between the CO brightness temperature and the kinetic temperature of the cloud. The carbon monoxide observations were carried out in collaboration with A. A. Penzias, R. W. Wilson and K. B. Jefferts, (these results will be discussed in more detail elsewhere) using a receiver developed at Bell Labs and the 36 foot telescope of the National Radio Astronomy Observatory located on Kitt Peak, Arizona, during February and May, 1971. The $H_2^{13}CO$ search was conducted with the 140 foot telescope of NRAO.

II. CARBON MONOXIDE OBSERVATIONS

The CO data was obtained using a 50 channel line receiver with a total bandwidth of 5 MHz and a channel width of 100 KHz. Some of the later observations were made using a 50 channel receiver with a 250 KHz width per channel. Frequency switching was used with the reference frequency displaced by 5 MHz. The $^{12}C^{16}O$ line at 115271.2 MHz was first detected in cloud H-1 (see Heiles, 1968) with an antenna temperature T_A = 1.9°K; figure 1b shows the observed emission line after a total of 90 minutes integration. A total of eleven dark clouds were observed and the line was found in eight cases with an upper limit (3 standard deviations) of T_A = 2°K in the three remaining clouds. The

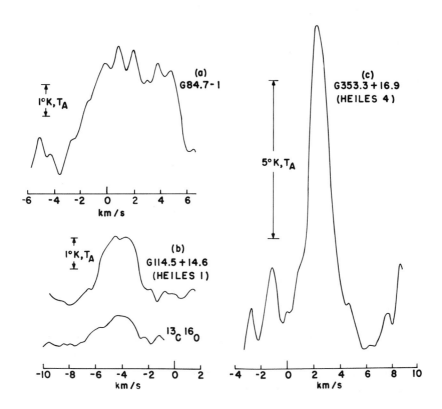

1. Observed line profiles of carbon monoxide emission in three dark interstellar clouds.

data is summarized in Table 1. Observations were made over extended regions covering approximately $1°$ in the three clouds H-2, H-4 and G84.7-1. In all cases the emission extended beyond the boundaries of the clouds as seen on the red Palomar sky survey prints. In particular for cloud 2 in the Taurus region the emission extended at least $30'$ beyond the sharp southern boundary into an area with a high star density, although this outer region still has about 2 magnitudes of extinction. Inside the boundaries of the dark clouds, the extinction $A_v \geqslant 5$.

TABLE 1
CARBON MONOXIDE EMISSION FROM DARK CLOUDS

Cloud	T_A		V_{LSR}(Km/sec)	ΔV(Km/sec)	T_{Kin}(°K)
G114.5+14.6 (H-1)	1.9°K	$^{12}C^{16}O$	-4.5	3.0	6.4
	0.8°K	$^{13}C^{16}O$	-4.7	2.9	
G174.6-13.6 (H-2)	2.1°K	$^{12}C^{16}O$	5.2	3.1	6.7
	0.8°K	$^{13}C^{16}O$	3.8	1.9	
G353.3+16.9 (H-4)	8.8°K	$^{12}C^{16}O$	2.0	3.1	18.3
	2.2°K	$^{13}C^{16}O$	2.4	<0.7	
G4.3+35.8 (L-134)	2.2°K	$^{12}C^{16}O$	2.9	2.0	6.9
	1.2°K	$^{13}C^{16}O$	1.6	<0.7	
G12.7	1.7°K	$^{12}C^{16}O$	13.5	8.8	6.0
	2.3°K	$^{12}C^{16}O$	33.6	6.0	7.2
G27.0+3	1.8°K	$^{12}C^{16}O$	3.7	7.3	6.2
G23.5+8	<2°K	$^{12}C^{16}O$			<6.5
G27.1+3	<2°K	$^{12}C^{16}O$			<6.5
G27.3+4	2°K	$^{12}C^{16}O$	9.6	6.5	6.5
G35.4+0	<2°K	$^{12}C^{16}O$			<6.5
G84.7-1	4°K	$^{12}C^{16}O$	2.0	10.4	9.9

The emission lines in the 11 clouds have a range in intensity from T_A = 1.7°K to T_A = 10°K with velocity widths from 2-10 Km/sec. Most of the features appear flat-topped and therefore heavily saturated, although much weaker than lines from clouds associated with HII regions, except for cloud H-4. A sample of the observed lines is presented in Figure 1. In all cases where data is available the $^{12}C^{16}O$ lines are broader than the 6 cm H_2CO lines. There appears to be very little change in velocity with location inside a single cloud. In the case of cloud H-4 there is however substantial structure in the intensity as a function of position. This is shown in Figure 2 along with a sketch of the cloud adopted from Heiles (1968).

The cloud with the broadest line, G84.7-1 is near the North American Nebula, which is in a complex region, consisting of emission nebulosity and dark clouds. The observed molecular region is much larger than the optical HII region.

The $^{13}C^{16}O$ line at 110201.4 MHz was searched for and found in the four well-studied clouds H-1, H-2, H-4 and L134. The isotope lines are surprisingly strong with $T(^{13}C^{16}O) / T(^{12}O^{16}O) = 0.25 - 0.55$ (see Table 1). However the ^{13}C isotope line widths are much narrower than the ^{12}C features especially for cloud H-4, where $\Delta V(^{13}C^{16}O) / \Delta V(^{12}C^{16}O) < 0.25$. ΔV is the full width at half maximum line intensity. The only exception to the narrow isotope lines is

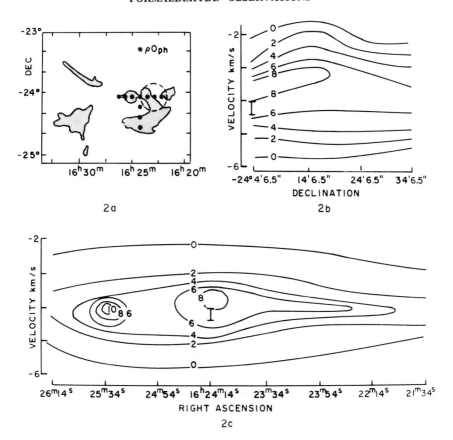

2. a) Sketch of cloud H-4 showing location of carbon monoxide observations.

b) Contours of antenna temperature in the velocity–declination plane.

c) Contours of antenna temperature in the velocity–right ascension plane. This cloud shows substantial structure in its CO emission, in contrast to the other clouds which were mapped.

the feature in cloud H-1, which is shown in Figure 1b. The high relative intensity and small width of the $^{13}C^{16}O$ lines are further evidence for strong saturation in the $^{12}C^{16}O$ line. In order to estimate the column density of carbon monoxide, an independent determination of the isotope ratio must be used. Strong saturation is evident in almost all interstellar carbon monoxide lines and the peak intensity ratios cannot be used for isotope measurements. It was therefore decided to search for the ^{13}C line of H_2CO at 6 cm in one of the dark clouds to obtain an estimate of $^{13}C/^{12}C$.

III. FORMALDEHYDE OBSERVATIONS

Cloud H-2 which has a strong *apparent* $^{13}C/^{12}C$ isotope ratio in carbon monoxide, T_A ($^{13}C^{16}O$) / T_A ($^{12}C^{16}O$) = 0.4, has been mapped by Palmer *et al.*, (1969) in the 6 cm (4830 MHz) line of formaldehyde. A search was conducted for the corresponding ^{13}C line (4593 MHz) at the position of maximum strength where the $H_2^{12}C^{16}O$ absorption line intensity T_L = 0.8°K. These observations were obtained using the 140 foot telescope of NRAO, a cooled parametric amplifier and the 384 channel autocorrelation line receiver with a bandwidth of 1.25 MHz. Frequency switching was used during a five-hour integration and the two difference signals were combined to produce a spectrum with a measured standard deviation per channel, σ = 0.01°K. No line was observed at the expected velocity. For an expected intrinsic velocity width of 0.4 Km/sec and using the measured hyperfine splitting in $H_2^{13}CO$ (Kirchhoff *et al.*, 1971) to determine the relative transition strengths, the column density ratio is ($H_2^{13}CO$ / ($H_2^{12}CO$) = O ± .016. The observed isotope abundance is therefore $^{12}C/^{13}C$ > 30 at the 2σ level and $^{12}C/^{13}C$ > 20 at the 3σ level. This is in sharp contrast to the apparent ratio of 2.5 from carbon monoxide and is further proof of the high degree of saturation in the CO emission line. The above argument assumes only that the 6 cm excitation temperature is the same for $H_2^{13}C^{16}O$ and $H^{12}C^{16}O$.

IV. DISCUSSION

If we adopt the limit of 30 set by the formaldehyde observations as the true isotope ratio $^{12}C/^{13}C$, then an observed relative intensity of 2 in carbon monoxide would indicate that the $^{12}C^{16}O$ line has an optical depth, τ > 15. However there is no reason to believe that the formaldehyde limit is the true value and we therefore adopt the terrestrial value of 89 which gives, for the clouds in Table 1, a carbon monoxide optical depth in the line center of τ = 22-49. The column density of $^{12}C^{16}O$ required to produce $\tau \sim 30$ and $T_A \sim 2$°K is $N \sim 10^{18} cm^{-2}$. The exact column density, which depends on the details of the excitation, will not differ substantially from $10^{18} cm^{-2}$.

The large optical depth and angular extent of CO emission in the dark clouds clearly means that the observed brightness temperature T_B is equal to the rotational excitation temperature T_{01}. This raises the important question of the relation between the excitation temperature and the gas kinetic temperature. The low observed brightness temperature clearly indicates that the excitation temperature is not highly anomalous and maser action is definitely ruled out. The excitation temperature is determined by the rates of collisional and radiative processes connecting the rotational levels of CO. For a simple rotational ladder in a diatomic molecule, like CO, it is possible to derive the conditions under which thermal equilibrium will prevail from a few basic theoretical considerations.

In the limiting case of *collision rates dominating all upward and downward transitions* into the J = O and J = 1 levels, *the excitation temperature must equal the kinetic temperature.* Quantitatively this corresponds to $C_{10} \gg A_{10}$ when only $\Delta J = \pm 1$ transitions are allowed, where C_{10} is the collision rate and A_{10} the spontaneous decay rate for J = 1→0. In the case of many quantum collisions, ($\Delta J \geqslant 2$) the thermodynamic limit is reached when $C_{10} \gg A_{21}$. This allows for any non-equilibrium effects which may result from coupling of the 0-1 transition to higher rotational levels.

The dominant collisional process expected in dark interstellar clouds is excitation by molecular hydrogen. The excitation rate per electron is much higher than for neutrals. However the electron concentration e/H_2 (Solomon and Werner, 1971) is far too small ($<10^{-4}$) for electrons to be important, particularly for carbon monoxide which has a small dipole moment. Assuming a typical kinetic cross section $\sigma = 2 \times 10^{-15} cm^2$ the hydrogen density necessary to thermalize carbon monoxide by collisions is:

$$\eta_{H_2} \gg \frac{A_{10}}{\sigma v} \qquad (1a)$$

which at a kinetic temperature of $10^{\circ}K$ (at $10^{\circ}K$ collisions to higher levels may be ignored) becomes:

$$\eta_{H_2} \gg 1 \times 10^3 \ cm^{-3} \qquad (1b)$$

The average density of hydrogen molecules expected in most dark clouds from the assumption of normal gas to dust ratios is also about 10^3 (see Solomon and Werner, 1971) which means that on the basis of pure collisions, the carbon monoxide is being heated but not necessarily fully thermalized. All of the above considerations apply only to an optically thin gas.

For large optical depths as is the case for carbon monoxide, the effect of radiative trapping will be to increase the upward transition rate and bring the excitation temperature into equilibrium with the gas kinetic temperature. Every J = 1→0 photon emitted as a result of a collision, will scatter in a random walk causing a large number of upward transitions until it either escapes from the cloud or is transferred into kinetic energy by downward collisions.

The mean number of scatterings experienced by a single photon generated inside a cloud, before it is destroyed by a downward collision, is $S = (A_{10} + C_{10}) / C_{10}$. The maximum number of scatterings is determined by the optical depth of the cloud and is equal to τ^2. The condition for thermalization of the J

= 1 state by trapping in a medium with collision rate C_{10} is the same as the requirement for production of a black body energy density in the 0 - 1 transition, and can be shown to be:

$$\tau \gg \left(\frac{A_{10} + C_{10}}{C_{10}} \right)^{1/2} \tag{2}$$

For the dark interstellar clouds, assuming $\eta_{H_2} = 10^3$ we have $C_{10} \sim A_{10} = 6 \times 10^{-8}$ and $(A_{10} + C_{10} / C_{10})^{1/2} \sim 1$. The optical depth derived in the previous section from the isotope intensity ratios is $\tau \sim 30 \gg 1$, and we can expect complete thermalization of the CO J = 0 - 1 transition.

The kinetic temperatures may then be derived in a straightforward manner from the antenna temperature T_A, the assumed microwave background temperature $T_{bg} = 2.8°K$ and the antenna efficiency $\eta = 0.60$. The equivalent antenna temperature of the microwave background radiation is:

$$T'_{bg} = \frac{\eta h\nu}{k \, (e^{h\nu/KT_{bg}} - 1)} \tag{3}$$

$$= 0.53°K$$

The kinetic temperature (see column 7, Table 1) is then given by $T_{Kin} = T_{01}$ where

$$T_{01} = \frac{h\nu}{k \, \ln \left\{ + \dfrac{\eta h\nu}{k(T_A + T'_{bg})} \right\}} \tag{4}$$

The kinetic temperature of eight of the clouds is the range $T_{Kin} = 6°K$ - $18°K$. The three clouds with upper limits probably also contain CO and have $T_{Kin} < 6°K$. Heiles (1969) has determined the kinetic temperature, using OH line intensity ratios, and found $T_{Kin} = 4.4$ and $5.4°K$ for clouds H-1 and H-2 in excellent agreement with the CO determination of 6.4 and $6.7°K$. The uncertainties in the instrumental calibration are sufficient to account for the differences in temperature. For cloud H-4, the OH measurements gave $T > 9.6°K$, which is also consistent with the CO value of $18°K$. The close agreement in kinetic temperature, derived by two completely different types of observation involving molecules with different transition strengths and abundances, is strong observational evidence that the OH,1665 MHz and 1667 MHz transitions, as well as the CO,115 GHz transition, are in thermal equilibrium.

The temperatures determined from the CO data are much lower than is usually assumed for interstellar clouds, but the expected grain temperatures inside a cloud, determined from a balance between emission and absorption of the incident radiation field, have been calculated to be in the range $T \sim 4°K$ - $6°K$ (Solomon and Wickramasinghe, 1969) and $T \sim 9°K$ - $15°K$ (Werner and Salpeter, 1969). Most of the dark clouds observed here have gas temperatures of $\sim 6°K$ and it may be that this low temperature is due to an equilibrium between the gas and grains.

The very low kinetic temperatures have important consequences for the dynamics of the clouds and for an understanding of all chemical and physical processes inside the clouds. One of the more interesting problems is the cooling of the $(J = 1, K$ doublet) 6 cm H_2CO line below the background radiation temperatures. The low kinetic temperature limits any collisional mechanism to the four levels of $J = 1$ and $J = 2$, which have $E/k \leqslant T$. The cooling mechanism, based on the H_2CO asymmetry, proposed by Townes and Cheung (1969) could operate at the densities expected in the clouds. However at $T \sim 6°K$, most of the collisions mixing the $J = 1$ doublet with the $J = 2$ doublet will be at less than twice the threshold evergy. Accurate collision cross sections are therefore required at very low kinetic energy. A recent paper by Thaddeus (1972) indicates that there will be no cooling by collisions with neutral particles at $T <$ $40°K$.

In obtaining the kinetic temperatures from the observations, we have tacitly assumed that the clouds have uniform density. For a cloud with a steep density gradient, the molecules on the outside (low density region) would be heated by radiation from the center. In this case, the heating by trapping would apply, but the observed excitation temperature would be lower than the kinetic temperature due to the dilution of the radiation field. For a spherical cloud the observed excitation temperatures would be approximately ½ of the kinetic temperature. The agreement between the OH and CO temperatures indicates however that the molecules at an optical depth $\tau \sim 1$ are completely thermalized, and we may accept the CO intensities as true indicators of the kinetic temperature. Further observations of OH and CO in a large number of clouds are required to confirm this result.

REFERENCES

Heiles, C. 1968, *Ap. J.*, **151**, 919.

Heiles, C. 1969, *Ap. J.*, **157**, 123.

Kirchoff, W. H.; Lovas, F. J. and Johnson, D. R. "Critical Review of Microwave Spectra," N.B.S.

Palmer, P.; Zuckerman, B.; Buhl, D. and Snyder, L. E. 1969, *Ap. J. (Letters)*, **156**, L147.

Penzias, A. A.; Jefferts, K. B. and Wilson, R. W. 1971, *Ap. J.*, **165**, 229.

Solomon, P. M. and Wickramasinghe, N. C., 1969, *Ap. J.*, **158**, 449.

Solomon, P. M.; Jefferts, K. B.; Penzias, A. A. and Wilson, R. W. *Ap. J.*, **163**, L53.

Solomon, P. M.; Scoville, N., Jefferts, K. B.; Penzias, A. A. and Wilson, R. W., 1972, to be published.

Thaddeus, P. 1972, *Ap. J.*, **173**, 317.

Werner, M. W. and Salpeter, E. E. 1969, *M.N.R.A.S.*, **145**, 249.

Townes, C. H. and Cheung, A. C. 1969, *Ap. J.*, **157**, L103.

OH and H_2O Microwave Emission from Infrared Stars

W. J. Wilson

The Aerospace Corporation
El Segundo, California

I. INTRODUCTION

In studies of OH and H_2O microwave emission sources associated with infrared (IR) stars, thirty-seven sources have been detected as of October 1971 and are listed in Table I (Wilson and Barrett 1972; Wilson *et al.* 1972). These sources show a variety of characteristics and it has been possible to identify three general types of OH/IR sources. The assumption that the OH and H_2O emission is closely associated with the IR stars is based upon OH and H_2O position measurements (Hardebeck 1971; Hardebeck and Wilson 1971; Schwartz 1971) and the observed facts that the OH and H_2O time variations and emission velocities are in agreement with the optical variations and stellar velocities. The correlation of the radio and optical velocities, one of the most interesting results from this study, is the basis of the expanding model which is proposed in III. From this model, it is possible to calculate rates of loss of stellar molecules to the interstellar medium. This calculation indicates that the molecules lost from OH/IR stars comprise only a minor portion of the interstellar material.

II TYPES OF OH/IR STARS

Based on their OH, H_2O, IR and visual characteristics it has been possible to divide the OH/IR stars into three general groups:

1. The majority of the microwave sources associated with IR stars have weak to moderately strong 1612-MHz OH emission which is concentrated in two narrow velocity ranges. The 1612-MHz emission is stronger than the main-line 1665-and 1667-MHz emission and it is essentially unpolarized. There is no detectable 1720-MHz OH emission and as yet no H_2O emission has been detected from these stars. In all the stars which have been studied and classified, it has been found that the OH sources are associated with late M-type Mira variables which have an excess IR radiation at 10μ, indicating the possibility of a circumstellar shell (Hyland *et al.* 1972). These sources have been designated as *Type II OH/IR stars* following the OH classification scheme introduced by Turner (1967). In addition to the Type II OH sources known to be associated with IR stars, there are eight OH sources listed in Table I which have OH emission characteristics similar to the Type II OH/IR stars but for which no IR star has yet been found.

2. There are seven OH sources in which the main-line 1665-and 1667-MHz OH emission is strongest and which also have associated H_2O emission. These stars are bright, nearby, late M-type Mira variables which do *not* show a large 10μ IR excess. These sources have been designated as *Type I OH/IR stars.*

Table 1

OH/IR STARS

STAR		R.A. 1950 DEC.				ℓ^{II}	b^{II}	REMARKS**
	H	M	S	D	M	D	D	
TYPE II		(1612-MHZ OH EMISSION)						
IRC 10011*	01	03	48	12	19.8	129	-50.1	CIT-3, M8, M
NML TAU	03	50	46	11	15.7	178	-31.4	IK TAU, M8E, M, 550
IRC 50137*	05	07	20	52	48.9	156	7.8	
IRC 40156	06	29	45	40	44.9	174	14.1	
IRC-20197	09	42	56	-21	48.1	256	23.3	
WX SER	15	25	32	19	44.1	30	53.5	IRC 20281, M8E, M, 425
IRC 30292	16	25	59	34	54.6	56	43.5	M9
OH1637-46	16	37	30	-46	13.9	338	0.1	OH338.5+.1, NO IR STAR DET.
OH1649-41*	16	49	52	-41	43.8	343	1.3	NO IR STAR DETECTED
OH1735-32*	17	35	58	-32	10.3	356	- 0.6	NO IR STAR DETECTED
IRC-10381	17	48	28	-08	00.7	19	9.5	
OH1821-12*	18	21	17	-12	28.0	19	0.4	NO IR STAR DETECTED
IRC-10434	18	30	30	-07	29.0	24	0.6	
W 41 *	18	31	27	-09	00.9	23	- 0.3	NO IR STAR DETECTED
IRC 10365	18	34	59	10	23.0	41	7.8	
IRC-10450	18	37	35	-05	45.8	27	- 0.1	
OH1837-05*	18	37	42	-05	00.6	27	0.2	NO IR STAR DET., DEC ≤2 MIN
W 43 A *	18	45	05	-01	48.3	31	0.0	NO IR STAR DET., DEC ≤2 MIN
OH1854+02*	18	54	56	02	07.7	36	- 0.3	NO IR STAR DET., DEC ≤2 MIN
R AQL *	19	03	58	08	09.2	42	0.4	IRC 10406, M8E, M, 300
IRC-20540	19	05	56	-22	19.2	15	-13.6	
RR AQL	19	54	58	-02	01.2	39	-15.6	IRC 00458, M7E, M, 394
IRC-10529	20	07	46	-06	24.7	36	-20.4	
OH2026+38*	20	26	40	38	57.0	78	0.2	ON-4, WEAK IR STAR DETECTED
UX CYG	20	53	00	30	13.4	74	- 9.4	IRC 30464, M6E, M, 560
V MIC	21	20	37	-40	55.2	01	-45.5	M6E, SRB, 347
IRC 40483	21	25	23	36	29.0	84	-10.2	
TYPE I		(1665/1667-MHZ OH AND H2O EMISSION)						
R HOR	02	52	12	-50	05.6	265	-57.4	M6E, M, 403
U ORI *	05	52	51	20	10.1	189	- 2.5	IRC 20127, M8E, M, 372
W HYA *	13	46	12	-28	07.1	318	32.8	IRC-30207, M8E, SRA, 382
S CRB *	15	19	21	31	32.8	49	57.2	IRC 30272, M7E, M, 361
U HER *	16	23	35	19	00.3	35	40.3	IRC 20298, M8E, M, 405
IRC-20424*	18	00	58	-20	19.2	10	0.8	
R CAS	23	55	53	51	06.6	115	-10.6	IRC 50484, M8E, M, 431
SUPERGIANTS		(1612-MHZ OH (STRONG) AND H2O EMISSION)						
VY C MA *	07	20	55	-25	40.4	239	- 5.1	IRC-30087, M3E, LC
VX SGR *	18	05	03	-22	14.0	8	- 1.0	IRC-20431, M4E, SRB, 732
NML CYG *	20	44	34	39	55.9	81	- 1.9	IRC 40448, M6

* OH INTERFEROMETER POSITIONS MEASURED BY HARDEBECK (1971) AND HARDEBECK AND WILSON (1971). ERRORS ≤ 0.25 MIN OF ARC, EXCEPT WHERE NOTED. OTHER POSITIONS ARE IR POSITIONS WITH ERRORS ≤ 1 MIN OF ARC.

** NAME, SP. TYPE, VAR. TYPE, PERIOD (DAYS).

3. There are three *Supergiant OH/IR stars,* VY CMa, VX Sgr and NML Cyg, which have strong, broad, unpolarized 1612-MHz OH emission and also have H_2O emission. The main-line OH emission is weaker than the 1612-MHz emission and it is significantly polarized. These supergiant stars are M-type irregular variables which have a large 10μ IR excess.

Not all of the sources fit neatly into the three categories since there are some sources which have properties of two groups. For example, the Type II stars, R Aql and NML Tau, have H_2O emission and IR characteristics similar to the main-line Type I OH/IR stars, even though their 1612-MHz emission is strongest. This may suggest that there is an evolution between the different types of stars.

III. EXPANDING MODEL AND CALCULATIONS OF MASS LOSS

Since the OH/IR stars have emission characteristics similar to known OH and H_2O maser emission sources it will be assumed that the OH and H_2O emission comes from masering molecules. (The brightness temperatures of the OH source in NML Cygnus and the H_2O source in VY CMa have been measured to be $\geqslant 10^{10}$°K which supports the maser assumption.) There have been two basic mechanisms proposed to explain the excitation of the OH and H_2O masers — excitation by radiation (IR or UV) and excitation by collisions with electrons and/or H atoms. The fact that the OH/IR stars have large IR fluxes and their temperatures are $\leqslant 2000$°K suggest that the IR radiation is a likely source for the excitation of the masers.

Litvak (1969) has calculated the conditions necessary for IR pumped OH masers and has found that cloud lengths of $\sim 2 \times 10^{15}$ cm and hydrogen densities of $\leqslant 10^9$ atoms cm^{-3} with kinetic temperatures of 100-1000°K are required to explain the observed properties. Both the long path lengths with little velocity change and the observed fact that the OH and H_2O emission velocities are essentially constant in time indicate that the OH and H_2O masering clouds must be located at a radius of $\sim 3 \times 10^{15}$ cm — which would be in the outer stellar atmosphere.

One of the most interesting results from the OH and H_2O measurements is the correlation of the OH, H_2O and optical absorption and emission velocities. Radial velocities for all the OH/IR stars with known optical velocities are summarized in Table 2 (Wilson and Barrett 1972, Wilson *et al.* 1972). These data suggest a correlation between the higher OH emission velocity and the stellar absorption line velocity. Since the absorption line velocity apparently represents the stellar velocity (Merrill 1940), the higher velocity OH cloud must be nearly stationary with respect to the star while the lower velocity OH cloud would be moving outward. In the cases of where H_2O emission is observed, the H_2O cloud would be between the two OH clouds and would be slowly expanding outward.

Table 2

HELIOCENTRIC VELOCITIES OF OH/IR STARS

STAR	OPTICAL EMS. KM/S	OH LOW VEL. KM/S	H_2O KM/S	OH HIGH VEL. KM/S	OPTICAL ABS. KM/S	ADD, TO CONVERT TO LSR
TYPE II						
NML TAU	41	30	--	63	63	-12.3
WX SER	-21	-17	--	- 2	- 7	16.0
R AQL	21	22 to 26	30 to 32	36	32	18.0
RR AQL	1	8	--	17	15	14.9
UX CYG	21	-29 to -20	--	--	- 6	16.0
TYPE I						
R HOR	46	44	--	47	56	-11.0
U ORI	-34	-29	-24	-23	-22	-12.8
W HYA	26 to 31	35	40	42	41 to 44	2.0
S CRB	-22	-21	-13	-13	-10	16.4
U HER	-44	-37	-33	-28	-28	18.1
R CAS	10	13	18	23	26	8.1
SUPERGIANTS						
VY CMA	10 to 17 37 to 75	4 to 27	33 to 57	49 to 80	57 to 80	-19.0
VX SGR	-34 to -26 11	-29 to -13	-10 to 3	- 1 to 16	12 -29 to -12	12.2
NML CYG	--	-44 to -16	-35	-11 to 9	7	16.5

Gertz and Woolf (1971) have proposed a model for M giants with circumstellar dust shells where dust grains begin to form in a stationary inner region, and at the edge of this region when the dust condenses, the grains are accelerated by radiation pressure. Adapting this model to the OH/IR stars requires that the stationary OH cloud be located in the inner region where the dust is forming and the lower velocity OH cloud be located in the outer region. The outer OH gas would be accelerated by the outflowing dust to the observed velocities of $\sim$7km sec[-1] for the Type I OH/IR stars, $\sim$25km sec[-1] for the Type II OH/IR stars, and $\sim$50km sec[-1] for the supergiant OH/IR stars. The H_2O cloud would be located just outside the stationary region.

The conclusion that material is flowing away from these stars at a velocities of $\geqslant$7km sec[-1] suggests that the OH/IR stars may be losing mass at a fairly high rate, since the escape velocity at a radius of 3×10^{15}cm for these $\sim$1 $M_\odot$ stars is only about 3km sec[-1]. Assuming a particle density of 10^6 to 10^7 hydrogen atoms cm[-3] at a radius of 3×10^{15} cm, based on the requirements of the IR pumping theory, and assuming an isotropic expansion velocity between 7 - 50km sec[-1] based on the observed velocity differences, the derived mass loss rate would be from 2×10^{-6} to 10^{-4} $M_\odot$ yr[-1]. Mass loss rates of 2×10^{-6} $M_\odot$ yr[-1] have been derived for many of the Type I OH/IR stars from IR observations (Gehrz and Woolf 1971) and these are in agreement with the values derived from the OH data for the Type I stars assuming a density of 10^6 hydrogen atoms cm^{-3}. The derived mass loss rates for the Type II and the supergiant OH/IR stars have values of $\sim 10^{-5} M_\odot$ yr^{-1} which are higher than the Type I OH/IR stars. This is in general agreement with the data of Gehrz and Woolf (1971) for supergiants and the longer period variables, although the actual mass loss rates appear to be larger.

It is possible to estimate the total amount of material the OH/IR stars have lost to the interstellar medium since the OH/IR stars have a spatial distribution similar to other late M-type long period variable stars (Hyland *et al.* 1972; Wilson *et al.* 1972; Blanco 1965). Assuming that the space density for Mira variables of 10^{-7} pc^{-3} (Oort and van Tulder 1942) has been nearly constant for 10^{10} years, and taking the observational evidence that only $\sim 5\%$ of the Mira variables have detectable microwave emission, which suggests that only 5% of the Mira variables have mass loss rates as large as 10^{-5} $M_\odot$ yr^{-1}, it is found that 0.025 hydrogen atom cm^{-3} in the interstellar medium originated from OH/IR stars. Since the density of the interstellar medium is ~ 1 hydrogen atom cm^{-3} (Spitzer 1968), it is concluded that the stars with detectable microwave emission do not contribute a significant amount of matter to the interstellar medium.

I thank P. R. Schwartz and J. A. Wilson for reviewing the manuscript.

REFERENCES

Blanco, V. M. 1965, *Distribution and Motions of Late-Type Giants, in Galactic Structure,* ed. A. Blaauw, M. Schmidt, Univ. Chicago Press, Chicago, p. 241.

Gehrz, R. D. and Woolf, N. J., 1971, *Ap. J.,* **165**, 285.

Hardebeck, E. G. 1971, *Ap. J.,* (In Press).

Hardebeck, E. G. and Wilson, W. J. 1971, *Ap. J.* (Letters), **169**, L123.

Hyland, A. R.; Becklin, E. E.; Frogel, J. A. and Neugebauer, G. 1972, *Astron. Astrophys.,* (in press).

Litvak, M. M. 1969, *Ap. J.,* **156**, 471.

Merrill, P. W. 1940, *Spectra of Long Period Variable Stars,* Univ. Chicago Press, Chicago.

Oort, H. D. and Van Tulder, J. J. M. 1942 *B.A.N.,* **9**, 327.

Schwartz, P. R. 1971, (private communication).

Spitzer, L. 1968, *Diffuse Matter in Space,* Interscience Pub., Wiley, New York.

Turner, B. E. 1967, Ph.D. Thesis, Astronomy Dept., Univ. of Calif. Berkeley.

Wilson, W. J. and Barrett, A. H. 1972, *Astron. Astrophys.,* (in press).

Wilson, W. J.; Schwartz, P. R.; Becklin, E. E.; Neugebauer, G. and Harvey, P. 1972, (in preparation).

Robinson: I've just received the following telegram:

"Successful detection of thioformaldehyde in absorption Sagittarius B2. Details as contained in IAU telegram sent September 30 summarised below: Sinclair, Ribes, Fourikis, Brown, Godfrey report detection in absorption of thioformaldehyde 2_{11} - 2_{12} transition in Sgr B2 at 3139.38 $\pm$ 0.03MHz. Antenna temperature of approximately 0.3° K. Velocity 56 $\pm$ 5 km/s. Width 20 km/s. Observed line to continuum ratio is 0.011, one over sixty of that observed for the 2_{11} - 2_{12} formaldehyde transition leading to molecular abundance ratio of 1 over 30. Terrestrial ratio of sulphur to oxygen 1 over 40."

Sinclair, Ribes, Fourikis, C.S.I.R.O. Radiophysics Lab, Sydney.

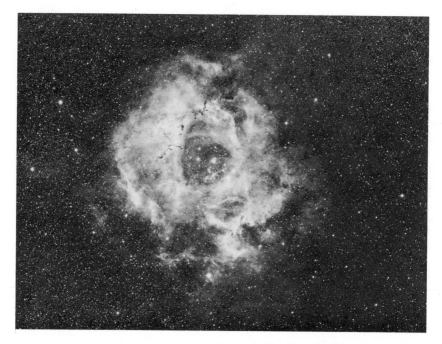

The Rosette Nebula, NGC 2237, in the constellation Monoceros. The small dark features are dense dust clouds. Photographed in red light by the 48-inch telescope. (Hale Observatories photograph)

Chemical and Rotational State
Stability of Interstellar Formamide

W. H. Flygare, R. C. Benson, H. L. Tigelaar
Noyes Chemical Laboratory
and
R. H. Rubin and G. W. Swenson, Jr.

University of Illinois Observatory
University of Illinois
Urbana, Illinois

I. INTRODUCTION

After the initial discoveries of NH_3, H_2O, and H_2CO in the interstellar medium many other molecules containing various combinations of H, C, O, and N atoms have been searched for. The initial molecules discovered involving H, C, and N atoms contain the unsaturated cyanide group $-C \equiv N$. These molecules are $H-C \equiv N$ (Snyder and Buhl 1971), $H_3C-C \equiv N$ (Solomon *et al.* 1971), and $H-C \equiv C-C \equiv N \, R'$ (Turner 1971). In contrast to this information, no saturated $R-NH$ nitrogen type molecules were found (R is $-CH_3$ or any other C and H atom group; R 'is -H, $-CH_3$ or some other combination).

The first molecule found which contained H, C, O, and N atoms was interstellar formamide, $O=C \overset{H}{\underset{H}{\langle}}$, which we recently reported (Rubin *et al.* 1971). Since this work we have conducted unsuccessful searches for two isomers of formamide; nitrosomethane, $H_3C-N=O$, and formoxime, $H_2C=N \overset{H}{\underset{.}{\sim}} O$ Although interstellar H-NCO (Snyder and Buhl 1972) has been observed we have failed to observe the less stable, but stronger (either absorbing or emitting), transitions in HCNO. The more stable polyatomic molecules, regarding UV and thermal rearrangement in the laboratory, are apparently more likely to be observed in the interstellar medium. In addition to the above comments on the chemical stability of isomeric forms in the interstellar medium we propose a model to explain the stronger emission from the $1_{10} \rightarrow 1_{11}$ K-doublet relative to the $2_{11} \rightarrow 2_{12}$ K-doublet in formamide. The emission observed in formamide as opposed to the normally observed absorption in formaldehyde is also explained by the model which involved the relative collisional and rotational radiative lifetimes.

II. FORMAMIDE

Formamide was identified by emission in Sgr B2 by the three strongest $\Delta F=0$ ^{14}N hyperfine transitions ($F=2 \rightarrow 2$, $F=3 \rightarrow 3$, and $F=2 \rightarrow 1$) in the $2_{11} \rightarrow 2_{12}$ K-doublet rotational transition. The $F=3 \rightarrow 3$ transition was blended with the H 112α transition (rest frequency of 4618.790 MHz). After resolving the $F=3 \rightarrow 3$ transition from the H 112α transition the rest frequencies of the three hyperfine components agreed very well with the corresponding laboratory measurements of $H_2^{14}NCOH$ $2_{11} \rightarrow 2_{12}$ ($F=2 \rightarrow 2$ = 4617.118 ± 0.020, $F=3 \rightarrow 3$ = 4618.970 ± 0.020, and $F=1 \rightarrow 1$ = 4619.988 ± 0.020 in units of MHz). The line temperature for the strongest of the hyperfine components ($F=3 \rightarrow 3$) was $T\ell =$ $(0.13 \pm 0.03)°K$. The column density of formamide molecules was found to be $N=2.2 \times 10^{16}$ cm^{-2}.

Evidence was also given for the existence of formamide in Sgr A. However, we found no evidence for formamide in W3 (continuum), W3 (OH), Ori A, Ori B, W49, W51, DR21, and the dust cloud L134.

After the above report on the discovery of formamide in Sgr B2 and Sgr A by the identification of the $2_{11} \rightarrow 2_{12}$ transition, Palmer, et al. have reported the observation of the lower frequency $1_{10} \rightarrow 1_{11}$ transition also in Sgr A and Sgr B2. The Tϱ for the $1_{10} \rightarrow 1_{11}$ (observed in emission) is considerably higher than the corresponding $2_{11} \rightarrow 2_{12}$ result. An analysis of the Tϱ for both transitions assuming Boltzmann statistics indicates that the lower frequency $1_{10} \rightarrow 1_{11}$ transition is emitting from energy levels which are at an excitation temperature considerably higher than the corresponding J=2 K-doublet energy levels.

III. COLLISIONAL EXCITATION AND ROTATIONAL RELAXATION MODEL

We have observed above that the J=1 and J=2 μ_a-dipole K=1 doublets in formamide are observed in emission but the relative populations of the J=1 and J=2 levels are not described by Boltzmann equilibrium. An energy level diagram for formamide is shown in Figure 1. The corresponding μ_a-dipole K-doublet transitions are observed in absorption in formaldehyde. Thus, the K-doublets in formamide are hotter than the background radiation and the K-doublets in formaldehyde are colder than the background. We will now propose a model to explain the non-equilibrium rotational emission from formamide in the interstellar medium. The model assumes that the H and H_2 concentration in the formamide cloud region exceeds the formamide concentration. This assumption is reasonable considering the observed column density for formamide. We also assume that the H_2-formamide collision time is short relative to the spontaneous rotational emission lifetimes. The rotational state lifetimes for a 5,000 MHz $\mu_a \cong 1$ Debye transition is $\tau_R \cong 10^9$ sec. The density of H_2 molecules to give a collisional time shorter than 10^9 sec is about $10^2/cm^3$. An energy level diagram showing the lower energy rotational states which are connected by μ_a and μ_b dipole transitions in formamide is shown in Figure 1. On the right of the diagram we note that the upper levels of the K-doublets have a higher excitation of angular momentum along the b axis than the corresponding lower level of the doublet.

If we now assume that the H atoms or H_2 molecules are translationally hotter than the formamide molecules, the collision will tend to excite molecules from the ground state (0_{00}) to the higher rotational states. Looking at the molecular diagram in Figure 2 we see that an H atom approaching the molecule along an axis parallel to the molecular c axis will lead (through a collision) to excitation of angular momentum along both the a and b axes. However, the predominant hard sphere collisional excitation will be to angular momentum along the b axis. This is evident from the position of the center of mass of the molecule and a hard sphere pictoral collisional cross section of the molecule; there are more

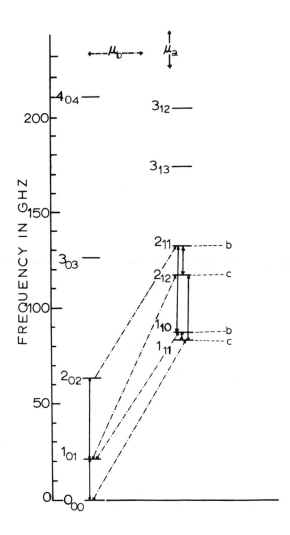

1. The energy levels of formamide. The double direction arrows on the levels indicate allowed μ_a and μ_b dipole transitions. Vertical relaxation between the K = 0, 1,2 . . . levels is allowed by μ_a-dipole transitions. Horizontal relaxation between the different K manifolds is allowed by the μ_b-dipole transitions in formamide. μ_b = 0 in formaldehyde which forbids horizontal relaxation between the K manifolds in formaldehyde.

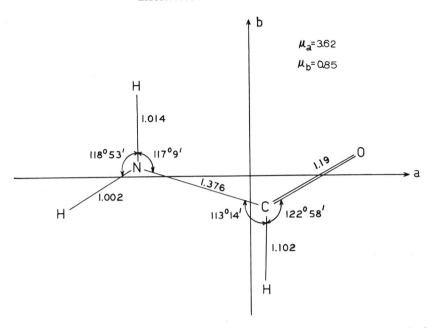

2. The formamide molecule showing the principal inertial a and b axes and the magnitudes of the dipole moments.

atoms extended away from the b axis than the a axis leading to a higher cross section for b axis excitation. Looking at b and a axis collisional excitation along with the above arguments indicate that the dominant collisional excitation of angular momentum from the ground state will be to states which have b axis excitation of angular momentum. In addition, it is evident from the energy spacings in Figure 1 and $1/\nu^3$ dependence on the radiative relaxations, that the lower levels of each K=1 doublet relax to the K = 0 levels (μ_b dipole) faster than the corresponding upper levels of the K doublets. These arguments show that the upper levels of the K-doublets are more populated, which explains the observed emission from the J=1 and J=2 K-doublets. Now, the reason for the higher J=1 excitation in formamide is apparently its proximity to the ground state. Thus, collisional excitation of one unit of angular momentum along the b axis in the $0_{00} \rightarrow 1_{10}$ process is more probable than the four units of angular momentum necessary for the $0_{00} \rightarrow 2_{11}$ excitation.

In the case of formaldehyde the μ_b dipole is zero and there is no allowed radiative relaxation between the K=0 and K=1 manifolds. Thus, the lower states in the K=1 doublets are metastable with respect to the lower energy K=0 levels. In addition to blocking the lower state radiative relaxation, Townes and Cheung (1969) have argued that collisional excitation will favor levels with the angular momentum along the c axis. Thus, according to Townes and Cheung the collisional excitation will favor the lower levels of the K-doublets (see Figure 1). The combination of collisional excitation of the lower levels and the inability of the K=1 doublets to relax into the K=0 manifold will lead to absorption of radiation in the K=1 doublets of formaldehyde. Both of these effects appear to work in the opposite direction in formamide leading to emission.

IV. OTHER ISOMERS

We have also searched, without success, for the $2_{11} \rightarrow 2_{12}$ transition in nitrosomethane ($CH_3N=0$), and the $1_{10} \rightarrow 1_{11}$ and $1_{01} \rightarrow 1_{000}$ transitions in formoxime ($CH_2=NOH$), indicating that these molecules are less abundant than formamide.

The result seem to indicate that a harsh UV environment will favor the isomer which is most stable to UV radiation. It is known that the following series of rearrangements occur in these molecules when subjected to UV radiation (Beak).

$$H_3C-N\equiv O \xrightarrow{u.v.} H_2C=N \overset{H}{\diagdown} O \longrightarrow H \diagup N \diagup C = O$$

These observations are also evident from recent searches for isocyanic acid (HNCO) and fulmimic acid (HCNO). HNCO has recently been observed in the interstellar medium. Since HCNO is a linear molecule with a large dipole moment, its transitions are very strong. It is less stable to UV radiation, however, and this may be the reason why we were unable to detect the J=1$\rightarrow$0 transition.

REFERENCES

Beak, P. (private communication).

Palmer, P.; Gottlieb, C. A.; Rickard L. J. and Zuckerman, B. 1971, *Bull. Am. Astron. Soc.,* **3**, 499.

Rubin, R. H.; Swenson, G. W.; Benson, R. C.; Tigelaar, H. L. and Flygare, W. H. 1971, *Ap. J. (Letters),* **169**, L39.

Snyder, L. E. and Buhl, D. 1971, *Ap. J. (Letters),* **163**, L47.

Snyder, L. E. and Buhl, D. 1972, *Ap.J.,* **177**, 619.

Solomon, P. M.; Jefferts, K. B.; Penzias, A. A. and Wilson, R. W. 1971, *Ap. J. (Letters),* **163**, L53.

Townes, C. H. and Cheung, A. C. 1969, *Ap. J. (Letters),* **157**, L103.

Turner, B. E. 1971, *Ap. J. (Letters),* **163**, L35.

Robinson: At Parkes, Ribes, Ables, and Godfrey have observed the $2_{12} - 2_{11}$ emission of formamide from Sgr B2. They detected the $1 \rightarrow 1$, $2 \rightarrow 2$, $2 \rightarrow 3$, $1 \rightarrow 2$, and $3 \rightarrow 2$ transitions, in good agreement with laboratory frequencies measured at Monash University by Godfrey and Crofts. The $3 \rightarrow 3$ transition is blended with the H112α recombination line. An attempt was made to remove the recombination line by subtracting the average of the H111α and H113α lines, but the H113α line is blended with the H142β line, and the removal cannot be achieved.

The 36-foot diameter radio telescope of the National Radio Astronomy Observatory, at Kitt Peak, Arizona – a site 6100 feet above sea level. This instrument operates at wavelengths longer than 1 mm. It has been used to discover about half of the molecules presently known to exist in interstellar clouds.

Detection of Acetaldehyde in Sagittarius

Carl A. Gottlieb

Harvard College Observatory
Cambridge, Massachusetts

We have detected acetaldehyde, CH_3CHO, at 1065 MHz (28 cm) in emission in Sgr A and Sgr B2 with the 140-foot telescope of the National Radio Astronomy Observatory. The observations were made on July 26 — August 2, 1971, by J. A. Ball and C. A. Gottlieb of the Harvard College Observatory, and A. E. Lilley and H. E. Radford of the Harvard College Observatory and the Smithsonian Astrophysical Observatory. At 28 cm the brightness temperature of the background continuum is large, yet acetaldehyde is observed in emission as is methyl alcohol (Ball *et al.* 1970) and formamide (Palmer *et al.* 1971). In the laboratory acetaldehyde and formaldehyde have similar chemical properties, and a comparison of the abundances of the two molecules should prove interesting, for it could indicate the extent of chemical evolution in our galaxy.

Figures 1 and 2 are the astronomical spectra taken with a receiver band width of 2.5 MHz and a spectral resolution after cosine weighting of 3 km/sec. The half

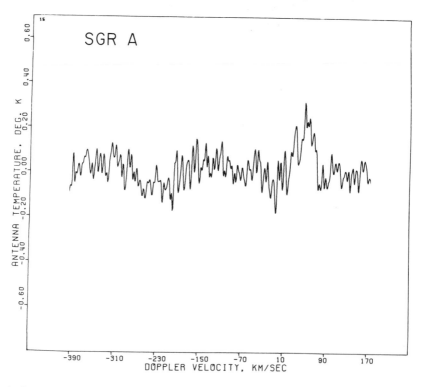

1. Spectrum of 28-cm line of acetaldehyde in Sgr A. The velocity scale is based on a line rest frequency of 1065.075 MHz, and is referenced to the Local Standard of Rest. The ON integration time was 447 min. The instrumental resolution was 11 kHz or 3 km/sec. A fourth order baseline has been removed for aesthetic reasons only, since the feature was clearly visible on the "raw spectrum".

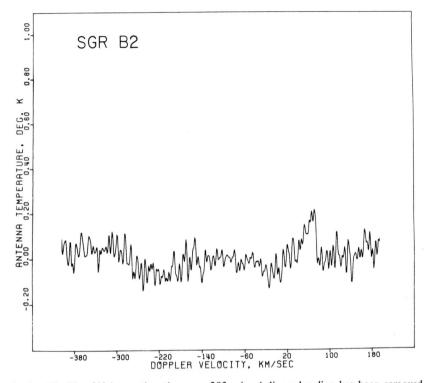

2. Sgr B2. The ON integration time was 283 min. A linear baseline has been removed. Otherwise all comments under Figure 1 apply.

Table 1

ACETALDEHYDE EMISSION FEATURES

Source	Pointing Position‡ (1950)						Continuum Antenna Temperature (°K)	Emission Features*		
	α			δ				T_L (°K)	full width to half power (km/sec)	$V_{LSR}^{\perp}$ (km/sec)
	h	m	s	°	'	"				
Sgr A	17	42	28	-28	58	30	215±72	0.3±0.1	40±8	46±10
Sgr B2	17	44	11	-28	22	30	94±24	0.2±0.1	24±5	58±10

$^{\perp}V_{LSR}$ is the Doppler velocity with respect to the local standard of rest and a line rest frequency of 1065.075 MHz. The velocity of the Sun with respect to the local standard of rest was taken to be 20 km/sec toward 18^h, +30° (1900).

*The error limits are our estimates of the peak errors due to noise. No correction has been made for instrumental resolution or for hyperfine splitting.

‡Because we measured neither the position nor the angular extent of the acetaldehyde emission, we give only the position toward which the antenna was pointed when we obtained the spectra shown in Figures 1 and 2.

power beam width of the antenna is approximately 28 arc min. All observations were made in the total power mode. The front end was an uncooled parametric amplifier whose system temperature on cold sky was $226 \pm 25°K$, and the spectral line receiver was the 413-channel one-bit autocorrelator. Table 1 summarizes the parameters of the astronomical spectra. All temperatures are

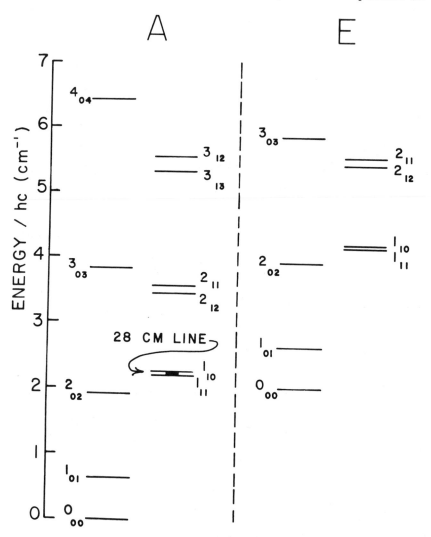

3. Lowest rotational energy levels of acetaldehyde. The E level is arbitrarily placed 2 cm[-1] higher than the A level. The first excited torsional levels start at 143 cm[-1] (Souter and Wood 1970).

referred to Virgo A, whose antenna temperature was taken as $68 \pm 14°K$.

The lowest rotational energy levels of acetaldehyde are shown in Figure 3. Internal rotation causes the end-over-end rotation levels to divide into levels of A and E symmetries (Kilb *et al.* 1957; Souter and Wood 1970). There are no allowed electric dipole transitions between the A and E branches since the dipole moment cannot depend on the internal rotation angle. The Einstein A coefficient for the 1_{11}-1_{10} transition is 4.6×10^{-11} sec^{-1}, and the radiative lifetime of the two states is $\sim 10^6$ sec.

The Doppler velocities are based on a measurement of the rest frequency made with a spectrometer similar to that used in the most recent measurements of the 834-MHz methyl alcohol line (Radford 1972). The main feature of this spectrometer is the large silver-plated aluminum cylindrical cavity that is resonant in the TE_{012} mode and whose active volume is 70 liters and loaded Q is 48,000.

The spectra shown in Figures 1 and 2 are identified with acetaldehyde, solely on the basis of the close agreement in velocity with other molecules observed in the galactic-center region, most notably methyl alcohol at 36 cm and formamide at 19 cm. The methyl alcohol velocities have been revised to 39 ± 5 km/sec in Sgr A and 54 ± 9 km/sec in Sgr B2 (Radford 1971). The formamide velocities are 47 ± 5 km/sec in Sgr A and 61 ± 5 km/sec in Sgr B2 (Palmer *et al.* 1971). We think that it is most meaningful to compare our assigned acetaldehyde features with those of methyl alcohol and formamide since these molecules are all observed in emission, the telescope beam widths are comparable, and the line widths are nearly the same.

Spectroscopic confirmation must await the detection of the 1_{11}-1_{10} transition in the E level that comes at 16 cm or another line (see Figure 3). Laboratory work is now in progress to determine the rest frequency of the 16-cm transition. Also, we are examining the excitation mechanisms of interstellar acetaldehyde. We hope to be able to make estimates of the column density so that we can compare the abundances of acetaldehyde and formaldehyde.

We thank John Payne and Tom Dunbrack for assistance with the telescope equipment and Douglas Pitman for assistance with the laboratory measurements. We are especially grateful to Professor Carl Heiles for many helpful comments and advice on observing techniques. This work was supported in part by the National Science Foundation through grant GP-19717.

REFERENCES

Ball, J. A.; Gottlieb, C. A.; Lilley, A. E. and Radford, H. E. 1970, *Ap. J. (Letters),* **162**, L203.

Kilb, R. W.; Lin, C. C. and Wilson, E. B. Jr. 1957, *J. Chem. Phys.,* **26**, 1695.

Palmer, P.; Gottlieb, C. A.; Rickard, L. J. and Zuckerman, B. 1971, *Bull. Am. Astron. Soc.* **3**, 499.

Radford, H. E. 1972, *Ap. J.* **174**, 207.

Souter, C. E. and Wood, J. L. 1970, *J. Chem. Phys.,* **52**, 674.

The Detection of a MM-Wave
Transition of Methlacetylene

David Buhl

National Radio Astronomy Observatory
Green Bank, West Virginia

Lewis E. Snyder

Department of Astronomy
University of Virginia
Charlottesville, Virginia

In May of 1971, we detected the $5_0 \rightarrow 4_0$ transition of CH_3CCH in emission in the galactic center source Sgr B2. This molecule is similar to cyanoacetylene and is a product of synthesis reactions producing complex hydrocarbons. The line was detected using the NRAO 36-foot telescope in Tucson, Arizona and a 3-mm mixer radiometer with a system temperature of $4000°K$. At this wavelength the antenna has a beamwidth of $70''$ and the spectral line receiver has a resolution of 2 MHz (Buhl and Snyder, 1971).

The line spectrum is shown in Figure 1. The center velocity is +60 km/sec with a velocity width of 20 km/sec. The antenna temperature is estimated to be $0.3 \pm 0.2K$. There is no evidence of the K=1 or K=2 transition which should be shifted 1.6 MHz (6 km/s) and 6.5 MHz (23 km/s) to the right. The K=1 line may be blended with the K=0 line.

Attempts have been made to search for the first two levels of this molecule without success. The $1_0 \rightarrow 0_0$ transition at 17.092 GHz was looked for by Dickinson and Papadopoulos using the Haystack antenna, and the $2_0 \rightarrow 1_0$ at 34.183 GHz was search for by Schwartz and Zuckerman using the Naval Research Laboratory antenna. The problem with detecting these levels is that the larger beamwidth at these wavelengths dilutes the line. In addition, the line strength is directly proportional to the transition frequency and the principal quantum number J. These two effects favor the mm-wave transitions of the molecule ($J > 2$).

The 5_0 level is 8 cm^{-1} above the ground state and would be populated at temperatures ≥ 10 K. Hence, for most reasonable interstellar temperatures the molecule will be excited up to at least J=5. If the excitation is approximately thermal the $5_0 - 4_0$ transition at 85.457 GHz will be about 15 times the intensity of the ground state $1_0 - 0_0$ transition. This is illustrated in Figure 2 where the line intensity relative to the ground state is plotted against excitation temperature for several transitions. The intensity increases as one goes to higher level transitions and toward higher excitation temperatures.

The line ratio for the $5_0 - 4_0$ transition (Figure 2) does not change much for temperatures above $20°K$. However, as the principal quantum number J increases, the slope of the curve also increases giving a greater change in line ratio with temperature. This means that the higher levels (J>6), in addition to being more intense, are more sensitive to changes in excitation temperature and therefore will be more accurate as interstellar thermometers.

The lower level transitions ($1_0 - 0_0$ and $2_0 - 1_0$) will be much weaker. The intensities predicted from Figure 2 would be $.02°K$ and $.06°K$, probably below the limit of sensitivity for current receivers. Thus the two negative results mentioned earlier are understandable.

The implication of these results is that the excitation of the methyl acetylene molecule is reasonably thermal. Any inversion of the ground state would be

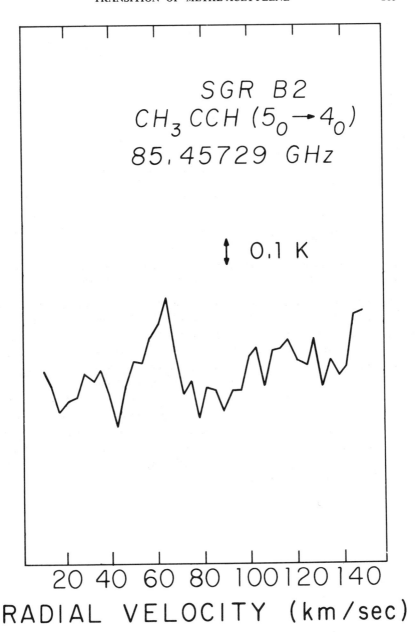

1. 3.5-mm line of methyl acetylene as seen in Sgr B2. Only the $5_0\text{-}4_0$ transition is evident; however, it could be blended with the $5_1\text{-}4_1$.

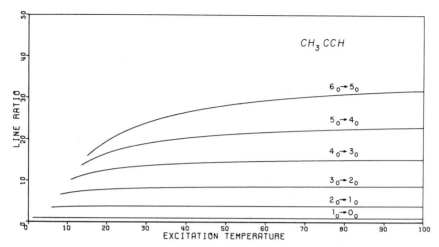

2. Predicted line strengths relative to the ground state for K=0 transitions of methyl-acetylene. Calculation is based on thermal excitation of the molecule.

easily detectable. Similar conclusions can be drawn from observations of the HNCO and OCS molecules (Buhl and Snyder 1972). The large organic molecules generally have quite a number of states below 20 cm^{-1} which will be excited. Thermal excitation should make the higher J transitions more intense. Measurements of a number of these transtions should give us a better picture of the excitation environment in which these molecules are imbedded.

The authors would like to thank Dr. E. K. Conklin and the 36-foot telescope group for aid with the observations. We also acknowledge the work of S. Weinreb, N. Albaugh, J. Edrich, D. Ross and E. Scheuetz on the NRAO 3mm line receiver. Finally we appreciate the contributions of C. Burrus, K. B. Jefferts, A. A. Penzias and R. W. Wilson of Bell Telephone Laboratories to the development of mm-wave receivers.

REFERENCES

Buhl, D. and Snyder, L. E., 1971, *Nature,* **232,** 161.
Buhl, D. and Snyder, L. E. 1972, *Proc. Liege Symp.* (in press).

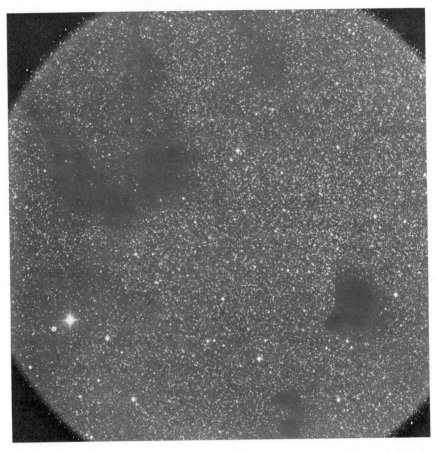

A photograph, taken by B. J. Bok, with a 90-in aperture telescope, of Barnard 72 (the snake-like obscuration in the upper left) and of Barnard 68 (the single globule in the lower left hand corner). These dark nebulae are good examples of small dust clouds.

Measurements of the $\lambda 6$cm H_2CO Excitation Temperature in Dark Clouds

Carl Heiles

Astronomy Department
University of California
Berkley, California

The equation of transfer, which provides the emergent brightness temperature, $T_{B(f)}$ (a function of frequency f), from a cloud of molecules having excitation temperature T_{ex} and optical depth $\tau_{(f)}$ in some microwave line, was given earlier today by Professor Barrett:

$$T_{B(f)} - T_c = (T_{ex} - T_c)\left(1 - \exp\left[-\tau_{(f)}\right]\right) \qquad (1)$$

where T_c is the continuum brightness temperature behind the cloud. For a frequency-switched system, often used for such observations, the quantity observed is the antenna temperature, equal to the left hand side of this equation multiplied by an appropriate factor denoted herein by G, which lies between 0 and 1 and accounts for telescope efficiency and size of the cloud as compared to the telescope's beam diameter.

Often a molecule has two or more transitions for which the excitation temperature T_{ex} is expected to be the same; for example, the six hyperfine components of the 6cm line of H_2CO. In general, these transitions will have different transition probabilities and statistical weight factors, and hence different optical depths in equation (1).

The second factor on the right hand side of equation (1) is equal to $\tau_{(f)}$ if $\tau_{(f)}$ is small compared to unity. In this case the intensities of the H_2CO hyperfine components are in the same ratio as the optical depths. However, if the optical depth becomes large the nonlinearity of this factor will change the relative intensities, an effect called saturation, thus allowing a determination of $\tau_{(f)}$ and T_{ex} individually.

Five of the six H_2CO hyperfine components have very nearly the same frequency, but the other lies about 18 kHz below the others. Since this is equivalent to a Doppler shift of only 1.1 km/sec, this component is easily distinguishable only for cold material having little macroscopic motion. The material in dust clouds satisfies this criterion. The apparent intensity of this component can thus be compared to that of the others and an attempt made to derive T_{ex}.

This Spring the Hat Creek 85-foot telescope, together with a two stage cooled parametric amplifier (system temperature about 55 $^\circ$K) and the 100-channel filter bank (filter width and spacing 2 kHz), was used to observe 44 positions in 8 dust clouds. The resulting profiles were fit with a function of the form

$$T_{A(f)} = K + \sum_{J=1}^{N} \left(T_{ex_j} - T_c \right)$$

$$\times \left\{ 1 - \exp - \left(A_j \sum_{i=1}^{6} a_i \exp - \left[(f - f_i - F_j)^2 / 2 \sigma_j^2 \right] \right) \right\} \quad (2)$$

using the least squares technique (Chauvenet, 1874). Here the index j sums over N clouds having velocity v_j, (apparent frequency F_j) excitation temperature T_{ex_j}, frequency dispersion σ_j, and central optical depth (of the strongest hyperfine component $2 \to 2$)A_j. The index i sums over the 6 hyperfine components; a_i are their relative intensities and f_i their relative frequencies (Tucker et al., 1970). This function, then, describes the set of six hyperfine components broadened by N Maxwellian velocity distributions, centered at different velocities, plus a constant K to account for a frequency-independent zero level error.

The signal-to-noise in this preliminary survey was insufficient to derive accurate optical depths or excitation temperatures because each point received only a few hours integration time. The data did show, however, that all of the profiles were extremely well-fit by equation (2). Most required only one velocity component (N=1); a few required two. From the initial list a small number of positions having strong lines were chosen for more detailed scrutiny.

These latter observations are currently underway and preliminary results for four positions became available just last night; the results are given in Table 1. The errors quoted are the mean errors as defined by Chauvenet (1874), and therefore do not represent strict bounds. An additional source of uncertainty, not included in Table 1, is the relative size of the cloud and the telescope beam, which influences the conversion from antenna temperature to brightness temperature (see discussion following equation (1)). This has been estimated by assuming the H_2CO to be concentrated in the apparent portion of the cloud as seen on the Palomar Sky Survey prints. This procedure is probably valid, given the linear dependence of H_2CO line intensity on optical extinction reported in this meeting by Kutner (see also Heiles and Turner, 1973). However, the H_2CO distribution must be mapped to ensure reliability and the present results should

TABLE 1

CHARACTERISTICS OF H_2CO IN DUST CLOUDS

α_{1950}	δ_{1950}	ℓ^{II}	b^{II}	τ^a	Δv^b (km/sec)	$(T_{ex} - T_c)/G^c$ (°K)	$T_{ex}{}^d$ (°K)
$04^h20^m35^s$	24°58'	172°1	-16°9	0.73 ± 0.14	0.51 ± 0.02	-0.92 ± 0.09	1.47
04 36 46	25 41	174.0	-13.7	0.64 0.12	0.87 0.03	-0.91 0.09	1.49
15 50 54	-04 31	4.1	35.8	0.52 0.18	0.62 0.04	-0.98 0.22	1.39
16 29 35	-24 24	353.9	15.8	0.96 0.25	0.85 0.04	-0.53 0.07	1.99

a. Optical depth at peak of strongest hyperfine component $(2 \rightarrow 2)$.
b. Full width at half intensity.
c. For definition of G, see discussion following equation (1). The beam efficiency has not
    been measured; it has been taken as 0.75.
d. T_c taken as 2.7 °K.
    Contains two velocity components; see text.

therefore be regarded as preliminary. Since the effect of this "beam dilution" would be to make the antenna temperature smaller than expected, the values of T_{ex} given in Table 1 should be regarded as upper limits. We suspect, however, that they will prove to be close to the actual values.

The first three of the four positions in Table 1 have essentially the same excitation temperature, 1.5°K. The fourth is near cloud 4 of Heiles (1969) and requires two velocity components for a good fit. Saturation occurs for only one of these components. This saturated component appears to have a higher excitation temperature than that found in the other 3 positions. If it is indeed different the H_2CO excitation temperature is influenced at least to some extent by local conditions.

Cloud 4, a well-studied object near, but not containing the fourth position given in Table 1, is unusual in three respects: first, its kinetic temperature is larger than that found in other dust clouds (as determined using the OH lines by Heiles, 1969, and using CO as reported at this meeting by Solomon). Second, this cloud lies near the star Antares, which has been shown to be a radio source at 3cm by Hjellming and Wade (1971); if its 2mm flux is large it may be radiatively pumping the H_2CO. Finally, this general region contains several hot stars and the surrounding optical radiation field is much more intense than that for a typical dust cloud — large enough to produce reflection nebulae. This may also imply an intense IR field, also able to radiatively pump the H_2CO (Litvak, presented at this meeting).

REFERENCES

Chauvenet, W. 1874, *A Manual of Spherical and Practical Astronomy*, **2**, (Philadelphia: J. B. Lippincott & Co.),p. 469.

Heiles, C. 1969, *Ap. J.* **157**, 123.

Heiles, C. and Turner, B. 1973, in preparation.

Hjellming, R. M. and Wade, C. M. 1971, *Science,* **173**, 1087.

Tucker, K. D.; Tomasevich, G. R. and Thaddeus, P. 1970, *Ap. J. (Letters),* **161**, L153.

A Study of Formaldehyde
6-CM Anomalous Absorption

Marc L. Kutner

Physics Department
Columbia University
New York, New York

I want to report a comparative study of 6-cm formaldehyde "anomalous" absorption and optical extinction in a dark   cloud in the Taurus complex. The radio observations were made with the NRAO 140-foot telescope in collaboration with Scoville, Solomon and Thaddeus. The purpose of this study was to investigate the dependence of the distribution of H_2CO, and the pumping mechanism responsible for anomalous absorption, on extinction, and ultimately on the total density.

Figure 1 shows the cloud studied. It lies in the Taurus complex of dark nebulae, and Heiles' Cloud 2, where anomalous absorption was discovered by

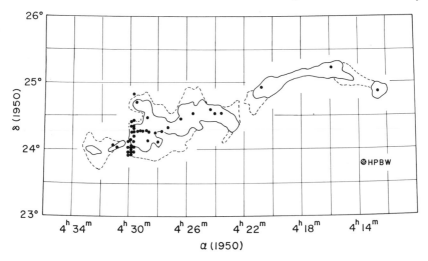

1. Map of the dark cloud. Dots indicate position of radio observations.

Palmer *et al.* (1969), is off the Figure to the upper left. The dotted and solid lines correspond roughly to the cloud outlines on the blue and red Palomar Sky Survey prints. One side of the cloud shows well defined edges, which is one reason for the selection of this cloud for study. The dots indicate the positions where radio observations were made, typical integration being ½ - 1 hr. At each of these positions the optical extinction in the red and blue was determined from the Palomar Sky Survey by the star counting method described by Bok (1956) and recently applied by Bok and Cordwell (1973). At each location stars were counted in a square roughly the size of the beam.

Figure 2 shows some of the observed spectra. There is a smooth velocity shift observed across the cloud, but no evidence of large velocity gradients. The bottom line profile, a synthesis of the spectra at three adjacent positions, shows some evidence of hyperfine structure.

Figure 3 is a comparison of the radio and optical data. The main thing to note

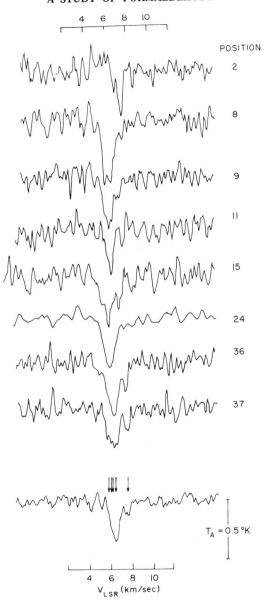

2. Some of the H₂CO spectra. The bottom profile is a synthesis of spectra taken at 3 positions.

is that there is a fairly tight correlation between the line strength and the extinction. The straight lines are fit to the data with slopes of about $0.1°$/mag in the blue and $0.15°$/mag in the red (results consistent with a $1/\lambda$ extinction law).

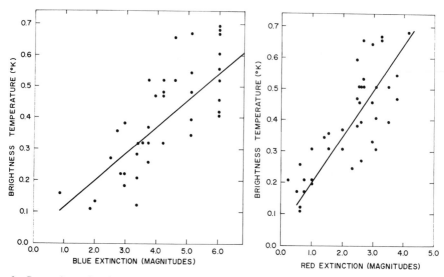

3. Comparison of extinction and radio observations.

At high extinction there is some error (that is near 5-6 magnitudes in the blue or 3-4 magnitudes in the red) due to the small number of stars counted in the square.

Assuming an excitation temperature of 1.7°K, we obtain from this data a maximum column density of ortho-formaldehyde molecules of

$$N_{H_2CO} \sim 2 \times 10^{13} \text{ cm}^{-2} \qquad (1)$$

and a formaldehyde to hydrogen ratio

$$N_{H_2CO}/N_{H_2} \sim 4 \times 10^{-9} \qquad (2)$$

in fairly good agreement with the ratio obtained from 2-mm formaldehyde observations in the Orion Nebula (Thaddeus *et al.* 1971) and the ratio obtained by Zuckerman *et al.* (1969) in their 6-cm H_2CO survey.

There are two general explanations for the observed correlation between line strength and extinction. The first is that it represents the distribution of the formaldehyde, the fading of the line at lower extinctions being produced simply by decrease in optical depth. The second is that the pumping mechanism responsible for anomalous absorption is density dependent and the fading of the line at lower extinction is due to the return of the excitation temperature to 2.7°K. Neither can be ruled out on the basis of our observations here, but the

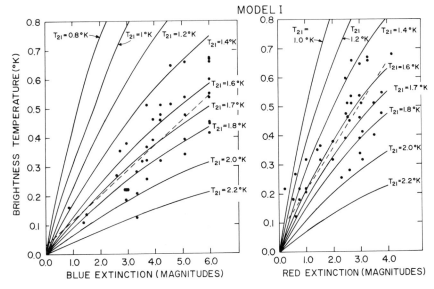

4. The results of Model I, showing the relationship between extinction and peak line temperature for various excitation temperatures.

results suggest how future studies of this type may be able to make such a distinction.

I have considered two basic models for the excitation, and have plotted the predicted relationship between the line strength and blue and red extinctions.

The first model is based on the assumption that the excitation temperature is constant throughout the cloud which would be the case if the anomalous cooling is due to deviations from a blackbody spectrum in the background radiation. We also assume that the formaldehyde is mixed with the dust and hydrogen so that N_{H_2CO}/N_{H_2} is constant.

Figure 4 shows the relationship between the peak line temperature and the extinction for various values of the excitation temperature. The data is consistent with this model and an excitation temperature of about $1.6°$ -$1.7°K$, an excitation temperature in good agreement with the results just presented by Heiles.

The second model is based on the assumption that the cooling is dependent on neutral particle density as would be the case if collisions are responsible. For this it was necessary to make an assumption about the dependence of the excitation temperature on the neutral particle density. I used the quantum mechanical calculation of Thaddeus (1972) and chose the case which showed the strongest cooling. It corresponds to a kinetic temperature of $60°K$ but provides a reasonable calculational model. I have also considered two cases of this model,

MODEL II a

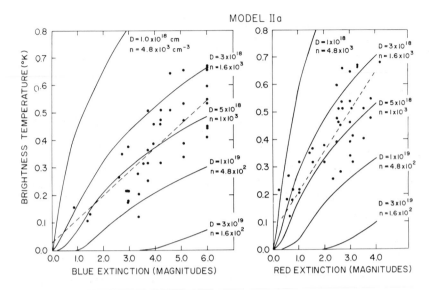

5. The results of Model IIa, showing the relationship between extinction and peak line temperature for a range of cloud thicknesses and densities.

with different assumptions on the distribution of formaldehyde. For Model IIa I have assumed that $N_{H_2CO} = 2 \times 10^{13}$ cm^{-2} all over the cloud that is. the formaldehyde is in a sheet evenly distributed over the cloud and for Model IIb I have assumed mixing of H_2CO and dust so that N_{H_2CO}/N_{H_2} is constant.

Figure 5 shows the relationship between line strength and extinction for Model IIa, the different curves corresponding to different assumed thicknesses of the cloud, and therefore different maximum densities of hydrogen. Of these curves the best fit to the data is for the case where the cloud thickness is 5×10^{18} cm, corresponding to a maximum hydrogen density of about 1000 cm^{-3}.

The important point to notice in this case is that for each curve there is a threshold for the appearance of the line, and the curve which agrees best with the data indicates that a line should not be seen in the regions of lower extinction. Since the case for the dependence of the excitation temperature on density was one which showed the strongest cooling, we may conclude that a more realistic case would push these thresholds to higher extinctions. Our observations do not show evidence for such thresholds, but the data is far from conclusive. A weakness of this model is the short lifetime of the H_2CO at lower extinction due to the photodissociation. As we take the models where the formaldehyde is increasingly mixed with the hydrogen the curves become steeper until we reach the case of Model IIb which is shown in Figure 6. Again this clearly shows the threshold effect. Another point to notice about the density dependent models is that they show a strong dependence on the assumed depth of the cloud, so we might expect that if different clouds were observed the slope would appear to be different. But the radiative cases should show no change unless we look at very dense regions where collisions will quench the cooling.

In conclusion, the most important result obtained from these observations is that there is apparently a rather tight correlation between the line brightness temperature and the optical extinction. On the basis of these observations alone it is not possible to determine the mechanism responsible for the anomalous cooling. However, this correlation between line temperature and extinction implies that further studies of this type, with a sample of a variety of clouds, and with special attention to regions of lower extinction, may provide a useful method for studying the question of the 6-cm pump.

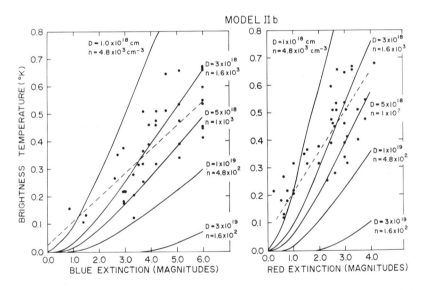

6. The results of Model IIb, showing the relationship between extinction and peak line temperature for a range of cloud thicknesses and densities.

REFERENCES

Bok, B. J. 1956, *Astron. J.* **61**, 309.

Bok, B. J. and Cordwell, C. S. 1973, *A Study of Dark Nebulae,* this volume.

Palmer, P.; Zuckerman, B.; Buhl, D. and Snyder, L. E. 1969, *Ap. J.,* **156**, L147.

Thaddeus, P. (1972), *Ap. J.,* **173**, 317.

Thaddeus, P.; Wilson, R. W.; Kutner, M.; Penzias, A. A. and Jefferts, K. B. 1971, *Ap. J.,* **168**, L59.

Zuckerman, B.; Palmer, P.; Snyder, L. E. and Buhl, D. 1969, *Ap. J.* **157**, L167.

The Kinematics of Molecular Clouds in the Nuclear Disc

N. Z. Scoville

Department of Astronomy
Columbia University
New York, New York

The most obvious characteristic of molecular gas in the galactic center is its clumpiness. Many of these clouds have high radial velocities (> 50 km/sec), large line widths (> 20 km/sec), large diameters (~ 30 pc), and tremendous masses (10^4 - $10^6 M_\odot$). These clouds are evidently very different from the interstellar clouds encountered in other regions of the galaxy, and it is interesting to try to explain why such clouds exist in the vicinity of the galactic center.

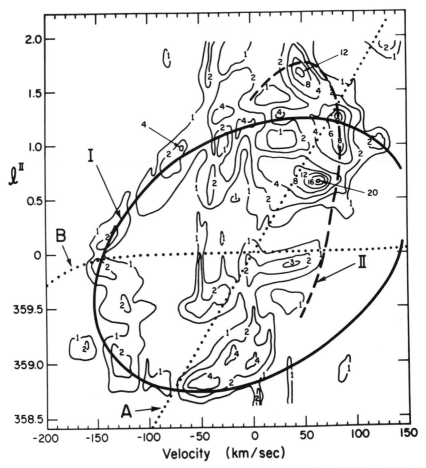

1. Contour diagram in the longitude-velocity plane of the average 6-cm H_2CO optical depth measured at $b^{II} = -2'$ (Scoville et al. 1972) and at $b^{II} = -12'$ (Scoville and Solomon 1972). Optical depth contour unit is 0.025. The solid (I) and dashed (II) curves show the longitude-velocity shape of two ring models discussed in the text. The dotted curves (A and B) show the minimum and maximum radial velocities expected from gas in the nuclear disc obeying the rotation law of Sanders and Lowinger (1972).

The large scale dynamics of these clouds can be studied best in the 18-cm OH and 6-cm H_2CO absorption lines and the 2.6-mm CO emission line. The information obtained from existing surveys at these wavelengths is very nicely complementary. Large angular coverage of the galactic center is provided by the OH survey of Robinson and McGee (1970), while the H_2CO observations of Scoville, Solomon, and Thaddeus (1972) give somewhat higher angular and velocity resolution in the most critical regions. Finally the CO observations (Penzias, Jefferts, and Wilson 1972) provide high resolution emission line data near the strongest cm-wavelength continuum sources which is where the absorption observations can be most misleading since they are biased toward gas lying in front of the sources.

Figure 1 shows the average of the 6-cm H_2CO optical depths at $b^{II} = -2'$ and $-12'$ plotted in a longitude-velocity diagram for longitudes close to the galactic center. From the observations at these two latitudes, it was found that the broad-line clouds appearing at positive velocities were strongest at $b^{II} = -2'$, while the negative-velocity clouds showed up most strongly at $-12'$. In order to represent most of the galactic center clouds observed in OH and H_2CO absorption in one longitude-velocity diagram this synthetic diagram was made by averaging the two separate sets of data.

The molecular clouds of Figure 1 may usefully be separated into three distinct groups: "local" clouds, 4-kpc arm clouds, and galactic center clouds. The local and 4-kpc arm clouds characteristically have velocities of 0 and -50 km/sec and show very narrow lines as compared with their 21-cm counterparts.

The unusually broad lines and high velocities of most of the other clouds in Figure 1 indicate that they are very different from the 4-kpc arm and local clouds. From now on I shall limit my discussion to these clouds: first introducing evidence that indicates they are in the nuclear disc, then showing that they may be joined up into one continuous feature, and finally presenting a kinematical model for this feature.

That most of these clouds belong to the nuclear disc is suggested by their being found to lie in a thin strip aligned with the galactic plane and not extending more than $2°$ in longitude from the galactic center. [There are a few clouds at higher longitudes (McGee *et al.* 1970, Scoville and Solomon 1972); it is not clear, however, if they are related to the much larger number of clouds observed closer to the center.] In addition, the large radial velocities of many of these clouds indicate they are not associated with the normal galactic spiral arms, which have nearly zero radial velocity in this direction. Finally, we note from Figure 1 that molecules of the 4-kpc arm are absorbing radiation from the strong thermal source Sgr B2 (at $\ell^{II} = 0°.7$); this source and its associated molecular cloud (seen at $\ell^{II} = 0°.7$ and $V = 62$ km/sec) must therefore lie behind the 4-kpc arm. This, together with the longitude, latitude, and velocity distributions of these broad-line clouds, suggests that they are in the rapidly rotating nuclear-disc at distances of several hundred parsecs from the center. Surprisingly, however, the velocities of these clouds show no obvious correspondence with the radial velocities expected from circular motion in the disc. (Circular motion corresponds to the region between curves A and B in Figure 1.) The large number of clouds with negative velocities at positive longitude immediately suggests that many of them possess very large radial motions either towards or away from the galactic center.

At this point we ask if the clouds are but components of what is basically a single feature. That this might be so is suggested by the continuity of the negative velocity clouds producing broad molecular absorption, which is evident in both 6-cm H_2CO (Figure 1) and in 18-cm OH (McGee 1970). The continuity of the denser positive velocity clouds is, however, much less evident, and, in fact, there appears to be a complete discontinuity between the positive velocity clouds at $\ell^{II} = 0°0$ and $0°7$, (Figure 1). This 'gap' could, however, be a result of

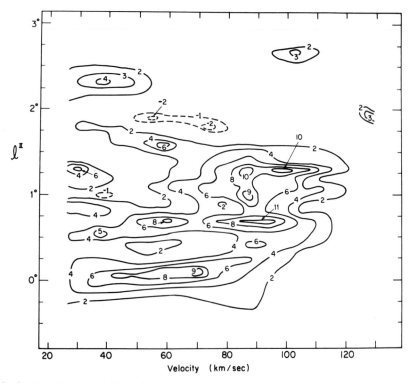

2. Contour diagram of CO emission in the longitude-velocity plane at $b^{II} = -2'$ from $\ell^{II} = 359°5$ to $2°8$ (Scoville *et al.* 1972a). Contour unit is $1 °K T_A$ ($\approx 1.7 °K T_B$).

the continuum source G 0.2 - 0.1, which accounts for most of the radiation in these longitudes, being in front of this gas. It is therefore of interest to study these positive velocity clouds in the 2.6-mm CO emission line (Figure 2). We see that in addition to emission corresponding to the deep absorption features at ℓ^{11} = 0°.0 and 0°.7, each of these clouds has a higher velocity emission component and there is also a very distinct bridge at $\sim$ 90 km/sec connecting these high velocity peaks. This bridge, in fact, is also seen in 6-cm absorption, although its low apparent optical depth does not show up in Figure 1. On this basis, I should like to suggest that most of the molecular clouds of the galactic center at both positive and negative velocities may be joined together into one continuous sequence. There apparently are non-uniformities in this sequence especially at positive velocities and perhaps a real gap at slightly negative longitudes and low positive velocities. Also at around ℓ^{11} = 1°.5 the sequence becomes confused and there are two possible conections between the negative and positive velocity clouds.

TABLE 1

Model Rings of Figure 1

MODEL	I^1	II^2
RADIUS[3] (pc)	218	305
ROTATIONAL VELOCITY (km/sec)	50	50
RADIAL VELOCITY[4] (km/sec)	±145	±70

NOTES TO TABLE 1.

1. Fit to negative velocity clouds in Figure 1.

2. Fit to positive velocity clouds in Figures 1 and 2.

3. The center of the ring is taken to be the galactic center.

4. Expansion or contraction velocity.

The elliptical shape of this sequence in the longitude-velocity diagram might be described kinematically as a ring that is rotating and is either expanding or contracting. The rotational motion accounts for the displacement of the ends of the sequence from zero velocity while the expansion or contraction of the ring accounts for the velocity separation of the two sides. Table 1 lists the parameters obtained by fitting model rings separately to the negative and positive velocity clouds (curves I and II of Figure 1). Perhaps the real form of the structure is a somewhat asymmetrical ring which varies between these two models along its circumference.

Whether the ring is expanding or contracting is a difficult question whose answer will require higher resolution absorption measurements. An initial analysis of the variations in 6-cm absorption intensity near the sources G 0.2 - 0.0 and Sgr B2 suggests quite strongly that the ring is expanding. In that case a suggested picture for this ring with possible locations of the continuum sources is shown in Figure 3. The low rotational velocity of the ring might be consistent with the notion that it is expanding and contains therefore much low angular momentum gas from closer to the center of the nuclear disc. However, it is then difficult, though probably not impossible, to explain why there is intense OH and H_2CO absorption at +40 km/sec in the direction of Sgr A since this gas

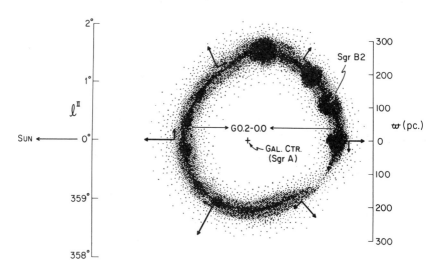

3. Sketch of rotating molecular ring near the galactic center as seen from the galactic plane. In this drawing the ring is taken to be expanding, although as discussed in the text, contraction can not be ruled out. The suggested locations of continuum radiation sources relative to this ring are also included. The label "Sgr A" refers to the non-thermal continuum component observed at $\ell^{II} = 0°.0$, $b^{II} = -0°.0$.

should lie behind the galactic center. If one makes the usual assumption that the non-thermal component in Sgr A is at the galactic center, then in order to form this absorption line, most of the other components in Sgr A must be beyond the galactic center and therefore unrelated to the non-thermal component. Thus, although the expansion model is tentatively adopted, contraction is not yet ruled out.

The total mass of gas in this ring is about $3 \times 10^6 M_\odot$, as estimated from the 6-cm observations, and the present kinetic energy observed in the radial motion of the ring is then $\sim 6 \times 10^{53}$ ergs.

The CO and H_2CO observations discussed here were made on the 36-foot and 140-foot telescopes of the National Radio Astronomy Observatory.*

*Operated by Associated Universities, Inc., under contract with the National Science Foundation.

REFERENCES

McGee, R. X. 1970, *Aust. J. Phys.*, **23**, 541.

McGee, R. X.; Brooks, J. W.; Batchelor, R. A. and Sinclair, M. W. 1970, *Aust. J. Phys.*, **23**, 777.

Robinson, B. J. and McGee, R. X. 1970, *Aust. J. Phys.*, **23**, 405.

Sanders, R. H. and Lowinger, T. L. 1972, *Astron. J.*, **77**, 292.

Scoville, N. Z. and Solomon, P. M. 1972 (unpublished).

Scoville, N. Z.; Solomon, P. M. and Thaddeus, P. 1972, *Ap. J.*, **172**, 335.

Solomon, P. M.; Scoville, N. Z.; Penzias, A. A.; Jefferts, K. B. and Wilson, R. W. 1972, *Ap. J.*, **178**, 125.

Goss: The HI mass of the 50 km/sec object in the direction of Sgr A has been determined with the Parkes interferometer to be $\sim 3 \times 10^9$ M$_\odot$, compared to the $10^5 - 10^6$ M$_\odot$ derived from the H_2CO observations. Presumably a large fraction of the mass is in the form of H_2.

Bok: All three previous speakers (Heiles, Kutner, Scoville) spoke of *clouds* in their beautiful presentations. May I suggest that it is very often inappropriate to think in terms of *sheets* or *filaments*. The Southern Coalsack is a fine example of an object which is first thought to be nearly spherical. Inspection of long-exposure photographs taken at Cerro Tololo shows many small-scale structure in the surface star counts, structure which may correspond to loosely connected sheets rather than a sphere. Magnetic alignment of bright and dark nebulae may be present.

Heiles: From a kinematic point of view, Cloud 2 may be composed of sheets as shown by recent observations of OH, H, and H_2CO microwave lines. The line profiles are double, with separations of about 0.7 km/sec and widths <0.7 km/sec. The velocity of each component is constant to ±0.2 km/sec over the cloud, which is nearly a square degree in area. These velocities are difficult to fit to a cloud which is as thick as it is wide. Hence, I suspect the cloud may be composed of sheets.

Annested: Are the sheets aligned along the magnetic field? If so, depletion of cooling elements as suggested here by Field, with consequent heating and expansion of an initial cloud could cause such structure. The magnetic pressure of a field, $\sim 3\mu$ G, would prohibit expansion across the field lines.

Isotope Abundances in Interstellar Molecules

B. Zuckerman

Astromony Program
University of Maryland
College Park, Maryland

My discussion will be divided into three brief sections. First, I will list a few of the astrophysical uses of isotopic abundances when studied in conjunction with interstellar molecules. This will be followed by a discussion of a few of the problems that arise when one tries to extract an isotopic adundance ratio from an observed ratio of antenna temperatures and finally, I will attempt to summarize what we have learned so far about interstellar isotopic abundances.

Table 1 gives terrestrial isotopic abundances and the molecules that can be used to obtain interstellar abundances.

Isotopic abundances determinations are important for studies of nucleosynthesis. But there are also many other astrophysical applications of isotope ratios when used in conjunction with interstellar molecules.

TABLE 1

TERRESTRIAL ISOTOPIC ABUNDANCES

Atom	Abundance (Most abundant isotope of a given element is set equal to unity)	Relevant Interstellar Molecules
^{12}C	1	CO, H_2CO
^{13}C	1/90	HCN
^{16}O	1	
^{17}O	1/2700	CO, OH
^{18}O	1/500	H_2CO
^{14}N	1	
^{15}N	1/270	NH_3, HCN
^{32}S	1	
^{33}S	1/125	CS
^{34}S	1/23	

1. In a given cloud, if one can determine the true isotopic abundance ratio for a certain element (say by observations of an optically thin molecule), then one can use this ratio to determine the optical depth and hence the abundance of optically thick molecules.
2. Then, by use of the various isotopic species of the optically thick molecule, one can probe different depths into the cloud and thus obtain information on variations in density, temperature and velocity field.
3. By comparing line shapes of different isotopic species of a given molecule in a particular cloud, one can say something about the nature of the mechanism that broadens the line. For example, we might distinguish between the case of a single cloud homogeneously stirred by turbulent motions or many smaller clouds moving every which way—the line profile determined by the velocity dispersion among the clouds.
4. Because nuclear processes produce a variety of isotopic ratios, it may be possible to say something about the origin of interstellar molecules from a study of isotopic abundances. For example, a high $^{13}C/^{12}C$ ratio might suggest that interstellar molecules originate in carbon star atmospheres.
5. By studying the less abundant isotopic species of OH and H_2O, it may be possible to gain useful information about maser phenomenon.

When one attempts to obtain an isotopic ratio from a ratio of antenna temperatures, there are a number of practical problems that arise. The most important of these is probably large optical depth. It may sometimes be possible to deduce the degree of saturation by use of a variety of molecules and elements. However, the best method is to find an optically thin molecule in each source. Because source size is generally needed to deduce optical depth, mm-wave observations or interferometric observations at cm wavelengths are especially useful.

A second problem is nonequilibrium pumping which may be different for two isotopic species of a given molecule. As an example, recall that the excitation temperature (T_{ex}) of anomalously cooled $H_2^{12}CO$ in dark clouds is $\sim 1.5°K$. Absorption optical depths are proportional to projected density divided by T_{ex}. If T_{ex} is different from $1.5°K$ for $H_2^{13}CO$ this must be accounted for in calculations of $^{12}C/^{13}C$ abundances deduced from observations of H_2CO. In the collisional cooling model of Townes and Cheung (1969) one expects T_{ex} to be the same for both $H_2^{12}CO$ and $H_2^{13}CO$. However, in the radiational pumping models of Litvak (1970) or Thaddeus (1972) T_{ex} might very well be different for the two isotopic species.

A final problem arises because of trapping of microwave or infrared radiation in the cloud. Different optical depths in two isotopic species implies different amounts of trapping of resonance radiation which can then result in different rotational or excitation temperatures. Then observations of only one transition in each isotopic species may be insufficient to deduce the true abundance ratio.

Now that we have some idea of the practical problems that arise in obtaining isotopic ratios, I will summarize what we have learned from a study of interstellar molecules. The $^{12}C/^{13}C$ ratio is perhaps most interesting since there is a substantial difference between the solar system value and the equilibrium value obtained in the CNO nuclear cycle. The most reliable determination of this ratio, in the sense that it is not affected by any of the above problems, was obtained (Bortolot and Thaddeus 1969) in the direction of ζ Oph from optical CH^+ lines. The $^{12}C/^{13}C$ ratio in this direction is $\sim$ terrestrial. Unfortunately, the optical lines are too weak to use in other directions. The radio molecules CO and HCN are still not too useful because of optical thickness and poor sensitivity problems. The best molecule to use at present is H_2CO (which, however, may be affected by differential pumping, see above). The study (Zuckerman *et al.* 1973) of optically thin sources (see Table 2) yields ratios that scatter about the terrestrial one. Near the galactic center we (Zuckerman *et al.* 1969) found that

TABLE 2

$^{12}C/^{13}C$ ABUNDANCES FROM OBSERVATIONS OF H_2CO

Source	$^{12}C/^{13}C$
W3	> 67
NGC 2024	$\gtrsim$ 84
W33N	105 ± 30
W43	47 $\pm \begin{smallmatrix} 30 \\ 10 \end{smallmatrix}$
W51	63 ± 20
4' North of W51	105 ± 30
DR21	47 $\pm \begin{smallmatrix} 30 \\ 10 \end{smallmatrix}$
Cas A	> 40

in Sgr A and Sgr B2 the apparent $^{12}C/^{13}C$ ratio was 11. The $H_2^{12}CO$ line was guessed to be optically thick so that the true isotopic abundances were unclear. Recent measurements of $H_2^{12}C^{18}O$ in Sgr B2 in Australia (Gardner *et al.* 1971) suggest that ^{13}C may be overabundant by a factor of 2 (although this determination is unlikely to be vitiated by effects of optical thickness, differential pumping may still be a problem). In Sgr A the Cal Tech interferometer has been used (Fomalont and Weliachew 1973) to show that $^{12}C/^{13}C = 25$. In the absence of small unresolved structure in this cloud (which might introduce an optical thickness problem) it appears unlikely that this ratio can be greater than 40. Thus in two clouds near the galactic center the $^{13}C/^{12}C$ ratio may be a factor of ~ 2 larger than terrestrial.

The interstellar $^{18}O/^{16}O$ ratio is still poorly determined. The best estimate is probably from observations of CO (Penzias *et al.* 1971) where the $^{18}O/^{13}C$ ratio in a few sources is apparently terrestrial. However to then deduce the $^{18}O/^{16}O$ ratio requires the assumptions that $^{13}C/^{12}C$ and $^{12}C/^{16}O$ are terrestrial. A preliminary upper limit on the $^{17}O/^{18}O$ ratio has been obtained from observations (Gottlieb *et al.* 1971) of OH in the galactic center. This upper limit (which is approximately equal to the terrestrial ratio) can be easily improved by further observations.

The nitrogen isotope ratio is still difficult to obtain and the best existing value is $^{14}N/^{15}N \geq 70$ based on the $NH_3(3,3)$ inversion transition in Sgr B2 (Zuckerman *et al.* 1971).

Preliminary values for the $^{32}S/^{34}S$ ratio have been obtained from observations (Turner *et al.* 1973) of the $J = 2 \rightarrow 1$ transition in W51, W3(OH) and Orion A. In each case the observed antenna temperature ratio was ~ 10 but unknown optical thickness in $C^{32}S$ renders interpretation impossible at this point.

In summary, we may say that, at present, there is no case where interstellar isotopic abundances have been shown to deviate from terrestrial with the possible exception of a slight overabundance of ^{13}C near the galactic center. In addition, there is one source (IRC +10216) which is not interstellar where, very likely, a deviation from a terrestrial ratio does exist. In this carbon star the intensity ratio of $H^{13}C^{14}N/H^{12}C^{15}N$ was observed (Morris *et al.* 1972) to be $\geq$ 6. If one assumes terrestrial ^{12}C, ^{13}C, ^{14}N, and ^{15}N abundances, this ratio should be ≤ 3. Thus in the absence of pumping and trapping effects that differ between the two molecules (perhaps due to near infrared radiation pumping the vibrational bands (Litvak 1972)), $[^{13}C]$ $[^{14}N]$ is overabundant relative to $[^{12}C]$ $[^{15}N]$. Since both ^{13}C and ^{14}N increase and ^{12}C and ^{15}N decrease in abundance in the CNO cycle this observation is consistent with a carbon star interpretation of IRC +10216.

REFERENCES

Bortolot, V. J., Jr. and Thaddeus, P. 1969, *Ap. J.,* **155**, L17.

Fomalont, E. B. and Weliachew, L. N. 1973, *Ap. J.,* in press.

Gardner, F. F.; Ribes, J. C. and Cooper, B. F. C. 1971, *IAU Telegram* , 2354.

Gottlieb, C. A.; Radford, H. E.; Ball, J. A. and Zuckerman, B. 1971, unpublished.

Litvak, M. M. 1970, *Ap. J.,* **160**, L133.

———————————— 1972, private communication.

Morris, M.; Zuckerman, B.; Palmer,P. and Turner, B. E. 1972, *Ap. J.,* **170**, L109.

Penzias, A. A.; Jefferts, K. B. and Wilson, R. W. 1971, *Ap. J.,* **165**, 229.

Thaddeus, P. 1972, *Ap. J.,* **173**, 317.

Townes, C. H. and Cheung, A. C. 1969, *Ap. J.,* **157**, L103.

Turner, B E.; Zuckerman, B.; Palmer, P. and Morris, M. 1973, in preparation.

Zuckerman, B.; Palmer, P.; Snyder, L. E. and Buhl, D. 1969, *Ap. J.,* **157**, L167.

Zuckerman, B.; Morris, M.; Turner, B. E. and Palmer, P. 1971, *Ap. J.,* **169**, L105.

Zuckerman, B.; Palmer, P.; Buhl, D. and Snyder, L. E. 1973, in preparation.

Session 3

Techniques

A photograph in red light of the HII region NGC 6611 in the constellation Serpens. Note the dark obscuring matter. (Hale Observatories photograph)

The ABC's of Frequency Calculations*

E. Bright Wilson

Mallinckrodt Chemical Laboratory
Harvard University
Cambridge, Massachusetts

The problem to be attacked here is how to predict the microwave spectrum of a molecule which has not, for one reason or another, been studied in the laboratory. It should be stated at once that it is not possible to guarantee at this time a very close prediction; it can however be close enough to be most useful as a preliminary to a laboratory investigation and, with luck, it can sometimes be useful astronomically.

Table I lists the well known energy terms which are usually important. Most spectra so far studied have involved mainly changes in

Table I.
Usual Energy Terms

Rigid rotation
Centrifugal distortion
Tunneling
Coriolis
Electron spin-rotation
Nuclear quadrupole coupling

rotational energy. Not too small molecules in low rotational states usually fit the simple rigid rotor model with Hamiltonian (Wollrab 1967, Townes and

$$H = \frac{\hbar^2}{2} \left(\frac{J_a^2}{I_a} + \frac{J_b^2}{I_b} + \frac{J_c^2}{I_c} \right)$$

Schawlow 1955, Allen and Cross 1963) where I_a, I_b, I_c are the principle moments of inertia ($I_a < I_b < I_c$) while J_a etc. are the operators (or matrices) for the components of angular momentum along the "molecule-fixed" axis a, b, c, measured in units of $\hbar$

The standard quantum mechanical procedures lead to determinantal equations for the energy in terms of the I's. Convenient computer programs exist for calculating energy levels and transition frequencies from values of I_a, I_b, I_c (Beaudet 1961). It then remains to estimate the latter. This required prediction of the molecular structure, since the atomic masses are known.

It is often possible to examine a number of related molecules and to transfer from them values of bond lengths and angles. A convenient compendium of molecules whose structures have been determined from microwave data is available (Starck 1967).

One point is worth noting however. Usually microwave data does not yield the true equilibrium structure but only an approximation thereto, because of the vibrational motion of the atoms. For prediction of spectra, the equilibrium structure is not even what is wanted; rather an "effective" structure which reproduces the ground state moments of inertia. However, many recent structures are what are known as r_s structures, believed to be nearer the equilibrium structures (Costain 1958). To correct them to get effective moments of inertia, a rough rule of thumb is to multiply the moments of inertia calculated from the r_s results by ~ 1.04.

If appropriate other molecules are not available, tables of bond lengths between specified atom pairs, and of bond angles can be consulted. (Sutton 1965)

In both these approaches, errors may be introduced. For example, conjugation or resonance (Pauling 1960) can have a considerable effect and may be present in the molecule of interest but not in the model compounds. Furthermore, for example, the C–C single bond is not strictly constant but depends in a linear way (approximately) on the number of attached atoms, falling from 1.53Å with six attached atoms (e. g. C_2H_6) to 1.38Å for only two attached atoms (e. g. H–C≡C–C≡C–H) (Costain and Stoicheff 1959). This has been explained variously as due to conjugation, (Pauling 1960) change of hybridization (Dewar and Schmeising 1959) or relief of steric hindrance (Bartell 1960). Similar effects show up with multiple bonds.

The substitution of several fluorine atoms (and to a lesser extent chlorines) tends to lower bond lengths to the atom to which they are attached (the data is mainly on fluorine on carbon; Pauling 1960). Finally, obvious crowding will produce distortions.

In the end, it is difficult to be sure of bond length predictions to better than 0.005 to 0.01Å, sometimes worse, and this means considerable uncertainty in the prediction of microwave lines, especially those of high angular momentum J, which are much more sensitive to the values of the moments of inertia.

Bond angles are more variable than bond lengths. Ideas of hybridization (Coulson 1952) have had a wide vogue as a guide here but do not seem to have been very fruitful with microwave structures (Myers and Gwinn 1952).

Planarity can often be predicted, especially where conjugation is likely to be present. A doubly bonded (or partially doubly bonded) carbon atom tends to hold its attached atoms and itself in one plane (Pauling 1960).

Carbon-carbon single bonds have the property that they act as axles about which the end groups can rotate, though not usually freely. Ordinarily, if there are six attached atoms or groups (as in ethane, C_6H_6) they take up a *conformation* which is staggered, i. e., the attached atoms on one end avoid those on the other. If there is a double bond to one carbon, there is usually a conformation as if one of the groups on the other carbon were attracted to the double bond (Herschbach and Krisher 1958) (this by no means implies that there is an actual attraction).

Rings may show competing forces. On the one hand, the bond angles within the ring would like to take their normal values. In a C_4 or C_5 saturated ring this would tend toward planarity. On the other hand the "barrier forces" about single bonds would like to stagger the substituents and this argues for a puckered ring. Some degree of puckering is common, but the extent varies from compound to compound (Laurie 1970).

There are of course more generalizations known to chemists but these may show the nature of some of the available rules.

The rigid rotor model works very well for sufficiently heavy molecules and low values of the angular momentum J. Thus, for example, the lowest 28 lines of $J < 6$ for trans-1-pentyne fit a rigid rotor to better than $0.5\,MHz$, or to about 1 part in 60,000 (Wodarczyk 1971). However, for very light molecules (e. g. H_2O!) or high J's, centrifugal distortion may become very important. Thus for oxygen difluoride, the transition $36_{4\ 32} \leftarrow 37_{3\ 35}$ is shifted by $7533\,MHz$ from its rigid rotor position (Pierce, DiCianni and Jackson 1963).

It is not easy to predict these shifts in advance although the theory is available and works very well. The first correction to H is of the form

$$H_{CD} = 1/8 \Sigma \tau_{\alpha\beta\gamma\delta} J_\alpha J_\beta J_\gamma J_\delta$$

(Wollrab 1967) with coefficients $\tau_{\alpha\beta\gamma\delta}$, which are known functions of the molecular masses, geometry and force constants. The force constants are seldom known well enough for polyatomic molecules to permit very useful predictions here—and this is naturally even more true of estimated force constants. This difficulty may introduce considerable uncertainty in any attempt to predict high J transitions, such as are likely to occur in the millimeter region. On the other hand if sufficient microwave frequencies have been measured, the three I's and five independent τ's can be fitted to them and then used to predict astronomically interesting transitions, often, but not always with considerable precision, using $H_{RR} + H_{CD}$ (Pierce, DiCianni and Jackson 1963; Kirchhoff 1971). It must of course also be remembered that the Ritz Combination Principle can sometimes be used for predictive purposes when the molecule possesses two or more non-vanishing components of the dipole moment, or even with only one component if appropriate lines are measurable. See Figure 1.

Most polyatomic molecules can exist in two or more equivalent equilibrium forms separated by some sort of potential energy barriers. For example a non-planar polyatomic molecule can always be inverted through its center of mass to a new, equivalent configuration. Often the energy barrier separating such configurations is so high that there is no need to consider the existence of the separate, equivalent forms. However, there are many cases where the barrier separating equivalent forms (not necessarily related by inversion through a

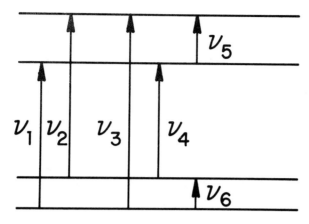

1. Illustration of Ritz Combination Principle. $\nu_5 = \nu_3 - \nu_1$.

center) is sufficiently low so that quantum mechanical tunneling becomes important and each rotational energy level is split into two (or more) levels. This splitting can range from negligible to huge so it can be either smaller or larger than the rotational spacing.

This splitting can influence the spectrum in different ways. If a component of dipole moment reverses on carrying the molecule from one configuration to the other, the observed transition (involving that component) will be as shown in Figure 2, so that the transition frequency will involve the tunneling separation as well as the change in rotational energy. Since the tunneling separation is a very sensitive function of the barrier height and shape, which in turn are difficult to predict, such transitions are not very accurately predictable for low barriers. Furthermore, the tunneling splitting may not be constant but can depend upon the rotational quantum numbers through interaction with centrifugal distortion.

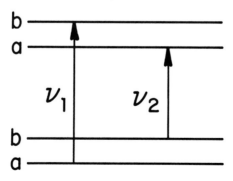

2. Transitions between levels split by tunneling. Case I.

Of course, if the barrier is known to be high, its exact value may be unimportant, because then the splitting may be negligible. Likewise if heavy atoms need to be moved, very low barriers must be involved in order for the splitting to be important.

If the dipole moment component of importance for the given rotational change is not altered on passing from one configuration to the other, the situation will be as in Figure 3. Here, if the splitting is independent of $J\tau$, it will not appear in the spectrum, which could then simulate a rigid rotor. However this is not always the case. First the a and b type states do have slightly different vibrational properties and hence the effective moments of inertia can be expected to be slightly different for the two. Second, centrifugal distortion may cause the splitting to depend noticeably on $J.\tau$. Third, the splitting may bring about near degeneracy between an a state of $J''\tau''$ and a b state $J'''\tau'''$ between which the coriolis coupling term can act. This near degeneracy will enhance the effect of the coupling and may show up as a doubling of the spectral lines, i. e. a frequency separation of ν_1 and ν_2 in Figure 3 (Scharpen 1968, Pickett 1972).

Ammonia is a notable example of the tunneling phenomena. In the ordinary inversion spectrum, $J' = J$ and the transition is between the a and b states of inversion. Ammonia is a symmetric rotor with mostly doubly degenerate rotational levels so the rotational quantum number change contributes no energy. However, centrifugal distortion strongly affects the inversion splitting so *transitions* of different J are spread over a wide range (Townes and Schawlow 1955).

Another common type of tunneling involves the internal rotation of a CH_3 group about the bond joining it to the rest of the molecule. There is normally a barrier of from a few hundred to a thousand wave numbers hindering the rotation from one to another of the three equivalent equilibrium orientations. Tunneling splits the three-fold degeneracy into a doubly degenerate (E) level and

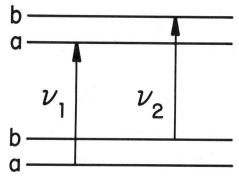

3. Transitions between levels split by tunneling. Case II.

a non-degenerate (A) level. Transitions do not occur between these types but coriolis coupling can be different and cause the rotational lines to be split into doublets (Lin and Swalen 1959). Acetaldehyde, CH_3CHO, is an example in which the doublets are mostly recognizable as such whereas in methyl alcohol, CH_3OH, the splitting is very large. Considerable numbers of molecules with doubling due to CH_3 tunneling have been studied in order to determine the potential barriers to internal rotation (Starck 1967).

The general phenomenon of tunneling will doubtless be encountered more and more often with larger molecules and it can be troublesome for predictions and even for the analysis of laboratory spectra. It has to be looked for with molecules with $-OH$, $-NH_2$ and $-NH$ groups, among others. Even heavier atoms (e. g. F) have shown the effect, but usually to a quite small extent.

Molecules with an odd number of electrons, or in triplet states can have important energy contributions from interactions of the magnetic moment of the electron with the magnetic moment generated by the rotation of the molecule and with those associated with nuclear spins (Townes and Schawlow 1955). A well-known example is provided by the OH radical.

The interaction of the nuclear electrical quadrupole moment of such nuclei as D, N, Cl, Br (in fact any nucleus with spin $> \frac{1}{2}$) with the electrical field gradient arising from the electrons is somewhat predictable (Wollrab 1967, Townes and Schawlow 1955). The absolute quadrupole moments are known for some nuclei but not for others (1969 Handbook of Chemistry and Physics); it is usually the relative values of eqQ (the product of the nuclear and electronic contributions) which matters, compared from one molecule to another. The field gradient parameter q is not easy to calculate accurately, even with advanced *a priori* methods, but rough values can be estimated from the arguments of Townes and Dailey (Townes and Schawlow 1955, Chapter 9). The internal relative spacings and relative intensities of the hyperfine components are accurately predictable (Wollrab 1967, Townes and Schawlow 1955) and can be most useful for identification of a transition. For example, with spin 3/2, as in [35]Cl and [37]Cl, certain Q branch transitions will appear as a quartet pattern. The ratio of the outer to inner frequency spacings determines the quantum number J (Azrak 1969).

In conclusion, the theory of the origins of molecular microwave transitions is adequate to account for and to fit parametrically the spectra of most simple molecules. This permits the accurate prediction of other transitions once a sufficient number have been observed and assigned (Kirchhoff 1971, Johnson, Kirchhoff and Lovas 1971). It is more difficult and much less accurate to predict the necessary parameter values from empirical rules about molecular structure and other properties, but rough estimates are possible. Finally, the theory can be useful in supporting the identification of an observed astronomical line, by predicting others (including isotopic lines) which should be observable, or by determining the fine structure expected, etc. However, there remain a much greater range of observable and controllable features in laboratory spectra so that the problem of certain identification is much easier. Identification could pose a considerable problem for astronomers if still larger molecules are encountered, because of the very dense spectra that can occur.** It seems very unlikely that a single coincidence in frequency to the uncertainty of the Doppler shift will be regarded as very strong evidence of the presence of a new chemical species.

*This paper is a not very literal transcription of the material presented under this title at the Symposium.

It is worth noting that a large number of molecules has been studied, many with very rich spectra, since the publication of the very useful NBS monograph 70, *Microwave Spectral Tables*. Other useful general sources include review articles in *Ann. Rev. Phys. Chem.*, **21, 20, 18, 15 and a tabulation of references organized by molecules by B. Starck, *Bibliographie Mikrowellenspektroskopischer Untersuchen an Molekulen* (Three periods) Physikalisches Institute der Univ. Freiburg/Br. 1963, 1966, 1970.

REFERENCES

Allen, H. C. Jr. and Cross, P. C. 1963, *Molecular Vib-Rotors* (New York: J. Wiley and Sons) Chapter 2.

Azrak, R. G., 1969 thesis, Harvard University, Cambridge, Massachusetts, pp.1-24, 25.

Bartell, L. S. 1960, *J. Chem. Phys.*, **32**, 827.

Beaudet, R. A. 1961, *Problems in Molecular Structure and Internal Rotation*, thesis, Harvard University, Cambridge, Massachusetts.

Costain, C. C. 1958, *J. Chem. Phys.*, **29**, 864.

Costain, C. C. and Stoicheff, B. P. 1959, *J. Chem. Phys.*, **30**, 777.

Coulson, C. A. 1952, *Valence* (Oxford: Clarendon Press), Chapter VIII.

Dewar, M. J. S. and Schmeising, H. N. 1959, *Tetrahedron*, **5**, 166.

Herschbach, D. R. and Krisher, L. C. 1958, *J. Chem. Phys.*, **28**, 728.

Johnson, D. R.; Kirchhoff, W. H. and Lovas, F. J. 1971, *Microwave Spectral Tables Supplement* (Washington: National Bureau of Standards) Sec. 1. (To be published)

Kirchhoff, W. H. 1972, *J. Mol. Spectry.*, **41**, 333.

Laurie, V. W. 1970, *Accounts of Chemical Research*, **3**, 331.

Lin, C. C. and Swalen, J. D. 1959, *Rev. Mod. Phys.*, **31**, 841-892.

Myers, R. J. and Gwinn, W. D. 1952, *J. Chem. Phys.*, **20**, 1420.

Pauling, L. 1960, *Nature of the Chemical Bond* (3rd ed.; Ithaca: Cornell University Press), Chapter 8.

Pickett, H. M. 1972, *J. Chem. Phys.*, **56**, 1715.

Pierce, L.; Di Cianni, N. and Jackson, R. H. 1963, *J. Chem. Phys.*, **38**, 730.

Scharpen, L. H. 1968, *J. Chem. Phys.*, **48**, 3552.

Starck, B. 1967 *Molecular Constants from Microwave Spectroscopy* (II-4 of Landolt-Bornstein) (Berlin: Springer-Verlag).

Sutton, L. E. 1965, *Tables of Interatomic Distances*, Special Publication 18 (London: The Chemical Society).

Townes, C. H. and Schawlow, A. L. 1955, (New York: McGraw-Hill Book Company), Chapter 4 (Chapter 9).

Wodarczyk, F. 1971, thesis, Harvard University, Cambridge, Massachusetts.

Wollrab, J. E. 1967, *Rotational Spectra and Molecular Structure* (New York: Academic Press) Chapter 2.

1969 Handbook of Chemistry and Physics (15th ed.; Cleveland: Chemical Rubber Company) P. e 75.

Litvak: How much help to the calculations would be laboratory observations (especially by lasers) in the far infrared?

Wilson: Grating measurements by present techniques are not as helpful as laser measurements will be as techniques become available. For many molecules, pure rotation spectra in the far infrared involves high J transitions, which unfortunately contain considerable centrifugal distortion of the molecular frame. Therefore, it is difficult to derive accurate constants for the ground state from these measurements. But large molecules have vibrations in the far infrared in the form of vibration–rotational transitions. From these transitions, we can determine accurate spectroscopic constants for ground states.

The XYZ's of Laboratory Frequency Measurements

David R. Lide, Jr.

National Bureau of Standards
Washington, District of Columbia

I. INTRODUCTION

The interpretation of microwave spectra observed in the laboratory is often critically dependent on having very accurate values of the frequencies of the observed transitions. In fitting a spectrum to a particular molecular model, one must be able, especially in the case of a very rich spectrum, to decide unambiguously whether an observed line agrees in frequency with the predictions of the model. For this reason there is a general tradition in the literature of microwave spectroscopy of reporting measured frequencies to a fairly high level of accuracy. However, the typical accuracy varies with frequency region and several other factors, and only in rare cases has the theoretically attainable accuracy been realized.

As a result of the observation of sharp-line microwave absorptions in the interstellar medium, certain laboratory spectra have achieved a new significance. The comparison of interstellar with laboratory spectra places demands on the latter which are somewhat different from those normally recognized by the molecular spectroscopist. This paper presents a brief summary of both present and past techniques for measuring the frequencies of microwave absorption lines in the laboratory. An effort is made to give some feeling for the reliability of laboratory frequency measurements and of predictions made from them. Useful sources of data on microwave frequencies and related quantities are also presented.

II. METHODS OF FREQUENCY MEASUREMENT

Microwave frequency measurements are much easier than the analogous wavelength measurements in optical spectroscopy. Laboratory microwave spectroscopy is normally carried out with essentially monochromatic radiation sources [klystrons or backward wave oscillators (BWO)] that are scanned smoothly over the frequency region of interest. It is only necessary to measure the frequency of the source as it crosses the peak of an absorption line.

A common way of doing this is shown in Figure 1. A small part of the output of the microwave oscillator is coupled into a mixer diode, into which is also fed the output of a very stable frequency standard at, say, 1 GHz. The diode serves both to generate harmonics of the standard frequency signal and to mix these harmonics with the source oscillator frequency. The resulting beat frequency is carried to a communications receiver, and the output of the receiver is displayed on one trace of a double-beam oscilloscope. The scope is swept by the same sawtooth that sweeps the source oscillator, and the absorption line is displayed on the other trace. Thus a "pip" or marker is produced whenever the difference frequency between the source and the harmonic of the frequency standard is

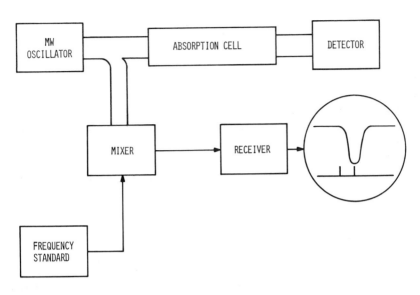

1. Typical system for microwave frequency measurements.

plus or minus the setting of the receiver. It is only necessary to tune the receiver until one of the markers coincides with the peak of the absorption line.

This simple and efficient scheme can be used throughout the lower microwave region. At high frequencies (above about 50 GHz), the harmonics of the 1-GHz standard tend to become weak and markers may be hard to generate. Here it is helpful to use an intermediate oscillator, say in the 5 to 10-GHz range, which is phase-locked to the frequency standard. Alternatively, if a variable oscillator is available in this range with sufficient short-term stability, its frequency may be monitored with a counter and the frequency standard eliminated entirely.

Frequency standards (or counters) with accuracy satisfactory for most laboratory microwave measurements on bulk gas samples — say, 1 part in 10^7 — are readily available. Likewise, there is little difficulty in measuring the difference frequency to an accuracy of 10 kHz or better with a good communications receiver. However, a potential source of error lies in the generally different time constants of the two channels displayed on the scope. This may produce a large displacement of the frequency marker relative to the absorption line signal, especially if the sweep speed is high. To avoid such errors, one should always reverse the direction of the sweep and carry out a duplicate measurement. If the amount of displacement is small, the average of the two measurements will give the correct frequency. By keeping the sweep speed relatively slow and carrying out this averaging procedure, the error introduced by different time delays in the two channels can easily be held below 10 kHz.

There is a current trend in laboratory spectroscopy toward the use of BWO sources which are phase-locked to a variable frequency oscillator. The microwave source may thus be scanned very slowly by varying the oscillator, and the spectrometer output displayed on a chart recorder. With this system frequency measurements are most easily made by continuously monitoring the variable oscillator with a frequency counter. The counter may be programmed to generate markers on the chart at suitable intervals. Measurement of the frequency of an absorption line then requires only a simple interpolation between markers.

III. MICROWAVE LINE WIDTHS

When microwave spectra are observed in bulk gas samples, several factors contribute to the width of an absorption line. At sufficiently high pressures the dominant broadening mechanism is intermolecular collisions. The line has a characteristic Lorentzian shape with a width which is directly proportional to gas pressure; the peak height is independent of pressure. As the pressure is lowered, however, other sources of broadening, including Doppler, wall-collision, modulation, and power saturation become significant. The peak height decreases and the line shape changes. A typical variation of an absorption line with pressure is illustrated in Figure 2.

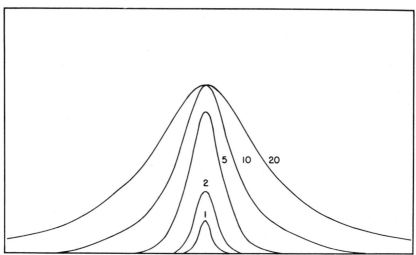

2. Typical variation in microwave absorption line as a function of pressure. The numbers indicate pressures in units of mTorr.

The details of the various line width contribution are discussed in standard texts, such as Townes and Schawlow (1955). The natural line width is much less than 1 Hz and can be completely ignored. The Doppler width (half-width at half-maximum intensity) resulting from thermal motion in the bulk gas is given by

$$\Delta \nu = (\nu/c) \, [(2kN_0T/M) \, \ell n2]^{\frac{1}{2}},$$

where ν is the peak frequency, c is the speed of light, k is the Boltzmann constant, N_0 is Avogadro's number, and M is the molecular weight. Broadening by collisions with the walls of the cell becomes significant when the mean free path is comparable to the cell dimensions. Another type of broadening is produced by the Stark or Zeeman modulation which is employed in most microwave spectrometers. This contribution to the line width is of the order of magnitude of the modulation frequency, which is usually in the range 5 - 100 kHz. Finally, if the microwave power level in the absorption cell is too high, the gas cannot dissipate the absorbed energy, and saturation broadening occurs. In principle, this source of broadening can be eliminated by keeping the microwave power low. However, in practice, the requirements of detector sensitivity impose a lower limit on the power level, and a trade-off between line width and signal-to-noise ratio is sometimes necessary.

The various line width contributions under typical conditions are summarized in Table I. We see that the Doppler width is generally larger than the wall-collision width, and that the contribution from modulation broadening may be the dominant factor, especially in the lower range of transition frequencies. In the usual practice, the sample pressure is adjusted such that the contribution from intermolecular collisions is roughly the same as the total contribution from the factors listed in Table I. Then the signal strength (peak height) is close to its maximum value and the line width is not far from its minimum limiting value. The majority of the microwave spectra reported in the literature were obtained under conditions which represent a compromise of this type between resolution and signal-to-noise.

The half-widths in typical laboratory spectra of bulk gas samples thus range from a few kHz in the very low microwave region to a few hundred kHz in the 100 GHz region. In the popular 20 - 40 GHz region, 100 kHz is an average figure.

A significant reduction in the limiting line width can be achieved by using molecular beam techniques to remove the Doppler broadening from thermal motion in the bulk gas. Three principal approaches can be mentioned. The molecular beam electric resonance method has been extended to the microwave region by several laboratories. This is particularly useful in studying molecules in

TABLE I.

Line width contributions from various sources (half-width at half maximum intensity, given in kHz) as a function of transition frequency.

Source	1 GHz	10 GHz	100 GHz
Doppler[a]	1	11	113
Wall collision[b]	0.4	4	10
Modulation[c]	5-10	10-100	10-100

[a] Assuming T = 300K and M = 30.
[b] Assuming convenient waveguide sizes at each frequency.
[c] Typical range for Stark modulation.

transient states or those which require high temperatures for vaporization. Half-widths of the order of 10 kHz have been obtained on transitions in the 20-GHz region (Green and Lew 1960). A summary of results on diatomic molecules is given by Lovas and Lide (1971).

Several spectrometers have been developed for the higher microwave frequency region which employ a molecular beam in a more-or-less conventional microwave cell (van Dijk and Dymanus 1969, Huiszoon and Dymanus 1966, Veazey and Gordy 1965). Half-widths in the range of 5 kHz at frequencies above 100 GHz have been obtained with such spectrometers. This technique is best suited to the high-frequency region, where absorption coefficients are very large.

Finally, beam-maser spectrometers (Thaddeus and Krisher 1961, Thaddeus, Krisher, and Loubser 1964) provide a means of achieving very narrow lines. Half-widths in the 2 - 5 kHz range have been reported. These spectrometers can be used throughout the microwave region.

IV. GENERALIZATIONS ON ACCURACY OF REPORTED FREQUENCIES

It was stated in the last section that laboratory microwave measurements on bulk-gas samples can usually be carried out on isolated lines under conditions where the half-widths are 100 kHz or less. Unless there are signal-to-noise limitations, it should be possible to measure the peak frequencies of such lines to a few kHz or better. However, the majority of the frequency measurements reported in the literature do not approach that level of accuracy; in fact, most published frequencies are quoted to only the nearest 0.01 MHz or 0.1 MHz. This is because most investigators have not felt the need to select the line center with the precision that is possible from a careful line-shape analysis. In addition, there is often insufficient control over sources of systematic error such as:

1. Overlapping by nearby lines
2. Partially resolved structure on the line
3. Overlapping by Stark components of other lines
4. Insufficient modulation
5. Time delay errors of the type discussed in Section I.

A critical evaluation of the accuracy of published microwave frequencies is not easy. One can only make use of the claims of the author and the experimental details which he gives to substantiate these claims; the rather subjective assessment of the practices in the particular laboratory; and the deviations between observed frequencies and those calculated by fitting to a particular model. In an evaluation of published data on about 2500 lines of 25 different molecules, W. H. Kirchhoff of the NBS Microwave Data Center concluded that about one-half of the frequencies could be trusted to an accuracy of ±0.10 MHz. The remainder were about equally divided between 0.2 MHz and 0.5 MHz accuracy. Only a few percent of the measurements were felt to be accurate to 0.05 MHz or better. These data came from relatively recent measurements; one should be even more cautious about the literature prior to 1960.

The typical accuracy of microwave measurements varies somewhat with the frequency region. At the lower end of the microwave region the accuracy tends to be higher (on a absolute scale) because line widths are reduced (see Table I). At the high-frequency extreme, the large Doppler contribution to the width and the generally difficult experimental conditions lead to a greater uncertainty in most measurements.

There are, of course, many exceptions to these generalizations. Certain important lines in all frequency regions have been known for some time to an accuracy of 0.01 or 0.02 MHz. Modern microwave systems make it relatively simple to measure any strong, isolated line to this level of accuracy. Nevertheless, a certain degree of caution is advisable in using frequencies reported in the literature unless there is a clear indication that all potential sources of error have been taken into consideration.

The accuracy is much higher for frequencies measured with molecular beam systems (see Section III). Such measurements are generally good to a few kHz, and in some cases uncertainties less than 1 kHz have been claimed.

V. A CAUTION ABOUT FINE STRUCTURE ON MICROWAVE LINES

There are several types of fine interactions which produce a structure on the rotational transitions observed in the microwave region. The most common is nuclear quadrupole hyperfine structure, which is generally observable in molecules containing a nucleus with spin greater than ½. Magnetic hyperfine structure, which occurs also for spin ½ nuclei, can be resolved in beam-type spectrometers. Fine structure is also produced by internal rotation, inversion, and other interactions between internal molecular motions and overall rotation. Non-uniform practices of reporting such fine structure in the literature sometimes leads to confusion.

While some authors report the observed frequency of each component of a multiplet pattern, others list only a mean or reduced frequency for the rotational transition. In the case of nuclear quadrupole hyperfine structure, a common practice is to tabulate the "hypothetical unsplit frequency" which would apply if the structure were completely collapsed. The shifts of individual components relative to this hypothetical frequency are often tabulated separately, but in some papers a calculation using the derived quadrupole parameters is necessary to regenerate the frequencies which were actually observed. In other cases, structure may be reported for certain transitions but not for others, because no effort was made to resolve every transition in the spectrum.

In using published data on spectra exhibiting hyperfine structure or other splittings, therefore, great care should be taken to determine which transitions showed structure (or would have done so if observed under sufficient resolution).

VI. PREDICTION OF MICROWAVE FREQUENCIES FROM MOLECULAR CONSTANTS

A published microwave spectrum generally contains only a small portion of all the absorption lines which the molecule exhibits. This practice has been acceptable because the primary objective of most studies is to obtain certain important molecular constants, and this can often be achieved by measuring only a small "window" in the total spectrum. Sometimes the strongest lines have been ignored, because the desired information could most easily be extracted from the transitions with low quantum numbers, which tend to be weaker. Thus a published list of microwave lines can rarely be relied upon to represent the complete spectrum over a wide frequency range.

When a set of molecular constants has been derived by fitting a limited set of observed frequencies to a given model, the accuracy of other transition frequencies predicted from these constants is obviously of great interest. The simplest case to consider is the fitting of a rotational spectrum to a rigid-rotor model. If transitions with low J values are used in the fit, the rigid-rotor constants obtained will be very close to the true values — close enough for most purposes for which these constants are used, such as calculating interatomic distances and angles. However, this does not mean that the constants will permit reliable predictions of unobserved transitions.

Even in diatomic molecules the influence of centrifugal distortion cannot be neglected. In NaF, for example, the J = 1←0 transition has been accurately measured by molecular beam techniques (Hollowell, Hebert, and Street 1964) as 26059.4831 ± 0.0030 MHz. If one had only this frequency available and were forced to use a rigid-rotor model to predict the J = 10←9 transition, the result would be 260,594.831 ± 0.03 MHz. The observed frequency of this transition (Veazey and Gordy 1965) is 260,457.106 MHz. The discrepancy of 137 MHz resulting from use of an inadequate model is far larger than any experimental uncertainties. Of course, if several transitions of a diatomic molecule have been accurately measured, the distortion constants may be determined, and predictions of additional transitions will be much more reliable.

A correction for centrifugal distortion can be made in diatomic molecules if the vibrational frequency ω_e is known. The leading distortion constant D_J is given by

$$D_J = 4B_e{}^3/\omega_e{}^2$$

Similarly, the distortion constant of a linear triatomic molecule may be calculated from the relation

$$D_J = 4B_e{}^3\,[(\zeta_{23}{}^2/\omega_1{}^2)$$
$$+ (\zeta_{12}{}^2/\omega_3{}^2)]\,,$$

where ω_1 and ω_3 are the totally symmetric vibrational fundamental frequencies, and ζ_{23} and ζ_{12} are Coriolis coupling constants. If sufficiently accurate vibrational data are available, one can often make a satisfactory correction for effects of centrifugal distortion on the predicted frequencies of these simple molecules.

Centrifugal distortion effects in asymmetric rotors can be much larger. Constants which fit a limited set of low -J transitions quite satisfactorily may give errors of thousands of MHz in predictions of higher transitions. Furthermore, it is often difficult to predict which transitions will be most sensitive to these errors. As a general rule, centrifugal distortion effects tend to be largest for asymmetric-rotor molecules of low molecular weight (and, therefore, large rotational constants).

There is an extensive literature on centrifugal distortion in asymmetric rotors; for a summary see Wollrab (1967) and Gordy and Cook (1970). A very thorough study has recently been made by Kirchhoff (1972). In this work an effort was made to obtain a realistic measure of the accuracy of predictions made with constants derived from fitting a partial spectrum. When a satisfactory model was chosen and appropriate techniques used for the data analysis, it was found that quite reliable predictions could be made. That is, the observed frequencies fell within the statistical limits of error given by the calculation. It should be noted that these calculated error limits depend very strongly on the transition which is being predicted, in a manner which is not always intuitively obvious. An example of the effect of model choice on the predictability of frequencies in SO_2 is given in Table II.

TABLE II

Prediction of rotational transitions in SO_2. The data set from which the constants were obtained included 97 transitions with $J \leqslant 53$ (but excluded the transitions listed here). Taken from Kirchhoff (1972).

Transition	With planarity constraint		Without planarity constraint	
	ν(obs)-ν(calc)	σ(calc)	ν(obs)-ν(calc)	σ(calc)
$29_{2\ 28}$-$28_{3\ 25}$	4.46 MHz	0.61	0.48	0.55
$31_{2\ 30}$-$30_{3\ 27}$	10.95	1.2	1.03	1.2
$35_{3\ 33}$-$34_{4\ 30}$	5.86	1.8	1.72	1.3
$37_{3\ 35}$-$36_{4\ 32}$	14.84	2.7	3.09	2.4
$51_{4\ 48}$-$50_{5\ 45}$	180.36	21.9	30.10	23.2

We may summarize the present situation as follows. In selected cases it has proved possible to analyze a partial spectrum of an asymmetric rotor in such a manner that reliable predictions of the full spectrum can be made — in the sense that one can predict both the frequency and its accuracy for any desired transition in any frequency range. However, great care must be exercised in the choice of model and in the computer techniques which are employed. In the future we can hope to see complete analyses of this type carried out on more molecules.

VII. POSITIVE IDENTIFICATION OF MICROWAVE LINES

In the analysis of laboratory spectra various tools are available for the assignment of specific quantum numbers to observed transitions. The Stark effect is an extremely powerful technique, and the Zeeman effect is useful in selected cases. Nuclear quadrupole hyperfine structure is an informative label for many microwave transitions. Relative intensities give a further boundary condition on assignments among a group of lines. Finally, the ability to fit a large set of observed transitions (expanded, if necessary, by additional measurements) to an appropriate model with high precision provides the ultimate test of the assignment of a spectrum.

Since most of these tools cannot be used for the identification of the spectra of interstellar molecules, heavy dependence must be placed on frequency coincidences with well-documented laboratory spectra (or with predictions based upon such spectra). The degree of confidence which one can place in such coincidences is a serious question. It is worth noting that NBS Monograph 70 lists one line per MHz on the average in the 20 to 30-GHz region. Thus there is a high probability that any unknown line will fall close to at least one line recorded in these tables.

A simple test may be carried out by generating a set of random frequencies and noting how many molecules have reported lines lying within a given interval around each frequency. Figure 3 shows a typical result, based on the tabulation of the 20 - 30 Gc region in Monograph 70, when the interval is chosen as ±1.00 MHz. It is seen that a possible identificaion exists in about 80 per cent of the cases, and that there is more than one choice 40 per cent of the time. However if the interval is reduced to 0.01 MHz, the probability of a match is reduced to a few percent, and multiple coincidences are very rare.

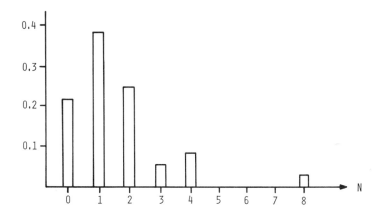

3. Fraction of cases (*ordinate*) in which N molecules had lines within ± 1.00 MHz of chosen frequency. This coincidence information is taken from lines in NBS Monograph 70 and randomly chosen frequencies in the 20 - 30 GHz region.

We may conclude, then, that when Monograph 70 is used as the basic data file, a coincidence within 1 MHz does not, by itself, provide very strong evidence for identification of an unknown line. When we further take into account that many lines are missing from Monograph 70 (even for those molecules which are included in the compilation), it is clear that great caution should be attached to this type of exercise.

REFERENCES

Gordy, W. and Cook, R. L. 1970, *Microwave Molecular Spectra* (New York: John Wiley and Sons).

Green, G. W. and Lew, H. 1960, *Can. J. Phys.,* **38**, 482.

Hollowell, C. D.; Hebert, A.J. and Street, K. 1964, *J. Chem. Phys.,* **41**,3540.

Huiszoon, C. and Dymanus, A. 1966, *Phys. L.,* **21**, 164.

Kirchhoff, W. H. 1972, *J. Mol. Spectry.* **41**, 333.

Lovas, F. J. and Lide, D. R. 1971, *Microwave Spectroscopy of High-Temperature Gases* (In *Advances in High Temperature Chemistry,* **3**, New York: Academic Press).

Thaddeus, P. and Krisher, L. C. 1961, *Rev. Sci. Instr.,* **32**, 1083.

Thaddeus, P.; Krisher, L. C. and Loubser, J. H. N. 1964, *J. Chem. Phys.,* **40**, 257.

Townes, C. H. and Schawlow, A. L. 1955, *Microwave Spectroscopy* (New York: McGraw Hill).

van Dijk, F. A. and Dymanus, A. 1969, *Chem. Phys. L.,* **4**, 170.

Veazey, S. E. and Gordy, W. 1965, *Phys. Rev.,* **138A**, 1303.

Wollrab, U. E. 1967, *Rotational Spectra and Molecular Structure* (New York: Academic Press).

Thaddeus: Whatever the hazards of identifying a molecule on the basis of a single line, it is only fair to point out that the situation is better in the radio region of the spectrum than in any other. Reasonable assignments have recently been made with observations made with a resolving power $\lambda/\Delta\lambda$ of 100 − 1000.

Penzias: A good fraction of the lines listed in the NBS tables are spectra of molecules containing rare isotopes, such as ^2H and ^{14}C. I would therefore think that any statistical argument based on these tables would over-estimate the probability of confusion by at least an order of magnitude.

SOURCES OF DATA ON MICROWAVE FREQUENCIES AND RELATED QUANTITIES

1. Direct spectral data (observed frequencies):
National Bureau of Standards Monograph 70:
Microwave Spectral Tables (Superintendent of Documents, U. S. Government Printing Office, Washington, D. C. 20402).
Vol. I   Diatomic Molecules (1964)
Vol. III   Polyatomic Molecules with Internal Rotation (1969)
Vol. IV   Polyatomic Molecules without Internal Rotation (1968)
Vol. V   Spectral Line Listing (1968)
Microwave Data Center (W. H. Kirchhoff, National Bureau of Standards, Washington, D. C. 20234).

2. Molecular constants derived from microwave (and other) spectra:
Starck, B. 1967, *Molecular Constants from Microwave Spectroscopy* (Landolt-Bornstein, N. S., Group II, **4**, Berlin: Springer Verlag).
Herzberg, G. 1966, *Electronic Spectra of Polyatomic Molecules* (Princeton: D. Van Nostrand Company).
Rosen, B. 1970, *Spectroscopic Data Relative to Diatomic Molecules* (International Tables of Selected Constants; Oxford: Pergamon Press).
Nelson, R. D.; Lide, D. R. and Maryott, A. A. 1967, *Selected Values of Electric Dipole Moments for Molecules in the Gas Phase* Nat'l. Stand. Ref. Data Ser., Nat'l Bur. Stds. (U. S.) **10**.

3. Useful bibliographies:
Favero, P. G. 1963, 1966, 1969, *Microwave Gas Spectroscopy Bibliography* (Laboratory of Radiofrequency Spectroscopy, University of Bologna, Italy).
Starck, B. 1963, 1965, 1970, *Bibliography of Microwave Spectroscopic Investigations of Molecules* (Structure Documentation Section, University of Ulm, Germany).
Lovas, F. J. and Lide, D. R. 1971, *Microwave Spectroscopy of High-Temperature Gases (In Advances in High Temperature Chemistry, 3, New York: Academic Press).

Centrifugal Distortion and Predictability
in the Microwave Spectrum of Formamide

Donald R. Johnson
William H. Kirchhoff

National Bureau of Standards
Washington, District of Columbia

The published microwave literature is currently under critical review at the National Bureau of Standards for information relating to the spectra of molecules which have been detected in the interstellar medium. In this paper, the approach which has been used for this review will be illustrated by a discussion of a typical example *viz.*, formamide.

Formamide (NH_2COH) serves as a good example for a number of reasons. Emission signals from the $2_{11} \leftarrow 2_{12}$ transition of interstellar formamide were reported early this year (Rubin *et al.* 1971). Since then other groups have reported a confirming observation of the $2_{11} \leftarrow 2_{12}$ and a new observation of the $1_{11} \leftarrow 1_{10}$ transition. Spectroscopically formamide is a good example of a small asymmetric rotor with dipole moment components along both the a and b axes. The nuclear spin of the nitrogen atom produces hyperfine splittings of the rotational energy levels which introduce additional complication to the spectrum.

The early spectroscopic work on formamide centered around the determination of the structure of the molecule. As a result, the published literature contains spectra from 10 isotopic forms of formamide but only 9 observed transitions for the most abundant isotopic form ($^{14}NH_2{}^{12}C^{16}OH$). Unfortunately none of these observed transitions were of immediate astronomical interest and only a few of them produced resolved hyperfine structure. From this limited set of observations it was not possible to accurately predict the astronomically interesting portions of the spectrum.

An extensive computer program has been written which performs a least squares analysis of both centrifugal distortion and hyperfine structure in molecules like formamide (Kirchhoff 1971). The present discussion will illustrate how this program has been used in conjunction with selective laboratory measurements to provide an accurately defined map of the entire spectrum of the normal isotopic species.

An accurate treatment of the rotational spectrum of formamide requires the determination of 9 rotational parameters: three of these (A,B,C) are associated with the square of the angular momentum operators (P^2), five τ's associated with P^4 and one H related to P^6. In addition to these rotational parameters, two quadrupole coupling constants are needed to describe the hyperfine splittings. In order to obtain statistically meaningful predictions of unmeasured transitions it was necessary to measure a number of transitions in addition to the measurements already appearing in the literature.

Figure 1 lists the unsplit center frequencies for the entire set of observed transitions for $^{14}NH_2{}^{12}C^{16}OH$. The original set of 9 transitions have been included as have the $1_{10} \leftarrow 1_{11}$ and $2_{11} \leftarrow 2_{12}$ transitions which were measured in other laboratories. The remainder of the transitions shown in this figure are measurements from our own laboratory made in the process of fixing the 11 parameters mentioned above.

TRANSITION		CALC FREQ
1(1, 0) -	1(1, 1)	1539.543
1(0, 1) -	0(0, 0)	21207.460
1(1, 1) -	0(0, 0)	82549.681
2(1, 1) -	2(1, 2)	4618.553
2(0, 2) -	1(0, 1)	42386.116
2(1, 2) -	1(1, 1)	40875.500
2(0, 2) -	1(1, 1)	-18956.106
4(1, 3) -	4(1, 4)	15391.972
4(0, 4) -	3(0, 3)	84542.447
4(1, 4) -	3(1, 3)	81693.565
4(2, 3) -	3(2, 2)	84807.915
4(3, 2) -	3(3, 1)	84889.120
4(3, 1) -	3(3, 0)	84891.113
4(0, 4) -	3(1, 3)	26922.925
5(1, 4) -	5(1, 5)	23081.205
5(0, 5) -	4(1, 4)	50693.730
6(1, 5) -	6(1, 6)	32297.249
7(1, 6) -	6(2, 5)	-16961.261
9(1, 9) -	8(2, 6)	-37260.811
9(1, 8) -	8(2, 7)	37567.480
10(1, 9) -	10(1,10)	84073.066
11(2,10) -	10(3, 7)	-82578.445
15(3,13) -	15(4,12)	-84384.739
18(3,16) -	17(4,13)	-51208.536
19(1,19) -	18(2,16)	-40939.390
19(2,18) -	18(3,15)	38827.820
20(1,20) -	19(2,17)	-54060.257
21(3,18) -	21(3,19)	37189.424
21(2,20) -	20(3,17)	53131.589
24(4,21) -	23(5,18)	-40665.778
25(2,24) -	24(3,21)	52890.419
29(5,25) -	28(6,22)	-54471.386
29(5,24) -	28(6,23)	-51087.541

1. Unsplit center frequencies for the entire set of observed transitions in $^{14}NH_2{}^{12}C^{16}OH$. Negative frequency implies that the energy of the first rotational level is below the second, reading from left to right.

A "bootstrapping" procedure has been developed for determining which transitions ought to be measured in order to be able to calculate the remainder of the spectrum with desired certainty. On the basis of a rigid rotor calculation, transitions with low quantum numbers and hence low centrifugal distortion corrections were predicted with sufficient certainty that assignment of these transitions to observed spectral features could be made unambiguously. For formamide, the set of $J = 4 \leftarrow 3$ transitions were measured in this first step. These transitions along with the original set of nine transitions were then fit to an eight parameter model which included five centrifugal distortion constants. Using the parameters obtained from the fit, absorption frequencies for the remainder of the spectrum were calculated. In addition, using standard deviations of each of the parameters and the correlation of these standard deviations between parameters, the standard deviation of each predicted transition frequency was calculated. Some transition frequencies had standard deviations which were on the order of a few hundred kHz. Measurement and inclusion in the fit of these transitions would not substantially improve the prediction of the remainder of the spectrum. Transitions with uncertainties on the order of tens of MHz were not measured because often more than one observed spectral line would fall within the uncertainty limits and hence assignment was not possible. Only those transitions with uncertainties on the order of a few MHz could be unambiguously assigned. These would be included in the fit and a new predicted spectrum would be obtained with smaller uncertainties. This procedure was repeated until the majority of the spectrum was predicted to better than 1 MHz at 95 per cent confidence. On the basis of the quality of the fit, decisions could be made on whether the model should be extended to include sixth order angular momentum terms or not. Using this procedure, a minimum number of transitions (twenty to thirty) were measured which could lead to the prediction of the entire spectrum of a single isotopic species in the ground vibrational state within suitable uncertainty limits.

TRANSITION UPPER STATE	LOWER STATE	OBSERVED FREQUENCY (EST. UNCERTAINTY)	CALCULATED UNSPLIT FREQUENCY + QUADRU- POLE SHIFTS (EST. UNCERTAINTY)	LINE STRENGTH + REL INTENS- ITY OF QUAD- RUPOLE COMP.	ENERGY LEVELS IN CM-1 UPPER STATE	LOWER STATE	REFERENCE
1(1, 0) -	1(1, 1)		1539.543 (.003)	[1.500]	2.805	2.754	
F = 0 -	F = 1	1541.018(0.02)	1.477 (.060)	[.111]			71C
F = 1 -	F = 0	1539.570(0.05)	.009 (.061)	[.111]			71C
F = 1 -	F = 1	1538.135(0.02)	-1.436 (.036)	[.083]			71C
F = 1 -	F = 2	1538.693(0.02)	-.858 (.027)	[.139]			71C
F = 2 -	F = 1	1539.295(0.02)	-.291 (.027)	[.139]			71C
F = 2 -	F = 2	1539.851(0.02)	.287 (.007)	[.417]			71C
1(0, 0) -	0(0, 0)		21207.450 (.016)	[1.000]	.707	.000	
F = 0 -	F = 1	21206.560(0.1)	-.945 (.077)	[.111]			57A
F = 1 -	F = 1	21207.960(0.1)	.473 (.039)	[.333]			57A
F = 2 -	F = 1	21207.430(0.1)	-.095 (.009)	[.556]			57A
1(1, 0) -	1(0, 1)		62881.765 (.054)	[1.500]	2.805	.707	
F = 0 -	F = 1		1.436 (.036)	[.111]			
F = 1 -	F = 0		-.009 (.061)	[.111]			
F = 1 -	F = 1		-1.477 (.060)	[.083]			
F = 1 -	F = 2		-.860 (.021)	[.139]			
F = 2 -	F = 1		-.282 (.035)	[.139]			
F = 2 -	F = 2		.285 (.012)	[.417]			
1(1, 1) -	0(0, 0)	82549.370(.17)*	82549.681 (.066)	[1.000]	2.754	.000	71B
F = 0 -	F = 1		-.964 (.053)	[.111]			
F = 1 -	F = 1		.482 (.026)	[.333]			
F = 2 -	F = 1		-.096 (.005)	[.556]			
2(1, 1) -	2(1, 2)		4618.553 (.009)	[.833]	4.271	4.117	
F = 1 -	F = 1	4619.988(0.02)	1.436 (.036)	[.150]			71A
F = 1 -	F = 2		-.473 (.039)	[.050]			
F = 2 -	F = 2		-.473 (.039)	[.050]			
F = 2 -	F = 2	4617.118(0.02)	-1.436 (.036)	[.231]			71A
F = 2 -	F = 3		-.209 (.028)	[.052]			
F = 3 -	F = 2		-.817 (.028)	[.052]			
F = 3 -	F = 3	4618.970(0.02)	.410 (.010)	[.415]			71A
2(0, 2) -	1(0, 1)	42386.070(**)	42386.116 (.031)	[2.000]	2.121	.707	60A
F = 1 -	F = 0		.455 (.039)	[.111]			
F = 1 -	F = 1		-.963 (.077)	[.083]			
F = 2 -	F = 1		.018 (.000)	[.250]			
F = 2 -	F = 2		.585 (.046)	[.083]			
F = 3 -	F = 2		-.046 (.003)	[.467]			
2(1, 2) -	1(1, 1)	40874.910(**)*	40875.500 (.029)	[1.500]	4.117	2.754	60A
F = 1 -	F = 1		.009 (.061)	[.111]			
F = 1 -	F = 1		-1.436 (.036)	[.083]			
F = 2 -	F = 1		.473 (.039)	[.250]			
F = 2 -	F = 2		1.051 (.026)	[.083]			
F = 3 -	F = 2		-.176 (.009)	[.467]			

2. Sample of the format of tabulated data as it will appear in the NBS critical review of the microwave spectrum of formamide.

CALC. UNSPLIT FREQUENCY*			(EST. UNCERTAINTY)
81693.564	4(1, 4)	- 3(1, 3)	(.057)
82549.681	1(1, 1)	- 0(0, 0)	(.066)
-82578.444	11(2,10)	- 10(3, 7)	(.085)
84073.065	10(1, 9)	- 10(1,10)	(.090)
-84384.738	16(3,13)	- 15(4,12)	(.108)
84542.447	4(0, 4)	- 3(0, 3)	(.059)
84779.935	17(2,15)	- 17(2,16)	(.131)
84807.914	4(2, 3)	- 3(2, 2)	(.059)
84889.119	4(3, 2)	- 3(3, 1)	(.074)
84891.113	4(3, 1)	- 3(3, 0)	(.074)
85093.391	4(2, 2)	- 3(2, 1)	(.059)
-86371.458	22(1,22)	- 21(2,19)	(.277)
86382.839	7(1, 6)	- 7(0, 7)	(.061)
87848.988	4(1, 3)	- 3(1, 2)	(.059)
-88225.562	10(2, 8)	- 9(3, 7)	(.102)
-92435.801	5(1, 5)	- 4(2, 2)	(.095)
-93093.414	16(3,14)	- 15(4,11)	(.109)
-93871.868	4(1, 3)	- 3(2, 2)	(.099)

3. Sample of the table of transitions ordered by frequency as it will appear in the NBS critical review of the microwave spectrum of formamide. Negative frequency implies that the energy of the first rotational level is below the second, reading from left to right.

In all, 22 new transitions were measured to up $J = 29$. As the quality of the predictions improved several of the earlier measurements were found to be sufficiently inaccurate that they were being predicted with greater accuracy than they had been measured and hence were excluded from the fit. Five of the original set of nine transitions abstracted from the early literature were eliminated by this procedure.

In the final analysis a set of 25 rotational transitions were fit to the 9 parameters with a standard deviation of 66 kHz. Quadrupole splittings were handled in much the same way with 42 resolved quadrupole components fit to 2 parameters with a standard deviation of 62 kHz.

The final format of tabulated data as it will appear in our review of formamide is shown in Figure 2. This is a sample page from the table which covers all transitions in $^{14}NH_2^{12}C^{16}OH$ from 0.5 GHz to 200 GHz with a total rotational energies ranging up to 200 cm^{-1}. The table is self explanatory except for the brackets with double asterisk which indicate that experimental uncertainties were unavailable and the raised asterisk indicates transitions not included in the fit.

As a convenience to the user a tabulation of transitions ordered by frequency has also been made. A sample from this table is shown in Figure 3.

The critical review of formamide carried out in collaboration with Dr. Frank Lovas of NBS is now complete and available in preprint form. Similar reviews are underway for several other interstellar molecules and will be available in preprint form as they are completed.

REFERENCES

Kirchhoff, W. H. 1972, *J. Mol. Spect.*, **41**, 333.

Rubin, R. H.; Swenson, G. W.; Benson, R. C.; Tigelaar, H. L. and Flygare, W. H. 1971, *I. A. U. Circular No. 2319*.

Robinson: At Monash University, P. D. Godfrey and J. G. Crofts have measured the $\Delta F = 0$ and $\Delta F = 1$ transitions of the formamide $2_{11} - 2_{12}$ state. Their measured frequencies are:

ΔF	ν(MHz)	ΔF	ν(MHz)
1 - 1	4619.99	1 - 2	4618.07
3 - 3	4618.99	3 - 2	4617.34
2 - 3	4618.36	2 - 2	4617.13

When these are compared to the NBS computed frequencies, the average difference is seen to be only 10 kHz.

Flygare: Once a thorough analysis of the rotational and centrifugal constants for the main isotopic species of a molecule is completed, the distortion constants can be scaled to other isotopic species. Thus, if the three rotational constants of other isotopic species are known, the scaled distortion constants can be used to predict new transitions for the isotopic species (with lower confidence than the main isotopic species). In any event, these predictions of isotopic species (in the absence of a true distortion analysis) would be useful in making predictions and defining search frequencies in molecular radio astronomy.

Johnson: Your comment is correct. We have made predictions such as these for $H_2{}^{12}CS$, $H_2{}^{13}CS$, and $H_2C^{34}S$ and get reasonable results.

E. B. Wilson: Raymond Azrak studied the ^{35}Cl and ^{37}Cl species of chloroethanol and found that a considerable improvement in fit could be achieved on the ^{37}Cl species by simply transforming the actual frequency correction (actual frequency minus rigid rotor) from the analyzed ^{35}Cl spectrum. A still better fit was found by taking over the actual centrifugal distortion constants τ. Of course, even better procedures are probably possible by scaling the τ's.

Thaddeus: It is of considerable value, in regards to optical depths of astrophysical lines, to have precise frequencies systematically measured for ^{13}C and ^{15}N isotopic species. But this information is often missing from the standard compilations.

Johnson: We are attempting to include information on all astronomically interesting species whenever such information is available in the literature. Please comment on any items which you feel are missing from our preprints so that we can make the published version as complete as possible.

Kirchhoff: The quoted uncertainties of 66 kHz for the fit of the formamide data represent a measure of the measurement error rather than model error. But, at this level of accuracy, the two types of error could be comparable.

Photograph of the Lagoon Nebula. Note the dense obscuring regions superimposed upon the background nebulosity. (Hale Observatories photograph)

Session 4

Theoretical Models of Molecular Excitation
and
Their Implications

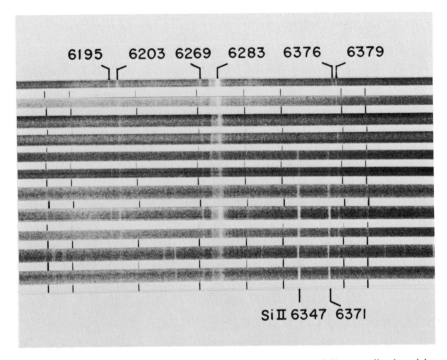

Spectrum of the interstellar diffuse band. The indicated lines of Si are stellar in origin. (Courtesy of G. Herbig)

Selective Predissociation as a Possible Mechanism for the Population Inversion of Interstellar H_2O

Takeshi Oka

National Research Council of Canada
Ottawa, Ontario

I. INTRODUCTION

The enormous intensity, variability and polarization of the interstellar H_2O emission line corresponding to the transition $6_{16} \rightarrow 5_{23}$ has indicated a maser action in interstellar space (Knowles *et al.* 1969). Some molecular process must be operating either to populate the 6_{16} level excessively or to depopulate the 5_{23} level to establish the required inverted population.

II. GROUPING OF ROTATIONAL LEVELS

It is unlikely (although not absolutely impossible) that either the 6_{16} or the 5_{23} level is anomalously populated compared with all other levels; there seems to be nothing special about these levels except that they happened to be very close in energy. Rather some process must either populate excessively a group of levels including the 6_{16} level (which we call Group-A levels) or depopulate a group of levels (Group-B levels) including the 5_{23} level. We divide many asymmetric rotor levels of H_2O into two groups: Group-A levels which are composed of all the $J_{0,J}$ and $J_{1,J}$ levels; and Group-B levels which are composed of all the other levels. The levels in Group-A are the levels with $K_c = J$ in which the H_2O molecule is rotating almost entirely around the C-axis. The justification for the grouping is seen when rotational relaxation by spontaneous emission is considered. Figure 1 summarizes the rotational levels of H_2O sorted according to the value of K_c. The pairs of levels $J_{0,J}$ and $J_{1,J}$ at the bottom of each column for a given K_c are the Group-A levels. It is seen that spontaneous emissions (indicated by lines connecting levels in Figure 1) starting from the $J_{0,J}$ and $J_{1,J}$ levels end at the $J\text{-}1_{1,J\text{-}1}$ and $J\text{-}1_{0,J\text{-}1}$ levels, respectively, which also belong to the Group-A levels. The only exceptions to this are the weaker transitions with $\Delta K_c = 3$ which include the maser transition. Therefore if once molecules are populated in Group-A levels with relatively high energy the only radiative transitions to Group-B levels are caused by the $\Delta K_c = 3$ transitions. In particular it may be noted, that the maser transition $6_{16} \rightarrow 5_{23}$ is the first way out of the Group-A levels. If some molecular process depopulates Group-B levels there will be a maser action. Once the maser action starts the increased induced emission will transfer molecules from Group-A levels to the Group-B levels via the maser transition.

The molecular process which depopulates Group-B levels must be one of the following: (i) the molecular formation process (ii) the pumping and relaxation processes by collision or radiation and (iii) the dissociation process. Normally the second process is used to explain various populational anomalies because it occurs many times during a lifetime of a molecule [see for example, Townes and Cheung (1969) or Litvak (1969)]. In this paper we wish to examine the

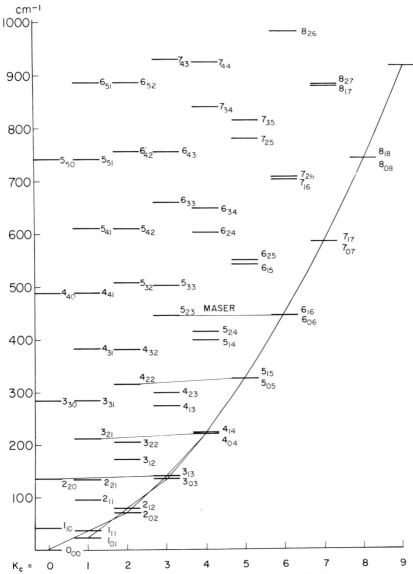

1. Rotational energy levels of H_2O molecules sorted out with K_c. The Group-A levels are composed of $J_{0,J}$ and $J_{1,J}$ levels which lie at the bottom of each column for a given K_c. The lines connecting various levels indicate spontaneous emission of the H_2O molecules. It should be noted that if molecules in Group-A levels relax from high evergy levels the maser transition $6_{16} \rightarrow 5_{23}$ is the first connection between the Group-A and Group-B levels. Therefore if some molecular process depletes molecules in the Group-B levels there will be an inverted population between the 6_{16} and the 5_{23} levels.

possibility of the third process, that of dissociation, as being a possible mechanism for the population anomaly. The adoption of it makes it necessary to take the view that molecules are formed and destroyed violently rather than the view that molecules live for many years protected by dust clouds. This may not be unreasonable in the region of the H_2O cloud where stellar formation or activity is expected.

III. SELECTIVE PREDISSOCIATION OF H_2O

Recently Johns (1971) has observed the vacuum ultraviolet spectrum of water in the region from 1240Å to 1070Å by using a high resolution vacuum spectrograph in Ottawa. Because of the large transition moments the H_2O molecule exhibits very rich Rydberg absorption over this range. Johns has observed that the widths of individual spectral lines are remarkably dependent on the rotational quantum numbers. In general the lines with the quantum number $K_a = 0$ in the upper electronic state are much sharper than the other lines. Figure 2 indicates two bands of the spectrum in which only the lines with $K_a = 0$ are visible; all the other lines are smeared out because of uncertainty broadening of the absorption lines caused by predissociation.

Such a dependence of predissociation on the rotational quantum numbers is not a rare occurrence and is explained from symmetry arguments [see page 458 of Herzberg (1966)]. In the present case, the discrete electronic state n from which the H_2O molecule dissociates into the continuous state i has such a symmetry that the product of their characters Γ_n and Γ_i belongs to the species corresponding to that of the rotation around the a-axis. The predissociation which is forbidden by electronic symmetry thus becomes allowed through mixing of the wavefunctions by rotation-electronic interaction.

The rotation-electronic interaction is essentially due to Coriolis force which is proportional to the angular momentum K_a; thus the levels with large K_a values are predissociated more than those with small K_a values. In particular for the $K_a = 0$ levels of electronic state in which the molecule has a linear structure the predissociation is forbidden and the molecule is stable. Because of the dipole selection rules, the stable rotational level J_{0J} in the excited electronic state are connected dominantly with some of the following levels in the ground electronic state; $J+1_{0,J+1}$, $J+1_{1,J+1}$, $J+1_{1,J}$, $J_{1,J}$, $J_{1,J-1}$, $J-1_{0,J-1}$, $J-1_{1,J-1}$, and $J-1_{1,J-2}$. The connection is dependent on the direction of the transition moment. Five out of these eight levels are the levels with $J = K_c$, that is, the Group-A levels.

The astrophysical implication of this is as follows. The H_2O molecules in the Group-B levels (with $K_c < J$) in the ground electronic state can absorb vacuum ultraviolet radiation over a much wider range of wavelength because of the broader linewidth than those in Group-A levels ($K_c = J$) which are connected to

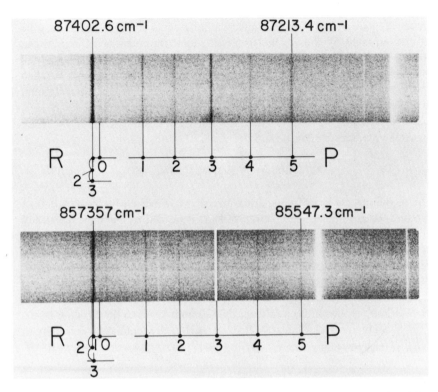

2. Two bands of the H_2O spectrum observed by Johns (1971) in the vacuum ultraviolet region. The remarkable simplicity of the spectral features indicate that only levels with $K_a = 0$ in the excited electronic state are stable but all the other levels are predissociated. This type of feature was observed over the entire region between 1240 and 1070Å. This gives the required mechanism of dissociating the H_2O molecules in the Group-B levels selectively.

the levels with $K_a = 0$. Another way of looking at the same thing is that the vacuum ultraviolet radiation corresponding to the sharp lines are absorbed on the surface of the H_2O cloud but the radiation corresponding to the diffuse lines penetrates to inside of the cloud and dissociate H_2O molecules in Group-B levels. An inverted population therefore will result between the 6_{16} and the 5_{23} levels.

The advantage of this mechanism over other infrared or far infrared pumping mechanisms is that vacuum ultraviolet radiation of very wide wavelength region can be used rather than resonant radiation in an extremely narrow region. The features of the vacuum ultraviolet spectrum observed by Johns are rather diversified depending on the properties of the vibronic wavefunctions in the upper state. We estimate the useful frequency range to be of the order of 100Å between 1240Å and 1000Å; the radiation between 1100Å and 1000Å may be absorbed strongly by the H_2 molecules. It should be mentioned that between 1800Å and 1240Å the predissociation of H_2O occurs indiscriminately on all rotational levels. Since the absorption in this range is weaker than that between 1240Å and 1000Å, we estimate that about one third of the total vacuum ultraviolet radiation absorbed by H_2O is used for selective predissociation.

In the following we consider the interstellar conditions which are necessary for such a process of work efficiently.

IV. SUPPLY OF H_2O MOLECULES

Since the mechanism of population inversion is based on the dissociation of molecules, some efficient production of H_2O is needed to maintain the steady state. The minimum for the rate of production is set by the time of spontaneous emission, 0.95 sec, for the $6_{16} \rightarrow 5_{05}$ transition since the molecules in the upper level of the maser transition are depleted at least by this rate. The pumping of molecules from lower rotational levels by far infrared radiation (except for a radiation trapping) or collisions does not change this number very much because they will also transfer molecules from the Group-A levels to the Group-B levels.

The vacuum ultraviolet radiation dissociates H_2O predominantly according to the following equation,

$$H_2O + h\nu \rightarrow H + OH \qquad (1)$$

Since the energy of the vacuum ultraviolet photons ($80,000 \sim 100,000$ cm^{-1}) is much higher than the dissociation energy of D(H-OH) $\sim 41,000$ cm^{-1} (Herzberg 1966), the resultant OH has very high energy. If the OH molecule is produced in the ground $^2\Pi$ state, the excess energy is 40,000 to 60,000 cm^{-1} whereas if the OH molecule is produced in the first excited $^2\Sigma^+$ state (which is about 33,000 cm^{-1} above the ground state), the excess energy is 6,000 to 26,000 cm^{-1}. We estimate that on the average 20,000 cm^{-1} of energy is transferred to the translational energy of H and OH. The OH will take 1/18 of the energy that is 1100 cm^{-1} or 1600°K. This hot OH will react efficiently to recover H_2O. The recovering reaction must be

$$OH + H_2 \rightarrow H_2O + H. \qquad (2)$$

The reverse reaction of Equation (1) is too slow to be efficient in interstellar space since the excess energy has to be taken away by photon emission. In the reaction of Equation (2) the excess energy is taken away by the resultant H atom and therefore the laboratory value for the rate constant,

$$k = 2.19 \times 10^{13} \exp(-5150/RT) \qquad (3)$$

can be used for estimating the rate in interstellar space, where k is given in cm^3 $mole^{-1}$ sec^{-1} and R is 1.987 cal $mole^{-1}$ deg^{-1} (Baulch, Drysdale and Lloyd 1968). If we assume the effective temperature to be $1000°K$, Equation (3) gives $k = 2.7 \times 10^{-12}$ molecules cm^{-3} sec^{-1}; the estimation of $1000°K$ was used in spite of the fact that translationally the H_2 molecules have lower temperature, because the internal energy of OH will supply the activation energy of 15 kcal $mole^{-1}$. This rate constant requires the density of H_2 to be 3.7×10^{11} cm^{-3} in order to recover the OH molecules in a second. If there are ten times more OH than H_2O in the area this number is reduced by a factor of 1/10. The rotational relaxation of the H_2O molecule induced by collisions with the H_2 molecules of this density has time of the order of 1 second; therefore if the ratio of OH to H_2O is more than 10 to 1, the selection of Group-A levels is not much spoiled by collision-induced rotational transitions.

V. SUPPLY OF VACUUM ULTRAVIOLET PHOTONS

The second requirement is, of course, abundant ultraviolet radiation; a star has to be near or in the H_2O cloud. Again in order to maintain the inverted population the dissociation has to occur in about 1 second. The time τ for induced absorption is given by

$$\tau = 3\hbar^2 / \pi \sum_i |\mu_{oi}|^2 U(\nu_i) \qquad (4)$$

where μ_{oi} is the dipole matrix element of the electronic transition from the ground state to the i-th excited state and $U(\nu_i)$ (erg cm^{-3} sec) is the energy density of the vacuum ultraviolet radiation at the frequency ν_i (Heitler 1953). Using $\mu_{oi} \sim 1$ Debye (which approximately reproduces the absorption intensity observed by Johns) and 10 electronic excited states, we find the necessary energy density to be of the order of 10^{19} erg cm^{-3} sec. The intensity of radiation is given by

$$I = \frac{c}{4\pi} U \qquad (5)$$

to be 2×10^{-10} erg cm^{-2} or 6×10^2 erg $cm^{-2} Å^{-1} sec^{-1}$. This means we need a star at a distance of 1 a.u. which gives vacuum ultraviolet radiation about five thousand times more than that from the sun including the Lyman a line.

Finally we consider the total number of photons. The number of microwave photons received from W49 for example is of the order of 6 photons cm^{-2} sec^{-1}. Using the distance of 14 kpc and assuming isotropic radiation, Townes and his collaborators estimated the number of microwave photons originally emitted from the H_2O cloud to be of the order of 6×10^{48} photons sec^{-1}, a tremendously large number. Since one ultraviolet photon corresponds to one microwave photon at most in the mechanism described in this paper, at least this large number of far ultraviolet photons are needed. The number of vacuum ultraviolet photons from the sun is short by factor of 10^8 to 10^9. In order to make up this number by increasing the brightness of the sun we need the temperature of the star to be of the order of 20,000°K, that is the temperature of a B0 star.

It should be pointed out that the mechanism described in this paper or some other mechanism might cause maser action in the inter-planetary space or outer-planetary space by using vacuum ultraviolet radiation from the sun: no oscillation but amplification. Slight amplification between us and the source will greatly reduce the required number of photons originally emitted from the H_2O cloud.

VI. CONCLUSION

Energy levels of the H_2O molecules have been classified into two groups: Group-A levels composed of all the $J_{0,J}$ and the $J_{1,J}$ levels and Group-B levels composed of other levels. It has been pointed out that the maser transition $6_{16} \rightarrow 5_{23}$ is the first transition to connect the Group-A levels with the Group-B levels by emission and therefore if some molecular process depletes molecules in the Group-B levels maser action results. Selective predissociation of H_2O by vacuum ultraviolet radiation observed by Johns (1971) has been considered as a possible process to deplete molecules in the Group-B levels. The necessary interstellar conditions for this mechanism are the presence of abundant OH (at least factor of 10 more than H_2O) and a bright star in the region of the H_2O cloud. A difficulty of this postulate is the photochemical problem on whether H_2 and OH can exist or can be recovered under such strong radiation.

Finally it is worth pointing out that a maser action is expected for the $5_{15} \rightarrow 4_{22}$ transition at 325 GHz because this transition is the first connection between the Group-A and Group-B levels for para-H_2O molecules.

I wish to thank J. W. C. Johns, G. Herzberg and A. E. Douglas for several helpful discussions and R. A. Back for a discussion of photochemistry.

REFERENCES

Baulch, D. L.; Drysdale, D. D. and Lloyd, A. C. 1968, *High Temperature Reaction Data,* Department of Physical Chemistry, The University, Leeds 2, England.

Heitler, W. 1953, *The Quantum Theory of Radiation* (London: Oxford University Press).

Herzberg, G. 1966, *Molecular Spectra and Molecular Structure III. Electronic Spectra and Electronic Structure of Polyatomic Molecules,* (Princeton: D. Van Nostrand Co., Inc.)

Johns, J. W. C. 1971, Unpublished.

Knowles, S. H.; Mayer, C. H.; Cheung, A. C.; Rank, D. M. and Townes, C. H. 1969, *Science,* **163,** 1055.

Litvak, M. M. 1969, *Ap. J. (Letters),* **156,** L471.

Townes, C. H. and Cheung, A. C. 1969, *Ap. J. (Letters),* **157,** L103.

Aannestad: Would it be possible to have the same kind of selective destruction of H_2O via the opposite reaction to the one you mention, that is,

$$H_2O + H \rightarrow OH + H \ ?$$

Oka: Not for this reaction! But for the reaction

$$H + OH \rightarrow H_2O + h\nu,$$

the product will be more A-type water than B-type.

W. J. Wilson: The cooler IR stars with H_2O emission require $\sim 10^{43}$ photons sec^{-1} —which is not likely to be available near these objects.

Oka: Many problems arise because of the enormous number of microwave photons observed astronomically, $\sim 6 \times 10^{48}$ sec^{-1} in some cases. I wonder if any maser action on this side of space—either interplanetary or extraplanetary—is possible. These masers can have a gain of $\sim 10^7$ before oscillation starts. Such a gain will reduce the number of original photons very much.

Litvak: Your first slide indicates that any mechanism capable of exciting the $J = 7$ and higher states involves cascades through the upper maser state, thereby leading to population inversion if there is adequate radiation trapping to lower levels. Won't OH and H_2 be photodissociated?

Oka: Yes. But because the spectra of OH and H_2 are much narrower than that of H_2O, the photons available for these dissociations are fewer than those available for dissociation of H_2O.

Masers and Optical Pumping

M. M. Litvak

Harvard College Observatory
Cambridge, Massachusetts

A maser in astronomy is usually an amplifier. As shown in Figure 1a, the microwave transition between the lowest and next energy level becomes a maser when the population in the upper (emitting) state exceeds that in the lower (absorbing) state. As in laboratory masers, the astronomical masers very likely depend upon the competition between collisional and radiative effects on the populations among the molecular states. The maser pump is the agent for exciting the overpopulation. Figure 1a shows three such states of a molecule. The straight arrow pointing from the lowest to the highest state represents collisional excitation (pumping), for example. The parities of the states are indicated on the right of the energy levels. This is an example of a "forbidden-type" transition. Following excitation of the highest state is fluorescence to the middle state, with the change of parity appropriate to electric-dipole transitions, the only radiation type considered here. It is easy to imagine for optical pumping that a fourth, even higher state, is excited, followed by two successive fluorescences. For non-equilibrium it is important that this fluorescent radiation leak out, otherwise collisional de-excitation will become more dominant the more this radiation is trapped in the cloud (Litvak 1969a, b). With only collisional de-excitation and collisional excitation, the populations come to equilibrium at the kinetic temperature of the colliding particles. Also, too large densities will lead to thermal equilibrium. As shown in Figure 1a, the lower state would become empty, entirely filling the middle state, except for the opposing rates (not shown) leading from the upper maser state eventually to the lower one. Figure 1c shows a four-level composite of the three-level scheme in terms of two successive microwave doublets, such as in ortho-formaldehyde, connected by millimeter-wave radiative transitions. If the collisional excitation from the second to the third level (indicated by the heavier straight arrow) is strong enough compared to the collisional excitation from the first to the fourth level, as proposed by Townes and Cheung (1969), then an anti-maser develops, *i.e.* an anomalous absorber arises because of excess population in the lower microwave levels (Litvak *et al.* 1966). In fact, anti-masers would occur for both doublets. However, collisional models have been disappointing. The hard-sphere collisions are supposed to give this asymmetry of collisional excitation. These forces are short-range compared to the electrostatic-type collisions such as dipole-induced dipole collisions between formaldehyde (having the electric dipole) and hydrogen (atom or molecule). These long-range collisions show no preference in exciting either level of the upper doublet and have much larger "forbidden-type" cross-sections, thereby greatly diluting the effect of the hard-sphere type of collisions. Dipole-induced dipole collisions involve "forbidden-type" transitions, while the dipole-quadrupole (on H_2) collisions with hydrogen molecules involve "dipole-type" transitions. Both will prevent collisions from giving anomalous absorption. In fact, if the kinetic temperature is

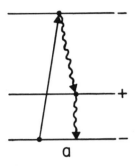
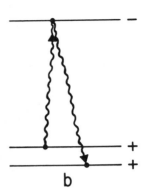

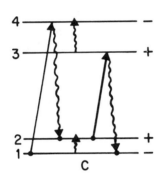

1. (a) Three-level maser scheme. Radiative processes (wavy arrows) obey electric dipole selection rules, collisions (straight arrow) may not. Parity of states indicated on right-hand side of levels. Collisions connecting other pairs of states maybe very strong, too.

(b) Two interlocking radiative transitions (having a common excited state and two hyperfine-split lower states. Unequal resonance-radiation trapping causes a transfer of population from one lower state to the other.

(c) Four-level maser scheme like the two rotational doublets in ortho-formaldehyde. Each set of parity states for a doublet resembles the three-level scheme of Figure 1a.

noticeably above the $3°K$ radiation temperature, there is a possibility of maser action – not anti-maser action – due to the two-temperature non-equilibrium (Solomon and Thaddeus 1969, Litvak 1970). Evidently, if collisions do not work optical anti-pumping might be the answer. By an accident of nature, perhaps the millimeter radiation for the higher frequency transition between the upper levels of each of the two doublets is somewhat stronger than for $3°K$ while the lower frequency radiation involving the lower levels of each of the doublets is not (Solomon and Thaddeus 1969). However, this pumping would probably favor maser *emission* in the upper doublet at the same time as anti-maser absorption in the lower doublet – a case not yet observed. Calculations indicate that for 6-cm absorption in Orion A, the continuum infrared radiation can cause anti-inversion in both doublets. Other cases, where no obvious sources of infrared radiation exist, perhaps a shock-heated layer of formaldehyde shines infrared resonance lines on cold molecules left far behind the shockwave. Anomalous absorption in these out-lying molecules is then observed. Larger angular sizes and microwave optical depths favors observation of these cold, slowly-moving molecules over the hot molecules. Details of the infrared pumping scheme are deferred till later. The formation of formaldehyde in the shockwave (Dean and Kistiakowsky 1971) by the reaction: $O + CH_3 \rightarrow CH_2O + H$, where CH_3 is derived from the methane reaction: $O + CH_4 \rightarrow CH_3 + OH$, seems highly-reasonable, according to calculations (Litvak 1971a). It is likely that the CH_4 is formed on grains, so that methane is about 10^{-5} times as abundant as hydrogen. The destruction process for CH_2O was assumed to be $O + CH_2O \rightarrow H + CO + OH$. Radiation cooling by O and H_2 was included in the calculations.

Figure 1b shows two interlocking fluorescences in place of the one in Figure 1a, because the middle level might actually be split by the hyperfine interaction. Because these hyperfine-split transitions usually have different line strengths, the radiative transport for the two radiations is usually unequal. A concrete example of this situation is shown in Figure 2, the ground state and two next-excited rotational levels for OH. The four transitions across the ground, $(^2\Pi_{3/2})$, hyperfine-split doublet are the famous ones observed in a variety of astronomical objects. These lower levels are connected to the next levels $(^2\Pi_{5/2}$ and $^2\Pi_{1/2})$ by far-infrared transitions, the relative line-strengths for which are shown in the Figure. Because the microwave optical depths in the ground state are quite large in order to explain the observed brightness temperatures, the implied optical depths involving absorption to the $^2\Pi_{5/2}$ states are very large, too large for important unequal trapping effects in the interlocking infrared transition having the $F = 2$ levels of *both* parity in $^2\Pi_{5/2}$ as common upper levels. The radiative connections are shown only for one set of parities, for the sake of clarity. The ones not shown have the same line-strengths but involve the opposite parities.

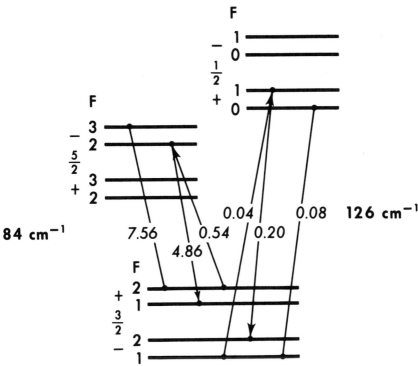

2. The three lowest rotational energy levels of hydroxyl. The interlocking far-infrared transitions between the ground and excited levels cause transfer of population in the ground hyperfine-split levels. The arrows (radiative transitions) indicate the direction of transfer for the case of external far-infrared radiation being selectively absorbed by the cloud. Volume excitation (by collisions or near-infrared pumping) of the excited state that is held in common by the interlocking transitions causes population transfer in the direction opposite to the arrows. Interlocking transitions between levels of the opposite parities exist but were not drawn for the sake of clarity. Thus, effects are fairly symmetric for upper and lower doublet levels.

The interlocking transitions involving the $F = 1$ levels in $^2\Pi_{1/2}$ have the most important effects for the larger microwave optical depths. If external 126 cm^{-1} radiation tries to penetrate the cloud from one side, the major pumping is shown in the direction of the arrows, for the part of the cloud farthest from the source but probably nearer to the observer. The transfer of population is from those labelled $F = 1$ to $F = 2$. These cause maser action at 1720 MHz. In contrast, if the $^2\Pi_{1/2}$ state is excited by some other means, such as by collisions or near-infrared absorption followed by cascading of the excitation to $^2\Pi_{1/2}$ state, the transfer of population is opposite to the arrows, giving rise to 1612-MHz emission, thereby possibly explaining the importance of this transition in infrared stars (Litvak 1969a). This particular pumping occurs because of the stronger trapping for the fluorescence with the larger line-strength. In the case of external far infrared, which must penetrate, it is because of the greater intensity in the transition with the weaker line-strength, at least in the farther half of the cloud. Anomalies in OH in the same dark clouds that show anomalous formaldehyde absorption (Turner and Heiles 1971), suggest the same infrared effects are occurring for both. The relatively thin OH shows excess 1720-MHz

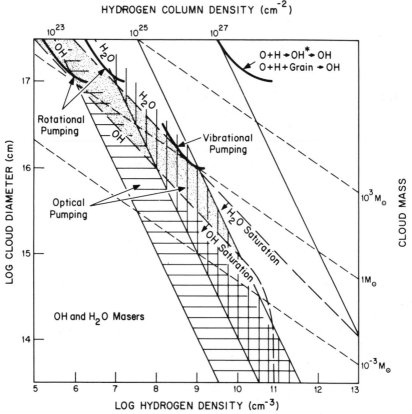

3. Cloud diameter *versus* cloud hydrogen density for obtaining maser emission like that of one emitting maser point in W49, according to pumping by excitation of rotational and vibrational levels via collisions, chemical reactions or optical means. Lines of constant hydrogen column density and cloud mass are shown. The speckled region is for gravitational collapse for clouds between $70°K$ to $200°K$ that are thin in the far-infrared region. Interferometer sizes of about 10^{13} to 10^{15} cm (probably enlarged by plasma-screen scattering) are observed.

emission, an expected result of infrared excitation of the $^2\Pi_{5/2}$ state. The clouds are too cold (5 - 10°K) for any significant collisional excitation.

Figure 3 shows heavily-drawn arcs that give the conditions of cloud diameter and hydrogen density that might be compatible with the strongest observed OH and H_2O maser emission points (Litvak 1970b). The type of states that are excited in the pumping are indicated. The temperature is assumed to be sufficiently large to be suitable. For example, rotational excitation involved OH that is slightly colder than 100°K while the H_2O is somewhat hotter than 100°K. The relative abundances of OH and H_2O to hydrogen are 10^{-6} and 10^{-5} respectively. Vibrational excitation involves nearly 1000°K, as does the pre-association $O + H \rightarrow OH^* \rightarrow OH$, process. This will be discussed more later on. More efficient chemical pumping fills the narrow region between vibrational pumping and pre-association or recombination on grains. The most optimistic rate of forming OH on grains was assumed. This is probably a factor of one-hundred faster than for pre-association at 1000°K. The speckled region denotes conditions of gravitational instability. It is interesting that the conditions for maser action lie near these, for about one solar mass. The large cloud sizes needed are indicative of a lack of adequate pumping if the interferometer sizes are interpreted as the real sizes. When corrected for some plasmascreen scattering which enlarges the spot sizes, the apparent sizes are probably close to 10^{13} cm (Litvak 1971b). A discussion of these sizes follows below. But, in Figure 3, the shaded areas show the allowed conditions for optical pumping, assuming a strong enough optical source at the center to provide the requisite number of photons/sec to account for the maser emission. This source is so intense that the pump rate can compete with high thermalizing-collision rates. The density is not likely to be uniform in these clouds, but rather perhaps like inverse-square with radius, much like the density profile shown by Dr. Bok earlier for a dark globule. Without absorption, the pump intensity follows the same radius dependence and can pump the whole cloud from center to outside. Of course, dust extinction for these hydrogen column densities (greater than 10^{23} cm^{-2}) might be a problem for ultraviolet and even infrared pumping. However, scattering is not detrimental since the source is inside the cloud, while most of the absorption, due to an icy-mantle, might be eliminated upon evaporation of the mantle by a passing shockwave or by the heating of the grains by the radiation from the shockwave that has passed and is collapsing into denser central regions. The collapsing shockwave might be expected for gravitationally-unstable clouds that are perturbed at their outer boundary by stellar winds or radiation pressure from nearby stars or by collisions with nearby clouds. The small refractory cores of the grains that remain might have negligible absorption, if they are silicates. The masers are saturated below the line of 1M$\odot$ and unsaturated above this line, approximately.

Galactic OH Sources

Source	Dist. (kpc)	Flux ($\times 10^3$ phot./m^2-s)	Freq. (MHz)	Lum. ($\times 10^{43}$ phot./s)
W3	3	20	1665	200
W49	14	30	1665	6,000
"	"	15	1667	3,000
W51	5	4	1665	100
"	"	1	1612	20
"	"	4	1720	100
W28	4	10	1720	200
NGC 6334	1	20	1665	20
" "	"	15	1667	10
Orion	0.5	2	1665	0.6
NML Cyg	0.5?	150	1612	40
VY CMa	0.5?	150	1612	40

A Comparison of Optical Pumps

Pump Object	Temperature (K)	Size (a.u.)	Maser Freq. (MHz)	Excitation ($\times 10^{43}$ phot./s)
O5 star (UV emitter)	50,000°	0.3	1665,1667	100
Protostar (UV shock)	4,000°	300	1665,1667	1,000
Near-infrared star	2,000°	10	1612	10
Protostar (IR shock)	1,000°	100	1612	100
Far-infrared nebula	100°	10,000	1720	1,000

4. Table of properties of some OH maser regions, their distance away, their microwave output and flux in photons per sec. Table of optical-pumping astronomical objects, their temperatures, diameters, and flux in photons per sec available for pumping (to be multiplied by the efficiency of pumping).

Figure 4 summarizes the microwave output of several HII region sources and the two unusual infrared stars, NML Cyg and VY CMa. For an optical pump to be adequate, under even the most efficient circumstances, *i.e.* of saturated maser action for which the maser rate is limited by the pump rate that replenishes the lost excitation, the optical photons/sec available must correspond to the maser emission. For the usual sources like ultraviolet from an O5 star or infrared from a cool star, the photon output is too small, especially for sources like W49. However, the large area sources, like the proto-stars with shockwaves that emit ultraviolet or infrared resonance lines, might be adequate. The type of source needed is dictated by the frequency that results from the pumping. For example, 1665-MHz is probably stronger than 1667-MHz, owing to the combined effect of ultraviolet and infrared lines emitted by the 4000°K shockwave. The amount of ultraviolet at 3100Å is fairly substantial despite its being on the tail of the blackbody curve. Excitation by electrons in the front excites the fluorescence of the ultraviolet OH state.

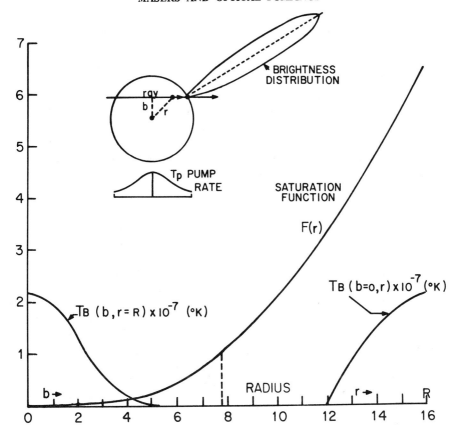

5. Spherical cloud (inset) emitting in a radially-peaked brightness distribution at a typical point on the amplifier surface. Brightness temperature as a function of b, the distance of the chord-ray from the center, at a point on the surface r = R (all radii in unsaturated optical depth units). Brightness temperature as a function of radius along a diameter b = 0, showing saturated amplification.

Figure 5 shows the results of a calculation of the output of a spherical maser cloud that is partially-saturated. The brightness distribution at a typical point is shown to be very directional. A blackbody would show isotropic emission at the point. Every point on the sphere shows the same pattern, thus the overall emission is still isotropic, but a very-long baseline interferometer would see a much smaller apparent size than the size of the amplifier. The brightness temperature profile shown at the bottom left indicates that the temperature drops to one-half of maximum for a value of $b \approx 2$ compared with a radius of R = 16, all lengths being in unsaturated microwave optical depth units and b being the distance of a ray to the center. The other temperature profile is for a ray through the center. The exponential rise of brightness temperature in the unsaturated central region is not discernible on this scale, which mainly shows the saturated amplification at the outer radii. Two cases were calculated for the non-linear transport equation, one with a given continuum temperature as a boundary condition, and the second with a negative excitation temperature of the same magnitude to represent the distributed volume-sources of spontaneous emission. The two different imputs to the amplifier lead to nearly identical results. The numerical example taken here corresponds to a pump rate T_p (r) (in temperature units) which is Lorentzian, with maximum value of $10^{5\,°}K$ and half-width at half-maximum of $h = 6\sqrt{2}$. It can be shown that a "hot spot" will appear even with constant T_p. The function F(r) is the saturation parameter, the ratio of microwave induced rate to pump rate, i.e. F(r) = $\int d\Omega T_B(r,\theta,\phi)/T_p(r)4\pi$, where Ω is the solid angle at a point in the amplifier, θ and ϕ being the associated polar angles. Actual rates in sec^{-1} are given by $\int d\Omega$ kT A/hν 4π, where A is the Einstein coefficient for spontaneous emission and hν is the photon energy of the transition. Saturation occurs when F is greater than unity, for this example for r > 8. The unsaturated core, or "hot spot", has a diameter of d = 16 in optical depth units. This agrees with an analytic formula giving $d \approx 2$ [F(0) (1/6 + h^{-2})]$^{-\frac{1}{2}} \approx 5$ F(0)$^{-\frac{1}{2}}$ with F(0) $\approx$ 0.1, the approximate value near the center (Litvak 1971a, b). A more complex calculation is underway that includes the effect of the frequency spectrum on the brightness distribution. Simple geometry shows that the angular width of the output brightness distribution (full angle at e^{-1} of maximum) is $\sqrt{2}$ d$^{\frac{1}{2}}$/R) while the size of the "hot spot" as viewed by an interferometer is given by $\sqrt{2d^{\frac{1}{2}}}/a$, in physical units, where a is the amplification coefficient (cm^{-1}) in the "hot spot" (Litvak 1971b). Because the astronomical brightness temperatures are much higher than in this example and because both the local amplification coefficient and pump-rate might have much higher values at the center compared to the edges, very small angular widths of the brightness distribution are expected, perhaps as small as 10^{-3}, compared to 0.35 here.

Starting with ultraviolet effects, Figure 6 shows the potential curves for OH as a function of distance between the O and H atoms. The ground state X$^2\Pi$ lies at the bottom. The A$^2\Sigma^+$ bound state is excited by 3100Å radiation for pumping purposes. Photodissociation, according to the Franck-Condon Principle, occurs by nearly vertical transitions. The transitions to unbound states that require a change of spin, as $^2\Pi \rightarrow ^4\Sigma^-$ or $^4\Pi$ are extremely weak. The $^2\Pi \rightarrow ^2\Sigma^-$ requires about 8 eV and 13 eV photons. Both of these transitions have weak oscillator strengths because they are forbidden in both the separated atoms and in the united atom, F. Some ultraviolet shielding by various species such as water vapor, carbon compounds, etc. is very likely to protect the OH, even without dust extinction. The unbound potential curves leading from ground

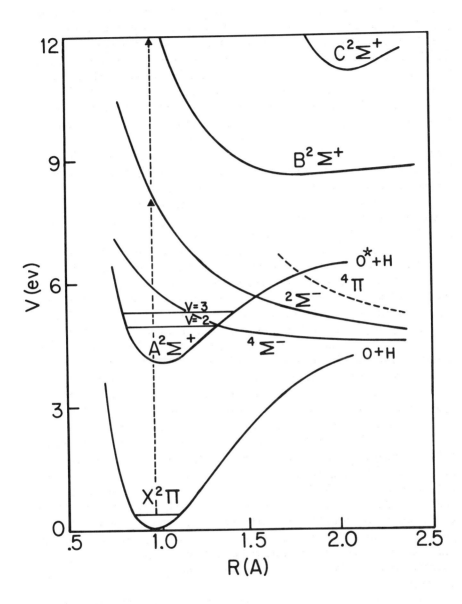

6. Potential energy curves for OH as a function of interatomic distance between O and H.

state O and H are given now, in increasing order of energy, as $^4\Sigma^-$, $^2\Sigma^-$ and $^4\Pi$. That the $^2\Sigma^-$ state is not the lowest, as suggested by Solomon (1968), makes the rate of pre-association extremely slow. First, kinetic temperatures much greater than $1000^\circ K$ would be needed to carry the interacting O and H atoms close to the crossing point of the $^2\Sigma^-$ curve with the bound A $^2\Sigma^+$ curve. The switching to the A state, followed by ultraviolet fluorescence, forms $X^2\Pi$ OH. Second, reaching the crossing of the $^4\Sigma^-$ and A $^2\Sigma^+$ curves still requires about $1000^\circ K$, temperatures for which there is no evidence. Moreover, the pre-association rate constant is probably at most $10^{-20} cm^3/sec$ if the pre-association (the inverse process) lifetime is about 10^{-6} secs, as estimated by Gaydon and Kopp (1971), because of the very weak interaction between $^2\Sigma^+$ and $^4\Sigma^-$ states. This rate is comparable to that measured at room temperature in lower A states. These rates are exceedingly slow and totally inadequate for maser pumping, as seen in Figure 3, even taking the rate as large as $10^{-17} cm^3/sec$, comparable to the maximum rate possible with atomic recombination via grains.

Figure 7 describes the infrared pumping of the water vapor maser. Collisions bring the rotational levels into thermal equilibrium up to about J=5 if the density is approximately $10^8 cm^{-3}$. The kinetic temperature is likely to be about $200^\circ K$, so that about one H_2O molecule per cc resides in either of the maser states, 6_{16} and 5_{23}, while the total H_2O abundance is about 10^{-4} or 10^{-5} times that of hydrogen. Because of the increasing Einstein A-coefficients and decreasing resonance radiation trapping as one considers higher rotational levels, the effective rotational temperature drops from the kinetic temperature to about $70^\circ K$ for the states just below the maser levels. It is to be noted that the 6_{16} level empties into only one lower level besides the maser 5_{23} level. The observed maser brightness temperatures require large microwave optical depths. Because the maser transition has a weak line-strength, the optical depths for the various far infrared transitions involving these levels must be very large. The resonance radiation trapping between the 6_{16}- and the 5_{05}-levels appears to make the lifetime of 6_{16} (upper maser level) comparable to that of 5_{23} (lower maser level) (Litvak 1969b). Collisions alone may not be adequate to pump these maser via collisional excitation of rotation or vibration, because collisions across the transition (always to be included) have a strong thermalizing effect. Infrared pumping via the vibration-rotation levels indicated by asterisks leads to rotational excitation which must de-excite to the 6_{16}-level, making population inversion possible. Opposing pump rates that fill the 5_{23}-level can be shown to be weaker.

Figure 8 explains the infrared pumping of ortho-formaldehyde that was mentioned earlier (Litvak 1970a). The total absorption line-strengths shown in the Figure are the same for the two doublet states, 1_{11} and 1_{10}. However, the absorption from the lower (1_{11}) state that excites the 0_{00}^* level does not

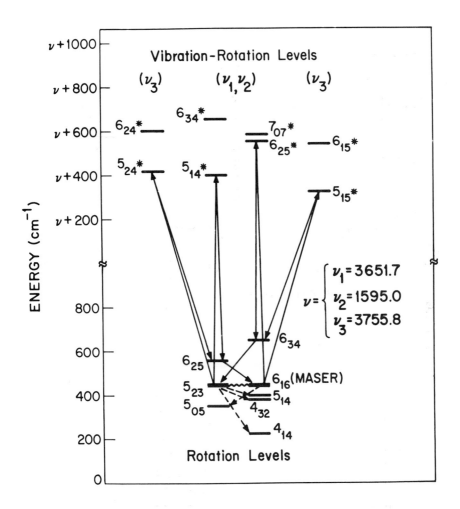

INFRARED PUMPING OF 22.2 GHz H_2O MASER

7. Rotation and vibration-rotation levels involved in infrared pumping of H_2O maser (22.2 GHz). The lower rotational levels (many not shown) are in nearly-thermal equilibrium because of collisions.

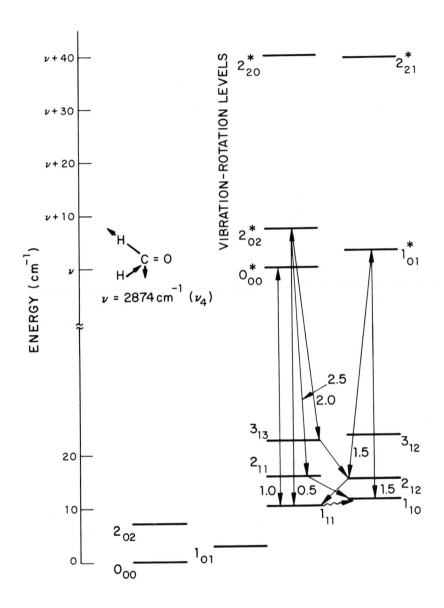

8. Lower rotation and vibration-rotation levels involved in infrared pumping of formaldehyde (ortho) anti-maser at 6-, 2- and 1-cm wavelengths (J = 1, 2 and 3 doublets, respectively).

transfer population out of the 1_{11}-level. The weaker absorption (by a factor of one-third) that transfers population from 1_{11} to 1_{10} (via excited states) cannot compete with the reverse process. Thus, strong anti-inversion is expected. Other vibration-rotation levels that might be excited by infrared pumping (such as the 2_{12} and 2_{11} levels) have a symmetric effect on the anti-maser, and only slightly decrease the efficiency of the pumping. Only the perpendicular infrared bands have a pumping effect. The asymmetric stretching mode shown here and two bending modes need to be considered. One bending mode (in-the-plane vibration) causes anti-inversion while the other (out-of-the-plane vibration) causes inversion. Collisional excitation of CH_2O-vibrational fluorescence in a shockfront might be expected to favor the modes causing anti-inversion. These hot molecules irradiate the cold molecules left far behind the shockfront. An expansion wave which is expected to follow the collapsing shock slows down the cold molecules to velocities of 1 km/sec or less (Litvak 1969a). It might be possible in some cases that ultraviolet pumping via the well-known perpendicular ultraviolet band of formaldehyde is also occuring, along with analogous ultraviolet pumping of OH anomalous absorption. Only very large optical depths in the infrared or ultraviolet would lead to maser action. Also, as can be seen in Figure 8, the next doublets (corresponding to J = 2 and 3) are also anti-inverted.

Figure 9 summarizes a method for predicting the non-equilibrium populations for the states in various complex molecules being formed in the galactic center and elsewhere. It is likely that steady-state conditions may be assumed, so that the rate of transferring population out of the ith state balances the rate into that state from all others. The population density in the ith state is n_i, the rate (sec^{-1}) of transferring population $i \to j$ is W_{ij}, while that from j to i is W_{ji}. Each de-excitation rate, assuming that E_{ij}, the energy difference, $E_i - E_j$, for the two states is positive, is given by the sum of three terms: 1) the spontaneous-emission rate given by the product of Einstein A-coefficient with the trapping transmission factor $L(\tau_{ji})$, 2) the stimulated-emission rate, given by the product of the usual Planck function for the external radiation having an effective temperature T_R (different radiation temperatures may be used for millimeter and submillimeter radiation), and the one-sided trapping transmission function $\frac{1}{2} L(\tau_{ji})$, if the cloud is illuminated over one face only, and 3) the collisional de-excitation rate. The corresponding excitation rate W_{ji} is the sum of two terms: 1) the absorption rate, equal to the stimulated emission rate, except for the ratio of the g-factors, the degeneracies of the two states, and 2) the collisional excitation rate, equal to the de-excitation rate times the Boltzmann factor for a kinetic temperature T_R, and times the ratio of the g-factors again. The trapping factors are simplest for the case of complete frequency redistribution upon each fluorescence and a two-stream slab geometry. A spherically-symmetric geometry has trapping factors that are nearly as

$$\frac{dn_i}{dt} = -n_i \sum_j{}' W_{ij} + \sum_j{}' n_j W_{ji} \approx 0$$

$$W_{ij} = A_{ij}\mathcal{L}(\tau_{ij}) + \frac{A_{ij}\,\frac{1}{2}\,L(\tau_{ij})}{\exp\,(E_{ij}/kT_R)-1} + w_{ij}$$

spont. emiss. stim. emiss. coll. de-exc.

$$W_{ji} = \frac{A_{ij}\,\frac{1}{2}\,L(\tau_{ij})\;g_i/g_j}{\exp\,(E_{ij}/kT_R)-1} + w_{ij}\,\exp\,(-E_{ij}/kT)\;g_i/g_j$$

absorption   coll. excitation

- resonance - radiation trapping transmission functions:

$$\mathcal{L}(\tau) = \frac{1}{2}\,[L(\tau) + L(\tau^\circ - \tau)]$$

$$L(\tau) = \int_{-\infty}^{\infty} e^{-\nu^2}\exp\,(-\tau e^{-\nu^2})\,d\nu/\sqrt{\pi}$$

(two-stream, complete frequency-redistribution)

- optical depth

$$\tau_{ij} \approx \frac{(2\pi)^2\mu_a{}^2}{3\hbar c}(N\ell\nu/\delta\nu)[S_{ij}\,(n_j/g_j - n_i/g_i)]$$

9. Steady-state rate equations for population and optical depths for multi-temperature non-equilibrium conditions in thick clouds.

convenient. The quantity τ_{ji}, the optical depth (for $j \to i$ at a point in the cloud), is proportional to the partial optical depth, $S_{ij}\,[n_j/g_j - n_i/g_i]$, where S_{ij} is the non-dimensional relative line-strength, such as found in Townes and Schawlow (1955), times the square of the ratio of permanent dipole-moment for the b- or c- components to the a-component, wherever these arise. The total optical depth across the cloud is τ°_{ji}. The square-bracket contains the population difference, usually calculated by assuming that the sum of all of the populations is unity. The total column density of molecules is applied as a factor in the optical depths. The cloud is divided into a convenient number of points at which the populations are simultaneously calculated for the given T_R, T_k, collision rates (for a given density), and total column density of molecules. Since the trapping depends upon the optical depths from the given point in the cloud to either end,

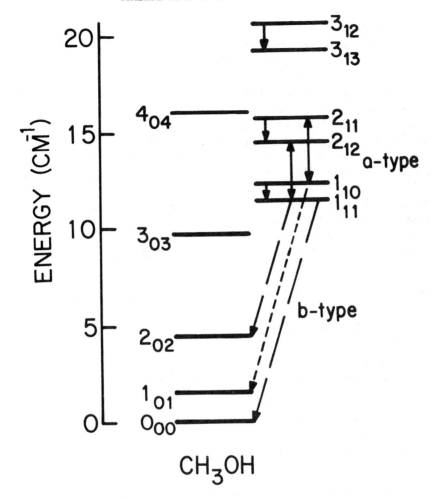

10. Energy level diagram for methanol-A. Emission from the K = 1 doublets (J = 1, 2 and 3) is possibly maser action.

the populations must be iterated upon to find a self-consistent solution of the rate equations. These are not too sensitive to the optical depths and a few computer iterations are adequate. The accuracy of using quasi-local solutions to the integral equations of radiative transfer is probably adequate but not completely explored (Litvak 1969a). As discussed earlier, collisions may obey dipole-type selection rules (case 2) or perhaps no particular selection rules (case 1). For simplicity, case 1 assumes the same *de-excitation* rate W for each transition. Case 2 assumes that each de-excitation rate is W S_{ij}/g_i. Roughly, W $\approx$ 10^{-9}n sec^{-1}, where n equals the hydrogen density. Actually, some

Einstein-coefficient A, usually for the most convenient millimeter-wave transition, is normalized to unity, so that all other rates are given with respect to this one. Usually W/A = 1 corresponds to approximately 10^3 hydrogens per cm^3, a convenient density.

Figure 10 is an energy level diagram for methanol-A. Emission in the three K = 1 -doublets (J = 1, 2, and 3) have probably been observed (Gottlieb 1971). Maser action is probably occurring, at least for the lowest doublet, because emission is seen against a background temperature of over $1000°K$ (Ball, Gottlieb, Lilley and Radford 1970). The b-type transitions to lower levels (0_{00}, 1_{01} and 2_{02}) are probably fairly important. These are absent in formaldehyde.

Figure 11 is a table of partial optical depths calculated for optically-thin methanol-A at T_k = 5, 10 and $20°K$, for a radiation temperature T_R = $2.7°K$ (appropriate for the predominately millimeter wavelengths). This table, for the sake of simplicity, lists in each box the range of W/A for which maser emission can occur (negative values of optical depth) and lists the maximum negative

Methanol-A

Relative Linestrength x Relative Population Inversion (T_R = 2.7°K)

	(no selection rules on collisions)			(dipole selection rules on collisions)		
	5°K	10°K	20°K	5°K	10°K	20°K
$1_{10}{\rightarrow}1_{11}$	W/A* = .01-10. -3(-4)	.001-100. -2(-4)	.001-100. -6(-4)	W/A† .1-100 -3(-4)	.01-100. -6(-4)	.01-1000. -8(-4)
$2_{11}{\rightarrow}2_{12}$	.1-100. -2(-4)	.01-100. -5(-4)	.01-100. -9(-4)	.1-100. -2(-4)	.1-100. -6(-4)	1.-1000. -6(-4)
$3_{12}{\rightarrow}3_{13}$	.1-100. -3(-5)	.01-100. -2(-4)	.01-100. -6(-4)	1.-100. -3(-5)	1.-100. -3(-4)	1.-1000. -5(-4)

*W/A is the ratio of collisional de-excitation rate (the same for all levels) to the Einstein-A coefficient for the $2_{12}{\rightarrow}1_{11}$ transition.

†The A is for the $2_{12}{\rightarrow}1_{11}$ transition. The collisional de-excitation rate is WS_{ij}/g_i for i→j. A = 2.5×10^{-6} sec^{-1}.

11. Partial optical depths (relative linestrength times fractional population inversion) for methanol-A for three kinetic temperatures and a $2.7°K$ radiation temperature.

value of the partial optical depth that is reached. Typically, the partial optical depth is a factor of ten smaller than its maximum value at each end of the range given. The numbers in the parentheses are the exponents of ten. Thus -3 (-4) denotes -3×10^{-4}, for example. We note that if the methanol-A clouds in Sgr A or B2 are about 6 min of arc, say, then the optical depth for the $J = 1$ doublet is considerably larger than for the $J = 3$ doublet, indicating a low kinetic temperature. Thus, a density of 10^3 cm^{-3}, a temperature of 5 - $10°K$ and a column-density of 3×10^{16} methanol per cm^2 is likely. However, the ratios of optical depths have not been definitely established because of the, as yet, unknown beam-filling factors, and future comparison with data will be made. Calculations on somewhat optically-thick methanol-E_1 indicates that maser action is likely to account for the emission observed in Orion A (Barrett,

Formamide and Formic Acid
Relative Linestrength x Relative Population Inversion

$T_R=2.7°K$	$T_K=5°K$	$10°K$	$20°K$	$5°K$	$10°K$	$20°K$
	Case 1. No Selection Rules on Collisions			Case 2. Dipole Selection Rules on Collisions		
$1_{10}\to1_{11}$	W/A†=.01-1.	.01-1.	.01-1.	W/A††= 1.	.1-10.	.1-10.
NH$_2$CHO	-4(-3)	-7(-3)	-8(-3)	-7(-4)	-2(-3)	-2(-3)
CHOOH	-2(-3)	-4(-3)	-4(-3)	-6(-4)	-2(-3)	-2(-3)
$2_{11}\to2_{12}$	.1	.01-1.	.01-1.	——	1.-10.	1.-10.
NH$_2$CHO	-3(-4)	-1(-3)	-2(-3)		-3(-4)	-6(-4)
CHOOH	——	——	-4(-5)*		-3(-4)	-4(-4)
$3_{12}\to3_{13}$	——	——	.1	——	——	10.
NH$_2$CHO			-2(-4)			-4(-5)
CHOOH			——			-7(-5)

† W/A = ratio of collisional de-excitation rate (same for all levels) to Einstein-A coefficient for $2_{11}\to1_{10}$. A=3.2 x 10^{-6} sec^{-1} (NH$_2$CHO) and 5.4 x 10^{-7} sec^{-1} (CHOOH).

†† $W_{ij} = W\, S_{ij}/g_i$ (de-excitation).

* For W/A = 1. only.

Note: Number in parenthesis is exponent of ten.

12. Partial optical depths for formamide and formic acid for the $K = 1$ − doublets ($J = 1, 2$ and 3).

Schwartz and Waters 1971), at conditions of lower temperatures and column densities of methanol than were calculated for thermal equilibrium conditions.

Figure 12 gives a similar table of partial optical depths for optically-thin formamide (NH_2CHO) and formic acid (CHOOH or HOCHO), which are similar in structure but show differences in maser action. While formamide is a likely maser in both $J = 1$ and 2 doublets, formic acid is not a likely maser in the $J = 2$ doublet if collision selection rules are not important. The hydrogen densities are probably close to 10^3 cm^{-3} and the kinetic temperatures are probably less than 20°K. The estimates of the column densities are low but inexact, being somewhere between 10^{12} and 10^{13} NH_2CHO or CHOOH molecules cm^{-2}. This type of calculation for two-temperature non-equilibrium is extremely important for understanding the molecular data.

A variety of maser and optical pumping cases have been discussed, ranging from direct optical pumping by external fluxes to internally-generated and trapped radiation with hyperfine splitting. There is a variety and efficiency obtained with optical pumping not to be found with collisions (which are often dominated by the simple electrostatic-type forces). Thus, different astronomical objects are expected (and observed) to exhibit different maser action for optical pumping and not for collisional pumping. There are known cases (dark dust clouds) for which the kinetic temperatures are too low for collisional excitation to be important, yet anomalies are seen in both formaldehyde and hydroxyl.

REFERENCES

Ball, J. A.; Gottlieb, C. A.; Lilley, A. E. and Radford, H. E. 1970, *Ap. J. (Letters)*, **162**, L203.

Barrett, A. H.; Schwartz, P. R. and Waters, J. W. 1971, *Ap. J. (Letters)*, **168**, L101.

Dean, A. M. and Kistiakowsky, G. B. 1971, *J. Chem. Phys.*, **54**, 1718.

Gottlieb, C. A. 1971, unpublished results.

Gaydon, A. G. and Kopp, I. 1971, *J. Phys. B: Atom. Molec. Phys.*, **4**, 752.

Litvak, M. M. 1968, ed. Y. Terzian, *Interstellar Ionized Hydrogen* (New York: W. A. Benjamin, Inc.) pp. 713-745.

————. 1969a, *Ap. J.*, **156**, 471.

————. 1969b, *Science*, **165**, 855.

————. 1970a, *Ap. J. (Letters)*, **160**, L133.

————. 1970b, I. A. U. Discussion, Brighton, England.

————. 1971a, Scottish Universities' Summer School in Physics, (New York: Academic Press), (in preparation).

————. 1971b, *Ap. J.*, **170**, 71.

Litvak, M. M.; McWhorter, A. L.; Meeks, M. L. and Zeiger, H. J. 1966, *Phys. Rev. Letters*, **17**, 821.

Solomon, P. M. 1968, *Nature*, **217**, 334.

Solomon, P. M. and Thaddeus, P. 1969, I. A. U. Meeting (cf. Litvak 1970a).

Townes, C. H. and Cheung, A. C. 1969, *Ap. J. (Letters)*, **157**, L103.

Townes, C. H. and Schawlow, A. L. 1955, *Microwave Spectroscopy* (New York: McGraw-Hill), pp. 557-612.

Turner, B. E. and Heiles, C. E. 1971, *Ap. J.*, **170**, 453.

Snyder: When H_2O is pumped by excitation of (ν_1, ν_2), for example, what percentage of the input energy ends up contributing to the $6_{16} - 5_{23}$ inversion? Can you estimate the efficiency?

Litvak: I don't have a good number, but perhaps a few per cent might be expected.

Wallerstein: What you refer to as a "proto-star shock" could be the same as a long period variable shock, and hence excite the 1665- and 1667-MHz hydroxyl lines in long period variables.

Litvak: I suggested this several years ago at the New York AAS meeting. The difference of OH (1612 MHz) velocities was attributed to a shock discontinuity.

Woolf: The typical infrared variation of the long period variables is a factor of 2. However, the collisional process is not thermal collisions. It is the acceleration of the envelope that arises from radiation pressure acting on the dust.

Litvak: I would think that collision rates might vary with the star's IR output, but perhaps sluggishly – either because optical depth effects might allow only very small temperature variations or because hydrodynamic effects are slow to propagate across the maser volume to change the density.

Sullivan: In several IR/H_2O/OH sources, the time variations in the infrared radiation (and indeed in many sources there are *no* IR variations) do not agree in character or in phase with the time variations in the OH and H_2O radiation. Thus it appears that if the OH or H_2O pump sources are in the infrared, these observational results are difficult to explain.

Litvak: As far as the data of Bechis and others are concerned, there are many IR stars which show continued correlation of the OH microwave emission with the IR output at several wavelengths.

Mira Variables as Black Clouds

N. Woolf

University of Minnesota
Minneapolis, Minnesota

I. ISHWARAS† – SUPER INTELLIGENT ORGANIZATIONS

Dyson (1960) has suggested that the development of an intelligent organism, organization, community or civilization is limited by its energy consumption. He therefore suggested that if intelligent life develops on planets around solar type stars, it should in its advanced phases be detectable by the middle infrared radiation that comes from its mechanism of thermalizing starlight. Hoyle (1957) in a fiction work has discussed an organization of this type. He suggests that an advanced life form with highly effective communication between the component parts should be considered as one individual. In his example this was a Black Cloud. We shall call such hypothetical individuals Ishwaras.†

Some of the assumptions by Dyson, *e.g.* that unintelligent multiplication of sub-units, and predatory behaviour towards the environment seem unlikely in a developed being. However, it would seem likely that such a being would operate through all the naturally occurring opportunities provided by the environment. Thus we are led to look for a place in which life could have developed, a place in which it has had many billions of years to evolve, a place which has developed an extended cool region utilizing a large fraction of the stellar energy, a place in which powerful modulable radio emissions arise naturally. Such conditions arise in the regions surrounding some Mira variable stars.

II. INFRARED OBSERVATIONS

At the University of Minnesota we have observed a few G dwarf stars in the 3-11 μ spectral band. This is to calibrate our infrared photometry by comparison with the known solar spectrum. It is one of three independent methods of calibrating our photometry, and in all cases we have found the radiation to be compatible with that expected directly from the photosphere. Thus there do not seem to be Dyson type civilizations (Sagan and Walker 1966) around these stars.

However, we have also observed certain other stars that have resulted from the natural development of stars like the sun, and have found all of them to be emitting substantial fractions of their total energy ($>$ 5 per cent) in the 10-20 μ region of the infrared. These stars are the Mira and SRa variable stars (Woolf and Ney 1969, Gehrz and Woolf 1971, Gillett, Merrill and Stein 1971) Five per cent of their energy output is 2×10^{28} watts. This figure is 10^{11} times larger than the rate at which solar energy strikes the earth.

The infrared radiation appears to be from a circumstellar dust shell, typically 10^{14} cm inner radius, and with a range of temperatures through it from several

†I am indebted to Dr. U. Arya for assistance in the selection of this word from Sanskrit. It is the ruler of extended though limited section of the Universe.

hundred °K downwards. Such a dust shell appears to arise naturally as part of the evolutionary expansion of the star. From the rates at which matter flows into and out of this shell it is inferred that the lifetime of this phase of stellar evolution is somewhat less than 10^6 years. Thus the possible energy consumption of an Ishwara is 10^7 times the total solar energy falling on the earth through its entire past history.

III. RADIO EMISSION

A number of these circumstellar regions also emit monochromatic radio emission near 18 cm (OH) and 1.35 cm (H_2O) (Schwartz and Barrett 1970, Wilson and Barrett 1970, Woolf 1972). The discussion by Woolf of the kinematic properties of the line emitting regions show that whereas the 1612 MHz OH emission and probably 1665 and 1667 MHz too, arises from very extended regions, the H_2O emission comes from the inner regions of the dust shell. The extraordinary high brightness temperatures associated with these emissions imply that maser amplification is taking place. Such amplification systems provide natural channels for intelligent modulation at the source. Such modulation can occur via injection of signals, and by modifying the velocity, velocity dispersion and excitation of the gas. Probably some combination of these processes is necessary.

IV. MIRA VARIABLES

Mira variables are a heterogenous group of stars, including objects relatively deficient in heavy elements, as well as solar composition objects. Typically they seem to be a population older than the sun, and with a moderately high velocity dispersion. Examples of such stars are the nearby variables W Hydrae, and R Aquilae. Both of these stars have H_2O emission. Their velocities with respect to the LSR are fairly large, 42 and 32 Km/sec respectively.

It can be inferred from the age of the stars, implied by the velocities, that these stars were originally G or late F dwarfs very similar to the sun. They cannot have been close binary stars, because the evolution of a binary star towards being a Mira star is truncated by mass exchange. Therefore, in the formation of these stars from interstellar clouds, conditions must have been similar to those during the formation of the sun, and a planetary system is likely to have originated around them. This planetary system will have been warmed by this central star for several billion years, just as ours has been, before the star started its evolution to higher luminosities and larger radii.

It would seem that if intelligent life had arisen on one of the planets, it would be advanced enough to be able to move outward during the stars evolution. In the only example of such a system known to us — our own — it seems that life will be adequately developed to cope with the expansion of the sun some billions of years from now. This assumes that these civilizations are not self destructive.

V. PREDICTED COMMUNICATION CHARACTERISTICS

The smallest H_2O emission regions will not have lengths much less than 10^{14} cm, thus signals will travel through in a time of about 1 hour. Modulation at frequencies much higher than 1 cycle per hour would seem to be excluded, largely by communication time across the cloud. Also there will be convective motions of the outer layers of the star that will modify the gas flow in periods of months. This will set a natural lower limit to artificial modulation of about 3 cycles per year. Thus artificial modulation may be expected at frequencies between these two. The modulation could involve change of frequency, change of amplitude, change of polarization or all three. Further, since there may be more than one line simultaneously excited (typically two lines), there may be a recognizable flow of information during a human lifetime.

The distance between Mira stars is quite large, and thus the communications we would be intercepting would be very much monologues. Under such circumstances one might hope that the intelligence considered the communication as also an art form. It would then not use an optimal modulation system whose result would resemble white noise, but would rather use a system that provided considerable redundancy and thus possibility of decoding. A study of communication use of the 1.35 cm line from several stars would be helpful in giving some statistical data on the existence of advanced communicative intelligences that are currently unavailable.

REFERENCES

Dyson, F. J. 1960, *Science,* **131,** 1667.

Gehrz, R. D. and Woolf, N. J. 1971, *Ap. J.,* **165,** 285.

Gillett, F. C.; Merrill, K. M. and Stein, W. A. 1971, *Ap. J.,* **164,** 83.

Hoyle, F. 1957, *The Black Cloud* (New York: Harper and Row).

Sagan, C. and Walker, R. G. 1966, *Ap. J.,* **144,** 1216.

Schwartz, P. R. and Barrett, A. H. 1970, *Kitt Peak Contribution No. 554,* 95.

Wilson, W. J. and Barrett, A. H. 1970, *Kitt Peak Contribution No. 554,* 77.

Woolf, N. J. and Ney, E. P. 1969, *Ap. J. (Letters),* **155,** L181.

Woolf, N. J. 1972, *Proc. Liege Symposium XVII,* in press.

W. J. Wilson: There is no evidence for any polarization pattern in the 1612-MHz OH/IR stars. For example, IRC +10011 has less than 3 per cent polarization at either velocity.

Woolf: Further studies of this polarization would be useful, because these calculations only predict the character of its wavelength dependence and not its result.

Goldreich: For which IR stars has VLB interferometry been done?

Woolf: NML Cygni has been studied in some detail. Observations of other stars are, I am told, being planned.

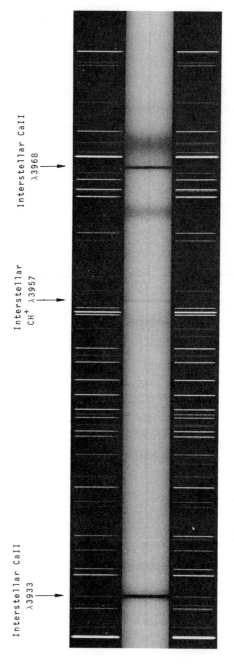

Interstellar CaII
λ3968

Interstellar
CH⁺ λ3957

Interstellar CaII
λ3933

Spectra (center) of the interstellar lines CaII and CH⁺ against the star 55 Cyg. (Courtesy of G. Herbig)

A Proposed New Method For Studying Interstellar Molecules

Michael J. Mumma
Bertram D. Donn

Astrochemistry Branch
Laboratory for Extraterrestrial Physics
NASA/Goddard Space Flight Center
Greenbelt, Maryland

Several mechanisms have been suggested to account for the formation of diatomic molecules in interstellar clouds. They include radiative association (Bates and Spitzer 1951, Bates 1951, Julienne and Krauss 1971), charge exchange and neutral atom exchange (Klemperer 1971), and formation on grain surfaces (Stecher and Williams 1968). As yet no observational data exist which can verify the occurrence of any of these mechanisms; estimates of their relative importance depend on estimates of recombination rate coefficients, atom-molecule exchange rates, and so forth. The purpose of this paper is to suggest a method of experimentally determining which of the proposed mechanisms are in fact occurring in interstellar clouds. The method utilizes the fact that most of the proposed creation mechanisms leave the newly formed molecule in a vibrationally and rotationally excited level of the ground electronic state or of a metastable electronically excited state.

Since the radiative lifetimes for rotation-vibration transitions are very short compared to the collision times in interstellar clouds for all molecules which have a changing dipole moment with v'', the newly formed molecules will cascade to the ground vibrational level before collisional quenching can occur. The cascade results in the characteristic infrared emission spectrum of the molecule. The fundamentals of known interstellar diatomics range from 1285 cm^{-1} for CS to 3735 cm^{-1} for OH (Herzberg 1950). There are several transmission windows in the earth's atmosphere in this wavelength range. If the infrared emission spectrum of a particular interstellar molecule is observed with a large telescope, then (in principle) the spectrum may be unfolded to yield the initial vibrational-rotational population of the state in question. The initial population may then be related to the possible formation processes. Under favorable conditions this could lead to identification of the formation mechanisms.

The ideas presented so far may be illustrated by the case of CO. In Figure 1 we see the potential energy curves for CO. There are two main sets of states, the singlet system and the triplet system, and intersystem combinations usually have very unfavorable matrix elements compared to transitions within each system (Herzberg 1950). Thus, whenever a molecule is created in the singlet system it remains in the singlet system and whenever a molecule is made in the triplet system it remains in the triplet system until it reaches the lowest triplet state $(a^3\pi)$ from which it undergoes an electric-dipole-forbidden transition to the ground singlet state $(X^1\Sigma^+)$. CO may be formed by radiative association of the Bates and Spitzer type through the $(A^1\pi - X^1\Sigma^+)$ transition and several transitions in the triplet system, and by inverse predissociation into either system (e.g. $B^1\Sigma^+ \to X^1\Sigma^+$). Thus we expect that whenever radiative association is important, CO will be formed in both the singlet and triplet systems. The lowest triplet state $(a^3\pi)$ is metastable and the lifetimes against radiating to the ground electronic state range from ~ 3 msec to 450 msec (James 1971). Typical transition probabilities for vibrational cascade for heteronuclear diatomics are ~ 100 sec^{-1} so vibrational cascade within the $a^3\pi$ state may compete favorably with the electronic transition to the ground state $(X^1\Sigma^+)$. The rotational distribution in each vibrational level of the ground state will be a complicated function of the direct transition probability in the singlet system and the cascade transition probability from the triplet system. The calculation of this vibration-rotation population would be difficult but we can say that it will differ significantly from the population which results from resonance fluorescence excitation of $CO(X^1\Sigma^+, J = O, v = O)$.

A second production mechanism is exchange of O with CH$^+$. In this case the CO can only be formed in the triplet system and not in the singlet. The reason is that the total electron spin must be conserved. Since the H$^+$ has no electron spin angular momentum and the spin of the initial system is one, then the CO must have spin of one. The reaction is exothermic by 7.5 eV which energetically can result in $CO(a'\ ^3\Sigma^+, v' \leqslant 3)$ or $CO(a^3\pi, v' \leqslant 6)$. This mechanism would produce electronic cascade in the triplet system resulting in emission of the Asundi Bands (6000-9000 Å) in addition to the infrared cascade in the $a^3\pi$ state.

Formation of CO on grains results in ground state CO if the molecules are desorbed by thermal processes. If the molecule desorbs as a consequence of the formation process it seems likely that it could be vibrationally excited. In either case it probably desorbs from the grain in the singlet state $(X^1\Sigma^+)$.

These three cases are summarized in Table I. It is clear that the absence of either singlet or triplet CO infrared emissions from a cloud where CO is known to exist argues against one or more of the production mechanisms. The list in Figure 2 is not meant to be exhaustive but only to indicate the possibilities which may exist.

We now address ourselves to the question of whether detection of these infrared emissions is feasible. For the purpose of estimating the surface brightness of a cloud we construct a model of a spherical dense cloud which is subject to photodissociation by the interstellar radiation field in the outer shell to a depth of ~ 2 magnitudes extinction. We use Stief's estimate (Stief *et al.* 1971) for the mean lifetimes against photodissociation, τ, under these conditions and we assume that a steady state exists, *i.e.*, destruction and formation of CO are in equilibrium. We further use Solomon's results (Solomon 1971) for the central column density in Orion ($\sim 6 \times 10^{19}$ cm^{-2}) attenuated by a factor (~ 60) to account for decreased column density at ~ 2 magnitudes extinction. Then the equilibrium condition is

$$\frac{dn_{CO}}{dt} = 0 = P - L \tag{1}$$

TABLE I

FORMATION MECHANISMS FOR TRIPLET vs. SINGLET CO

Radiative Association: Singlets and Triplets

$$C(^3P) + O(^3P) \rightarrow \begin{array}{l} CO(a^3\pi) + h\nu \\ CO(X^1\Sigma^+) + h\nu \end{array}$$

Exchange Interaction: Triplets

$$CH^+ \,(^1\Sigma^+) + O(^3P) \rightarrow CO(a^3\pi) + H^+ (singlet)$$

$$\Delta E = D_o^o(C-O) - D_o^o(CH^+) = 7.5 \text{ eV}$$

Formation on Grains: Probably Singlet

$$C + O + Grain \rightarrow CO + Grain$$

where P is the production rate per cm^3 and $L = \dfrac{n_{CO}}{\tau}$. The column brightness, B (photons/cm^2/sec/4π steradian), is given by

$$B = \frac{N_{CO}}{\tau} \sim 10^{18}/3 \times 10^{12} \sim 3 \times 10^5 \qquad (2)$$

where N_{CO} is the column density in the shell. We separately estimate the production rate from radiative recombination and neglect the other formation mechanisms

$$P = \Gamma n_o n_c \sim 10^{-16} \,(2 \times 10^{-7}\, n_H{}^2), \qquad (3)$$
$$\sim 2 \times 10^{-15} \text{ cm}^{-3} \text{ sec,}$$

where we have used the maximum possible value for the radiative recombination rate, Γ, (Julienne 1971) and we adopt $n_H = 10^4$ in the shell. If the outer shell corresponds to a depth of a few parsecs, then the column brightness due to radiative association is

$$B' \sim 2 \times 10^{-15} \,(10^{19}) \sim 2 \times 10^4 \qquad (4)$$
$$ph/cm^2/sec/4\pi,$$

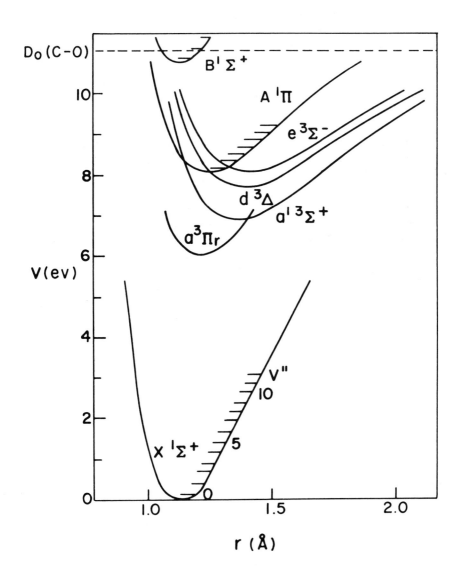

1. The potential energy diagram for CO, showing the singlet states and some of the triplet states relevant to the production mechanisms discussed in the text.

which is in reasonable agreement with the loss rate considering that we have neglected other formation mechanisms.

This represents a radiated power of 1×10^{-19} watts/cm^2 at the surface of an earth-based telescope with a detector acceptance cone of 10 arc minutes, *i.e.*,

$$\text{Power} \sim 3 \times 10^5 (10^{-5} \text{ steradians}) (4 \times 10^{-20} \text{ joules/photon})$$
$$\sim 1.2 \times 10^{-19} \text{ watts/cm}^2$$

This power level is about two-to-three orders of magnitude below the noise level of a typical state-of-the art detector with this large an acceptance angle. Based on these calculations the method appears to have a limited chance of success at this time, but the rewards of a successful search would be so great that we feel the effort should be made. Significant improvements in the sensitivity of infrared detectors should be forthcoming in the near future.

The method may be extended to the stabilizing ultraviolet radiation (radiative association and exchange interaction) with a resulting improvement in the detector noise level; however the ultra-violet emissions do not exhibit the "gain effect" of the infrared emissions which is a direct result of the vibrational cascade. The intensities of the ultraviolet features will be diminished from the calculated infrared intensities by the appropriate branching ratio for formation.

REFERENCES

Bates, D. R. and Spitzer, L. 1951, *Ap. J.*, **113**, 441.

Bates, D. R. 1951, *M. N. R. A. S.*, **3**, 303.

Julienne, P. S.; Krauss, M. and Donn, B. D. 1971, *Ap. J.*, **170**, 65.

Klemperer, W. 1971, *Highlights in Astronomy*, ed. De Jager (Dordrecht: Reidel), 443.

Herzberg, G. 1950, *Molecular Spectra and Molecular Structure, I. Spectra of Diatomic Molecules (2nd ed; Princeton: Van Nostrand), Table 39.*

James, T. C. 1971, *J. Chem. Phys.*, **55**, 4118.

Julienne, P. 1971, private communication.

Stecher, T. P. and Williams, D. 1968, *Ap. J.*, **146**, 88.

Stief, L. J.; Donn, B. D.; Glicker, S.; Gentieu, E. P. and Mentall, J. E. 1972, *Ap. J.*, **171**, 21.

Solomon, P. 1971, private communication.

Session 5

Astrochemistry

Chemistry in Interstellar Space

Bertram Donn

Astrochemistry Branch
Laboratory for Extraterrestrial Physics
NASA/Goddard Space Flight Center
Greenbelt, Maryland

It should no longer be necessary to justify citing interstellar chemistry as a full fledged field of research. It is, however, very necessary to delineate the characteristic of the subject and to emphasize the contrast and the relationship between chemistry in interstellar space and in the terrestrial laboratory. That is the purpose of this brief introduction.

Interstellar space is characterized by: extremely low densities and consequent large deviations from thermodynamic equilibrium; a mixture of the elements in the ratios of their cosmic abundances; in many regions a relatively very high ultraviolet flux; and temperatures between about $5°K$ and $200°K$. In this connection it is appropriate to repeat Harold Urey's remark that the apparently unusual conditions under which chemistry occurs in space are really the ordinary conditions for chemical phenomena in nature. It is the general circumstances common to terrestrial chemistry that are rare and exotic; a strong tendency towards thermodynamic equilibrium; stable molecules; temperatures about $300°K$; and pressures near atmospheric. This viewpoint may be emphasized by the fact that the earth is the only place where liquid water or indeed any liquids are known to occur.

One must be careful not to adopt an earth-centered approach to cosmic chemistry. It is equally important not to disregard relevant chemical theory and experiment. What must be done is to devise appropriate procedures for applying experiment and theory to interstellar chemistry.

When considering chemical processes associated with interstellar molecules perhaps the first thing to note is the short lifetime, on an astronomical scale, of molecules in the interstellar radiation field. Polyatomic molecules decompose in less than about one hundred years as Stief's analysis in this session shows and consequently cannot travel from their place of origin unless protected (Stief *et al.*, 1972). This requires that the observed gaseous molecules are formed in the regions where they are found.

The characteristics of those regions which determine the chemistry in the clouds is the next step to examine. In an HI region with $n(H) \approx 10^2$ as Herbig (1968) obtained for the cloud in front of Zeta Ophiuchi, times between collisions of H atoms with each other are 10^8 sec. For hydrogen collisions with other species this must be multiplied by their abundance relative to hydrogen and becomes 10^3 times greater for the total abundant reactive atoms, carbon, nitrogen and oxygen.

The more interesting regions are those dense, heavily obscured areas where the complex array of molecules have been observed. Total densities may exceed $10^6 cm^{-3}$ (Barrett *et al.*, 1971; Solomon *et al.*, 1971). The mean free path for atom-molecule collisions and the time between collisions now become about 10^{10} cm and 10^5 sec, respectively. The latter interval far exceeds radiative lifetimes for electronic and permitted vibrational transitions. Forbidden vibrational transitions, *e.g.* H_2, can reach 10^7 sec (Herzberg 1950). Infrared rotational lifetimes can also become long, *e.g.*, in CN for k = 1, $\tau \sim 10^5$ sec; k = 10, τ = 100 sec; and k = 30, τ = 3 sec (Arpigny 1965). For H_2, j =2→0 transition, τ = 5 x 10^{10} sec (Spitzer 1949, Herzberg 1969). Similar results are expected for other homopolar molecules. Consequently in dense clouds, vibrationally excited homopolar molecules, or rotationally excited molecules in their lowest states are possible. In general, excited molecules with appreciable internal energy are not expected. Excited molecular hydrogen is a particularly interesting possibility as, *e.g.*, Werner and Harwit (1968), have pointed out. A means of excitation of molecules in regions of high density is not readily apparent.

The mechanism and rates of gas phase reactions between simple molecules are becoming well understood. Detailed experimental and theoretical results are available in a number of cases. Although the specific systems themselves tend not to be of astrophysical interest, the types of reactions are very pertinent, particularly with regard to the non-equilibrium character of the interstellar medium. The present status of the subject has been discussed in two reviews by J. C. Polanyi (1971, 1972) where extensive references may also be found and in his paper in this section.

Exchange reactions have been proposed on several occasions (Carroll and Salpeter 1966, Klemperer 1971) as a means of producing diatomic molecules that have a low yield by two body radiative association. Small energy barriers can seriously inhibit reactions if the temperature is sufficiently low. As temperatures in interstellar clouds are low, between about $200°K$ and $5°K$ (Solomon *et al.*, 1971; Barrett *et al.*, 1971; Heiles 1969) careful attention must be given to possible energy barriers.

The inhibiting effect of small activation energies in interstellar clouds is seen in Table 1. For combinations of temperature and activation energy above and to the right of the dividing line, reaction rates are reduced by at least a factor of one hundred.

The Table also shows that at room temperature activation energies of $1°K$ cal/mole or less are difficult to measure. Determinations of activation energies based on reaction rate measurements as a function of temperature cannot generally do better than about ± 0.5 kcal/mole (Trotman-Dickenson and Milne 1967; Kwei *et al.*, 1970). Energy barriers as small as 0.2 kcal/mole have sometimes been determined in molecular beam experiments (Table 2). A summary of measured activation energies or barrier heights for exothermic exchange reactions appears in Table 2. It is to be noted that even strongly exothermic reactions may have appreciable energy barriers to reaction.

None of the reactions studied have direct astrophysical interest. Reactions between radicals which are of importance for interstellar chemistry are considerably more difficult to study experimentally. The best prospects for

TABLE 1

Inhibiting Factor of Activitation Energy on Rate Constant ($\log_{10}$)

T(°K)	Activation	Energy		
	0.05	0.1	0.5	1.0 kcal/mole
	0.002	0.004	0.02	0.04 eV
5	$\bar{2}$	$\bar{5}.6$	$\overline{22}$	$\overline{44}$
10	$\bar{2}.9$	$\bar{3}.2$	$\overline{11}.5$	$\overline{22}$
20	$\bar{1}.5$	$\bar{2}.9$	$\bar{6}.5$	$\overline{11}.5$
50		$\bar{1}.5$	$\bar{2}$	$\bar{5}.6$
100			$\bar{2}.9$	$\bar{3}.2$
300				$\bar{1}.3$

obtaining data for such species may be theoretical calculations of potential energy surfaces for reactions of interstellar interest.

The direct application of such data to interstellar chemistry is hindered by the non-equilibrium conditions of interstellar molecules. The translational energy will have a near Maxwellian distribution (Spitzer 1968). However, as shown above, vibrational and rotational energy distributions will generally deviate considerably from a Boltzmann distribution because the radiative transition probability is much shorter than collision frequencies. Formaldehyde is an interesting example in which the excitation temperature is reduced below both radiative black body temperatures or collisional kinetic temperatures (Zuckerman *et al.*, 1970). In the case of water (Sullivan 1971) and the hydroxyl radical (Robinson and McGee 1967) high excitation temperatures and inversion of energy states occur, leading to maser action.

The usual treatment of reaction rates assumes thermodynamic equilibrium and a Boltzmann energy distribution for each degree of freedom associated with the molecule. Rate constants in the literature are weighted averages over all energy states. A number of recent experimental investigations have delineated the dependence of the reaction probability upon particular energy modes of reactant molecules; translation, vibration or rotation. Shock tube experiments are described by Bauer and his associates (Lewis and Bauer 1968). Spectroscopic

TABLE 2

Activation Energies of Exothermic Reaction

Reaction	Heat of Reaction k cal/mole	Activation Energy k cal/mole	Method
$H + Cl_2 \rightarrow Cl + Cl$	45	2.5	Temp. Dependence[a]
$F + H_2 \rightarrow HF + H$	31.3	1.7	Temp. Dependence[a]
$H + Br_2 \rightarrow HB_r + B_r$	41.1	0.9	Temp. Dependence[a]
$H + HBr \rightarrow H_2 + B_r$	16.5	0.9	Temp. Dependence[a]
$O + OH \rightarrow O_2 + H$	16.5	0 ± 0.5	Temp. Dependence[b]
$K + HI \sim KI + H$	6.1	0.2 ± 1	Mol. Beam[c]
$K + HBr \rightarrow KBr + H$	4.5	0.15	Mol. Beam[d]
$K + RI \rightarrow KI + R$	~25	~0.5	Mol. Beam[e]

$$R + CH_3, \ C_2H_5, \ C_3H_7, \ C_4H_9$$

a. Thrush, B. A. (1965), Prog. Reaction Kinetics, v. 3

b. Baulch, D. L., Drysdale, D. D. and Loyd (1968) High Temperature Reaction Rate Date No. 2, Dept. of Physical Chemistry, Leeds University, England.

c. Ackerman, M. A., E. F. Green, A. C. Moursound, and J. Ross (1964), J. Chem. Phys. 41, 1183.

d. Beck, D., E. F. Green, and J. Ross (1962), J. Chem. Phys. 37,2895.

e. Kwei, G. H., J. A. Norris, D. R. Herschbach, J. Chem. Phys. 52, 1317.

studies of infrared chemiluminescence in Polanyi's laboratory have given detailed information (Polanyi and Tardy 1969; Anlauf et al., 1969). A molecular beam investigation of the problem has been reported by Jaffe and Anderson (1969). Quite generally, vibrational excitation is required for an appreciable reaction rate. The review by Polanyi (1971a) discusses several aspects of non-equilibrium chemical processes.

Although no experimental results are available for astrophysically significant reactions, the specific dependence of the rate on energy mode must be known before interstellar rates can be calculated. An example is the reaction

$$O + H_2 \rightarrow OH + H$$

which has been proposed on several occasions (Carroll and Salpeter 1966; Heiles 1968; Field *et al.*, 1968) as a source of interstellar OH. It is nearly thermally neutral, $\Delta H = 2$ kcal/mole $= 0.09$ eV, and has an activation energy of 9.45 kcal/mole (Baulch *et al.*, 1968). A detailed study of the reaction dynamics of this reaction is required before a reliable interstellar rate constant can be calculated. The actual value may be significantly smaller than the result calculated from laboratory rate parameters.

There are many problems with molecule formation by gas phase reaction which has led to frequent proposals for surface catalyzed processes. Because of the high obscuration, and resultant large grain density, surface catalyzed reactions have considerable appeal as a mechanism of molecule formation. The gas kinetics of gas-surface interactions are readily calculated for particular grain models. Determination of reaction rates are another matter entirely. In the low density clouds, for a given hydrogen atom, the time interval between atom-grain collision for a given molecule is 10^{15} sec, whereas successive impacts on a grain occur at intervals of 10^5 sec. In the dense clouds we will assume that both the gas to solid ratio and the grain characteristics remain unchanged, although the validity of the assumption is not obvious when the density becomes several orders of magnitude higher. For total gas densities of 10^6 cm^{-3} the hydrogen molecule-grain collision interval drops to $\sim 10^{12}$ sec and the collision interval on grains becomes 10^{-2} sec. The grain surface ($a = 3 \times 10^{-5}$ cm) contains 10^7 sites and for a sticking probability of unity, would become completely covered with hydrogen in $\sim 10^5$ sec. Except for carbon monoxide, time intervals for the polyatomic molecules would be about 10^6 times longer. The rate of coverage depends only on gas density and physico-chemical properties of the surface, primarily sticking coefficients and not upon grain size. However, depending upon the mechanism of surface reaction, reaction rates may be size dependent. The sticking probability and surface reaction rates will depend upon surface composition, structure and degree of contamination. Both experimentally and theoretically, the study of surface reactions is less advanced than the present status of gas phase reactions. I am not saying that surface reactions may not be important for interstellar chemistry but rather they present more difficult problems to analyze. The results are consequently even more uncertain. Future progress depends upon making attempts to relate terrestrial theory and experiment to interstellar chemistry.

Recent developments in techniques for studying surface phenomena open new approaches to clarifying this phase of the problem. Present ultrahigh vacuum systems yield operating pressures below 10^{-10} Torr (Ehrlich 1963) which corresponds to less than 10^7 mole/cm^3. Experiments on surface phenomena are regularly done at 10^{-9} -10^{-10} Torr and are described in recent issues of the *Journal of Vacuum Science and Technology*. These experiments are within a factor of ten in density of processes occurring in dense clouds. Among other methods for studying surface phenomena at low pressures and temperatures are: (1) flash desorption (Ehrlich 1963); (2) field emission microscopy (Ehrlich 1963); (3) molecular beams; (4) low energy electron diffraction, LEED: and (5) Auger Spectroscopy. Descriptions and applications of these methods as well as other techniques for studying surface phenomena may be found in the proceedings of the National Symposium on Vacuum Science and Technology as published in the *Journal of Vacuum Science and Technology* for 1969, 1970 and 1971. Unfortunately, for the interstellar chemist, because these methods work best on conductors and conductors are important catalyst surfaces, studies have concentrated on metals or semi-conductors. It becomes the responsibility of the student of interstellar chemistry to interest experimenters in astrophysically significant surfaces and reactants. This applies to gas phase reactions as well as surface reactions.

REFERENCES

Aulauf, K. G.; Maylotte, D. H.; Polanyi, J. C. and Bernstein, R. B. 1969, *J. Chem. Phys.*, **51**, 5716.

Arpigny, C. 1965, *Mem. Acad. Roy. Belgique, Classe. Sci.*, **35**, 1.

Barrett, A. H.; Schwartz, P. R. and Waters, J. W. 1971, *Ap. J.*, **168**, L101.

Baulch, D. L.; Drysdale, D. D. and Lloyd, A. C. 1968, *High Temp. Reaction Rate Data, No. 2* (Leeds: Dept. of Phys. Chem., Leeds University).

Carroll, T. O. and Salpeter, E. E. 1966, *Ap. J.*, **143**, 609.

Ehrlich, G. (1963) in *Advances in Catalysis*, **14**, ed. by D. D. Eley, H. Pines and P. B. Wwisz (New York: Academic Press), p. 255.

Field, G. B.; Aannestad, P. A. and Solomon, P. 1968, *Nature*, **217**, 435.

Heiles, C. 1968, *Ap. J.*, **151**, 919.

————. 1969, *Ap. J.*, **157**, 123.

Herbig, G. H. 1968, *Zt. Astrophys.*, **68**, 243.

Herzberg, G. 1950, *Molecular Spectra and Molecular Structure* (New York: Van Nostrand), p. 279.

————. 1969, *Mem. Soc. Roy. Sci. Liege, Ser. V*, **17**, 121.

Jaffe, S. B. and Anderson, J. B., 1969, *J. Chem. Phys.*, **51**, 1057.

Klemperer, W. 1971, *Highlights in Astronomy*, ed. J. De Jager (Dordrecht: Holland), p. 443.

Lewis, D. and Bauer, S. H. 1968, *J. Am. Chem. Soc.*, **90**, 5390.

Polanyi, J. C. 1971, *J. App. Opt.*, **10**, 1717.

————. 1972, *Acc. Chem. Res.*, April.

Polanyi, J. C. and Tardy, D. C. 1969, *J. Chem. Phys.*, **51**, 5717.

Proceedings of National Symposium on Vacuum Science and Technology 1969, 1970, 1971, *J. Vac. Sci. Tech.* **6**, *No. 1*, **7**, No. 1; **8**, No. 1.

Robinson, B. T. and McGee, F. F. 1967, *An. Rev. Astron. Astrophys.*, **5**, 183.

Solomon, P. M.; Jefferts, K. B.; Penzias, A. A. and Wilson, R. W. 1971, *Ap. J.*, **168**, L107.

Spitzer, L. 1949, *Ap. J.*, **138**, 337.

————. 1968, *Diffuse Matter in Space* (New York: Interscience), p. 103.

Stief, L.; Donn, B.; Glicker, S.; Gentieu, E. P. and Mentall, J. 1972, *Ap. J.*, **171**, 21.

Sullivan, W. T., III. 1971, *Ap. J.*, **166**, 321.

Trotman-Dickenson, A. F. and Milne, G. S. 1967, *Tables of Bimolecular Gas Reactions, National Bureau of Standards, National Standard Reference Data Series No. 9* (Washington: U. S. Govt. Printing Office).

Werner, M. W. and Harwit, M. 1968, *Ap. J.*, **154**, 881.

Zuckerman, B.; Buhl, D.; Palmer, P. and Snyder, L. E. 1970, *Ap. J.*, **160**, 485.

Photochemistry of Interstellar Molecules

L. J. Stief

Astrochemistry Branch
Laboratory for Extraterrestrial Physics
NASA/Goddard Space Flight Center
Greenbelt, Maryland

I. INTRODUCTION

Within the last three years five diatomic and some fourteen polyatomic molecules have been discovered in interstellar space, mainly by radio astronomy. Prior to this, only a few diatomic species were observed, mainly by optical astronomy in the visible region. This unexpected increase in both the number and complexity of interstellar molecules is evidence of significant chemical phenomena associated with the interstellar medium. While the interaction of many scientific disciplines is required, *e.g.* astronomy, astrophysics, spectroscopy, photochemistry, chemical kinetics, surface chemistry, *etc.,* I wish to discuss here the role of photochemistry in determining the lifetime of interstellar molecules.

The photochemistry and lifetimes have previously been discussed (Stief, Donn, Glicker, Gentieu, and Mentall 1972) for the molecules H_2CO, NH_3, H_2O, CH_4 and CO. This has been extended to include new molecules detected since the first calculations were made (OCS, $CH_3C \equiv CH$), one molecule (NO) which has been the subject of an extensive negative search, and two other molecules (C_2H_2 and C_6H_6) which might be expected to differ from the usual pattern.

For an interstellar molecule the lifetime against photodecomposition depends upon three factors: the absorption cross section, the quantum yield or probability of dissociation following photon absorption, and the interstellar radiation field. The first two factors plus the mode of photodecomposition or primary photochemical process constitute the photochemistry of the molecule and this will be discussed in the next section. The radiation field adopted will be discussed briefly in Section III while Section IV will present the calculations on lifetime against photodecomposition. Finally, some discussion will be given of the implications of these results.

II. PHOTOCHEMISTRY OF INTERSTELLAR MOLECULES

The quantitative spectroscopy and photochemistry of the ten molecules considered here have been investigated in the laboratory to varying degrees. References for the absorption cross sections used in the lifetime calculation are collected in Table I. The photochemistry of the molecules H_2CO, NH_3, H_2O, CH_4 and CO as it applies to the interstellar medium has been discussed previously (Stief *et al.,* 1972) and will only be summarized here. The important primary processes for the first four molecules are formation of atomic hydrogen and formation of molecular hydrogen; the relative importance of these decomposition channels is usually strongly dependent on wavelength. Presently available evidence indicates that the quantum yield of dissociation or decomposition probability is unity for H_2CO, NH_3, H_2O and CH_4.

Carbon monoxide is unique in that no laboratory studies of its photodecomposition have been reported for wavelengths below the decomposition threshold at 1115 Å. The only spin allowed process energetically possible with interstellar photons ($\lambda > 912$ Å) is formation of carbon and oxygen atoms in their ^3P ground states. If long-lived excited states are involved below 1115 Å as have been shown to be at longer wavelengths, the dissociation probability could be significantly less than unity.

The photochemistry of the remaining five molecules will now be considered separately in more detail.

A. Carbonyl Sulfide

The photochemical decomposition of OCS may be considered in terms of the primary processes:

$$OCS + h\nu \rightarrow CO + S \qquad (1)$$
$$\rightarrow CS + O \qquad (2)$$

From the studies of Sidhu, Csizmadia, Strausz, and Gunning (1966) at 2537 and 2288 Å, the quantum yield for process (1) is 0.90 with formation of $S(^1D)$ predominant over $S(^3P)$. The preferential formation of $S(^1D)$ would be expected on the basis of spin conservation.

Formation of atomic oxygen via processes (2) becomes energetically possible at 1753 Å for $O(^3P)$ and 1360 Å for $O(^1D)$. Calvert and Pitts (1966) cite qualitative evidence for the occurrence of (2) at very short wavelengths. The process $OCS \rightarrow C + O + S$ is not energetically possible for interstellar photons ($\lambda > 912$ Å). The ionization potential of OCS is 11.2 eV (Matsunaga and Watanabe 1967) which corresponds to an ionization threshold of 1110 Å. The ionization efficiency is about 50 per cent in the region below 1100 Å.

B. Methyl Acetylene

Photolysis in the near ultraviolet continuum (2000 - 1600 Å) occurs via the primary process

$$CH_3C \equiv CH + h\nu \rightarrow CH_2C \equiv CH + H \qquad (3)$$

Ramsay and Thistlethwaite (1966) have observed the propargyl radical ($CH_2C \equiv$

CH) in absorption during flash photolysis of methyl acetylene and the products observed by Galli, Harteck and Reeves (1967) at 2062 Å are consistent with process (3). No quantitative data on the quantum yield of decomposition is available for absorption in the continuum.

In the banded region below 1600 Å, Stief, De Carlo and Payne (1971) found evidence at 1470 Å for processes leading to atomic hydrogen ($\Phi \geqslant 0.40$) and molecular hydrogen ($\Phi = 0.15$). Although several processes probably occurred, it is important to note here that the primary quantum yield of dissociation must be 0.65 or larger. At 1236 Å, Payne and Stief (1971) found that molecular hydrogen formation occurred to the exclusion of atomic hydrogen formation. An important process was found to be

$$CH_3 C \equiv CH + h\nu \rightarrow C_3 + 2H_2 \qquad (4)$$

which occurs via the short-lived $C_3 H_2$ radical.

Interstellar photons ($\lambda > 912$ Å) are capable of ionizing methyl acetylene since the ionization potential of 10.36 eV (Nakayana and Watanabe 1964) corresponds to $\lambda = 1196$ Å. The ionization efficiency may be estimated as 50 - 75 per cent below 1175 Å.

C. Nitric Oxide

The photodissociation of nitric oxide has been discussed (Calvert and Pitts 1966) in terms of the following primary processes:

$$NO + h\nu \rightarrow N + O \qquad (5)$$

$$\rightarrow NO^+ + e \qquad (6)$$

McNesby and Okabe (1964) discuss the evidence that predissociation begins above $v' = 7$ of the $B^2 \Pi$ state, corresponding to 6.5 eV or 1910 Å. Hence, only interstellar photons below 1910 Å are capable of dissociating NO. Ionization occurs below 1338 Å based on Dressler and Miescher's (1965) value for the ionization potential. The data of Watanabe, Matsunaga and Sakai (1967) indicate that the ionization efficiencies are mostly in the range 40 - 80 per cent for $\lambda = 912$ to 1300 Å.

There is little direct information available on the primary quantum yield. From the low pressure data of Leiga and Taylor (1965), the product quantum yields for photolysis of NO at 1236 Å, 1470 Å and the region 1550 - 1650 Å are at least not inconsistent with primary quantum yields in the range 0.7 to 0.9.

D. Acetylene

The photochemistry of acetylene may be considered in terms of the following primary processes:

$$C_2H_2 + h\nu \rightarrow C_2H + H \qquad (7)$$

$$\rightarrow C_2 + H_2 \qquad (8)$$

$$\rightarrow C_2H_2^* \qquad (9)$$

The data of Zelikoff and Aschenbrand (1956) at 1849 Å lead to a value of approximately 0.2 for the quantum yield of process 7. Stief, De Carlo, and Mataloni (1965) suggest that the quantum yield of process 8 is of the order of 0.1 at both 1470 and 1236 Å. In the laboratory, the long-lived excited acetylene molecule formed in process 9 reacts with excess acetylene to form a variety of products (Takita, Mori, and Tanaka 1969); this has little application to the photodecomposition of interstellar molecules. The observation at low resolution of a broad, structureless emission in the region 4000 to 6000 Å during photolysis of C_2H_2 at 1236 Å was tentatively identified by Stief, De Carlo, and Mataloni (1965) with the Swann bands of C_2. The recent observation by Becker, Haaks, and Schurgers (1971) of this same broad emission feature under high resolution in both the 1236 Å photolysis of C_2H_2 and in the acetylene-oxygen atomic flame rules out a diatomic molecule as the emitter. Arguments are presented that C_2H_2 may be the emitting species. If this is correct, then at low pressures process (9) followed by emission from some low-lying excited state does not lead to dissociation of C_2H_2 and the total primary quantum yield may not exceed 0.2 at all wavelengths studied.

The ionization potential of C_2H_2 is 11.4 eV (Nakayama and Watanabe 1964, Diebler and Reese 1964) and thus photons of $\lambda < 1087$ Å are capable of ionizing acetylene. Below 1000 Å, the ionization efficiency is approximately 75 per cent (Metzger and Cook 1964).

E. Benzene

Benzene does not undergo any significant photochemical reactions in the first absorption band (2800 to 2100 Å), although important photophysical processes do occur. We therefore restrict our attention to wavelengths below 2100 Å. Although several recent photochemical studies have been performed in this spectral region, the complexity of the system is such that detailed knowledge of primary processes is still lacking. Shindo and Lipsky (1966) and Foote, Mallon, and Pitts (1966) have shown that the primary quantum yield of dissociation at 1849 Å extrapolates at low pressure (0.1 Torr) to approximately unity. The quantitative experiments of Hentz and Rzad (1967) at 1 Torr pressure show that a minimum value for Φ is 0.30 and 0.15 at 1470 and 1236 Å respectively. Photoionization occurred at the latter wavelength since the ionization potential of benzene is 9.24 eV (Watanabe, Nakayama, and Mottl 1962) corresponding to an ionization threshold of 1341 Å. Below about 1250 Å, the average ionization efficiency is 50 per cent (Person 1965).

III. THE INTERSTELLAR RADIATION FIELD

The component of the interstellar radiation field we will be concerned with is the total dilute starlight coming from all stars. Short wavelength radiation from hot stars can ionize the atomic hydrogen in their immediate vicinity. We are not concerned here with these relatively small regions dominated by ionized H but with the larger remaining HI regions where hydrogen is neutral. A more thorough discussion of the interstellar radiation field adopted for these calculations has been given previously (Stief *et al.*, 1972) and will only be summarized here. For unobscured regions we adopt the results of Habing (1968) who obtains an ultraviolet energy density U between 30 and 50 x 10^{-18} ergs cm^{-3} $Å^{-1}$ for wavelengths between 1000 and 2200 Å. Lambrecht and Zimmermann's (1955) results for wavelength to 3646 Å are sufficiently consistent with this value that we have adopted a constant U = 40 x 10^{-18} ergs cm^{-3} . $Å^{-1}$ for $\lambda > 912$ Å. The uncertainty of the calculations and the small effect on the lifetimes do not warrant the allowing for any wavelength dependence.

To determine the radiation field in obscuring clouds, we combine the adopted constant flux with an interstellar extinction curve covering the visible and ultraviolet regions. For the visible we used Johnson's (1965) curve for the Perseus region and for the ultraviolet, Stecher's (1969) curve determined from rocket observations of Persei. Figure 1 is a plot of transmissivity as a function of wavelength for various optical depths in a cloud. These transmissivities may be multiplied by the adopted uniform interstellar radiation field to obtain the radiation field as a function of cloud depth.

IV. LIFETIME OF INTERSTELLAR MOLECULES

The probability P that a molecule is decomposed by light in interstellar space free of obscuring clouds is

$$P = \frac{U\Phi}{h} \quad \sigma_\lambda \, \lambda \, d\lambda = \frac{U\Phi}{h} \langle \sigma\lambda \rangle \qquad (10)$$

where h = Planck's constant, U is the energy density, Φ is the primary quantum yield of dissociation, σ_λ is the absorption cross section, and the limits of integration are from 912 Å to the photodissociation threshold. From the list of interstellar molecules observed to date, we have chosen six for which sufficient data is available: NH_3, H_2O, H_2CO, CO, COS and $CH_3C \equiv CH$. Calculations have also been made for four additional molecules which are of interest for a variety of reasons: CH_4, NO, C_2H_2 and C_6H_6. Methane is of interest since there are indications it may have been detected by absorption in the infrared (Herzberg 1968). Nitric oxide has been extensively searched for in some 67

Table I. References for absorption cross sections

Molecule	Reference
OCS	Cook and Ogawa (1969); Matsunaga and Watanabe (1967); Sidhu et al (1966).
H_2CO	Gentieu and Mentall (1970); Calvert and Pitts (1966).
$CH_3C \equiv CH$	Nakayama and Watanabe (1964)
NH_3	Watanabe and Sood (1965); Samson and Myer (1969).
CH_4	Metzger and Cook (1964); Samson and Myer(1969).
H_2O	Metzger and Cook (1964); Samson and Myer (1969).
NO	Watanabe, Matsunaga and Sakai (1967); Samson and Myer (1969).
C_6H_6	Bunch, Cook, Ogawa and Ehler (1958); Marmo (1968).
C_2H_2	Metzger and Cook (1964); Nakayama and Watanabe (1964).
CO	Cook, Metzger and Ogawa (1965); Myer and Samson (1970).

sources (Turner, Heiles, and Scharlemann 1970) with only negative results to date. It has been suggested that this may be due to an unusually short photodissociation lifetime for NO. Acetylene and benzene may have long lifetimes if processes other than bond rupture occur (*e.g.*, internal conversion, fluorescence, phospheresence, *etc.*). Benzene is also of interest as the simplest aromatic molecule since all the organic molecules observed so far have been aliphatic.

The references for the absorption cross sections for the ten molecules under consideration are summarized in Table I. For all the molecules except NH_3, NO and C_6H_6 it was necessary to interpolate the data over relatively small wavelength intervals not covered by the data available; frequently this was in the region 1000 to 1050 Å. For most of the molecules considered, we have taken the quantum yield of dissociation to be unity. This is to be expected for those molecules with essentially continuous absorption and is consistent with available laboratory data. The exceptions are CO, C_2H_2 and C_6H_6. At this stage we will carry Φ as an unevaluated factor for these three molecules.

From equation 10 and using the laboratory data for σ, the procedure was to plot $\sigma\lambda$ versus λ and obtain $\langle\sigma\lambda\rangle$ from the area under the curve. Both of these were done numerically on a computer with areas determined by the spline interpolation method (Scudder 1971, Thompson 1970). Table II gives the lifetimes in unobscured regions for the molecules in question.

It is obvious that lifetimes will be orders of magnitude longer in clouds where there is high obscuration of interstellar radiation. In order to examine the effect of increasing depth of the cloud, we have calculated the lifetimes of the ten molecules discussed above as a function of increasing optical thickness. The

Table II. Lifetimes of Interstellar Molecules in Unobscured Regions.

Molecule	τ (years)	Molecule	τ (years)
COS	10	H_2O	65
H_2CO	30	NO	100
$CH_3C = CH$	30	C_6H_6	5/Φ
NH_3	40	C_2H_2	20/Φ
CH_4	40	CO	100/Φ

Φ = primary quantum yield of dissociation.

latter is measured in terms of the extinction in the visual, A_v, as shown in Figure 1 as a plot of transmissivity T_λ versus λ for steps of 1.0 in A_v. The dissociation probability for an obscured region is given by

$$P = \frac{U\Phi}{h} \quad T_\lambda \, \sigma_\lambda \, \lambda \, d\lambda = \frac{U\Phi}{h} \langle T\sigma\lambda \rangle \qquad (11)$$

Similar to the previous calculation, the value of $\langle T\sigma\lambda \rangle$ was obtained via plotting and numerical integration using a computer. Figures 2 and 3 show lifetimes against photodecomposition with increasing obscuration, the latter measured in terms of extinction in the visual (A_v).

V. IMPLICATIONS OF THE LIFETIME CALCULATIONS

The results in Table II suggest that in clear interstellar regions, most of the molecules under consideration have lifetimes against photodissociation of the order of 100 years or less. That these lifetimes are extremely short on the galactic time scale is evident from the following. If one assigns to the molecules velocities comparable to those of the clouds as a whole, that is velocities of the order of 10 to 100 km/sec, the total distance traveled by a molecule in its lifetime of 100 years $(3 \times 10^9$ sec) is only 3×10^{15} to 3×10^{16} cm or 0.001 to 0.01 pc. Since cloud diameters are typically of the order 0.1 to 10 pc and intercloud distances even larger, it is evident that the average polyatomic molecule can not travel a significant distance in unobscured regions without being subject to destruction by interstellar photons. Thus polyatomic molecules can exist only in dense clouds which protect them from the full interstellar radiation field. This is consistent with observations. Further, these molecules can never have been exposed to the unobscured radiation from the time of formation until protected in clouds. This requirement imposes a severe restriction on possible mechanisms of formation. It implies that polyatomic molecules were formed or released in the gas phase in the cloud where they now occur.

The carbon monoxide lifetime will be 10^2 to 10^3 years if the quantum yield Φ for decomposition is 1 to 0.1. An alternate estimate for CO may be made by considering the dissociation continua only, for which $\Phi = 1$, and ignoring the possible contribution from predissociation in the banded region. From the continua estimated by Cook, Metzger, and Ogawa (1965) from their own data, we calculate $\tau = 10^3$ years. This may be longer than the true value if predissociation contributes significantly or shorter if the apparent dissociation continua has been overestimated due to insufficient resolution of the closely spaced bands in the spectrum. Nevertheless, it is evident that carbon monoxide

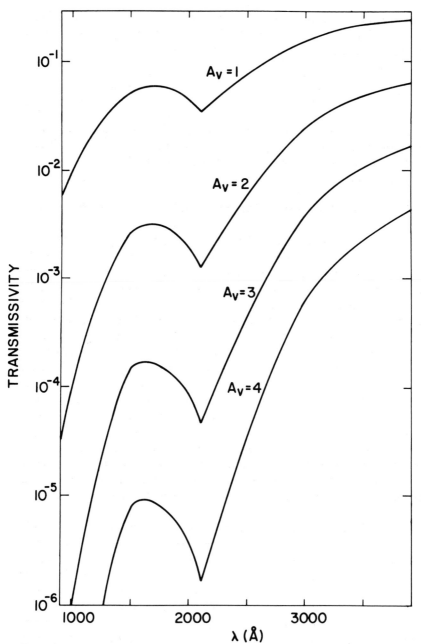

1. Transmissivity in cloud as a function of wavelength. A_v is the cloud extinction in magnitudes at 5500 Å

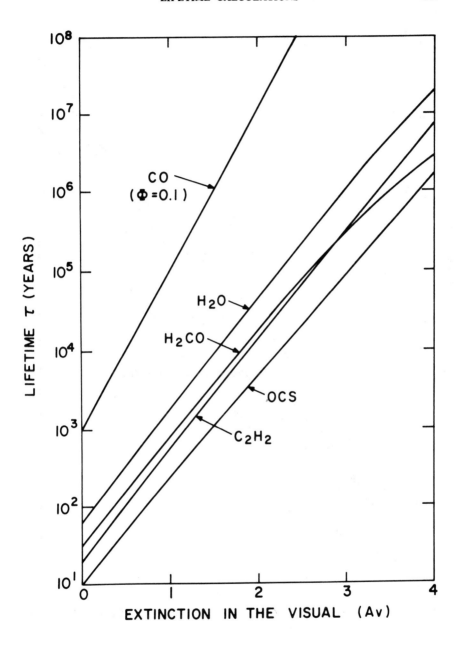

2. Lifetimes of molecules in clouds.

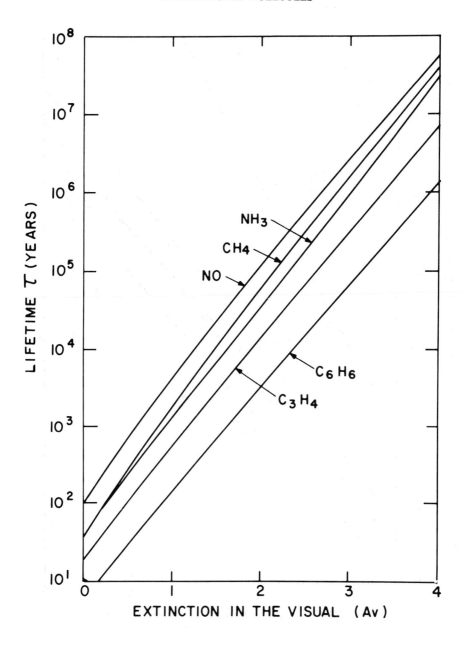

3. Lifetimes of molecules in clouds.

requires considerably less protection than formaldehyde and the other interstellar molecules. This property can readily explain the observation (Wilson, Jefferts, and Penzias 1970) that the size of the carbon monoxide cloud is at least an order of magnitude larger than that of a typical formaldehyde cloud. The smaller quantum yield and the estimate from the dissociation continua both lead to a lifetime comparable to that for other diatomics.

The benzene and acetylene lifetimes are at least 5 and 20 years respectively if the quantum yields are unity. While it is likely that Φ is in fact somewhat less than unity for both these molecules, it is doubtful if Φ will be sufficiently small to increase τ much above the typical value of 50 to 100 years. The lifetime for COS appears to be somewhat shorter than the typical value, although it is not certain if the difference is significant.

Figures 2 and 3 show, as expected, that the lifetimes of molecules in clouds with a few magnitudes extinction become the order of 10^6 years. Contrary to the lifetime of less than one hundred years estimated in the cloud-free regions, lifetimes in clouds of moderate opacity are significant compared to the time scale of the clouds themselves.

In addition to photodissociation, the lifetimes calculated here may include a contribution from photoionization. From the ionization potentials of the ten molecules being considered, it is concluded that interstellar photons ($\lambda > 912$ Å) are capable of producing the parent ion for all molecules except CO. Ionization will in general be less important than bond rupture since it occurs over a narrower wavelength range and since ionization efficiencies are less than unity for polyatomic molecules. Ionization will be even less important in clouds due to attenuation of the short wavelength radiation.

In addition to photodissociation and any photoionization that occurs, molecules can also be destroyed by interaction with high energy radiation (X and γ-rays) and energetic particles. These processes are much less effective than destruction by ultraviolet radiation. However, when the ultraviolet is highly attenuated in clouds, the energetic radiation and particles will persist and become relatively more important. The ultimate lifetimes in clouds may depend upon these processes and may be shorter than those shown in Figures 2 and 3. These processes require further study.

The author is indebted to Dr. Bertram Donn for his patient tutoring in astronomical problems, especially those concerning the interstellar medium. The present paper owes much to stimulating discussion with him. Thanks are also due to Thorton Pinder for his help with compiling the data on absorption cross sections and to William Mish for his expert guidance and assistance in the computer calculations.

REFERENCES

Becker, K. H.; Haaks, D. and Schurgers, M. 1971, (private communication).

Bunch, S. M.; Cook, G. R.; Ogawa, M. and Ehler, A. W. 1958, *J. Chem. Phys.*, **28**, 740.

Calvert, J. G. and Pitts, J. N. 1966, *Photochemistry* (New York: John Wiley and Sons, Inc.).

Cook, G. R.; Metzger, P. H. and Ogawa, M. 1965, *Can. J. Phys.*, **43**, 1706.

Cook, G. R. and Ogawa, M. 1969, *J. Chem. Phys.*, **51**, 647.

Dibeler, V. H. and Reese, R. M. 1964, *J. Chem. Phys.*, **40**, 2034.

Dressler, K. and Miescher, E. 1965, *Ap. J.*, **141**, 1266.

Galli, A.; Harteck, P. and Reeves, R. R., Jr. 1967, *J. Phys. Chem.*, **71**, 2719.

Foote, J. K.; Mallon, M. H. and Pitts, J. N. 1966, *J. Amer. Chem. Soc.*, **88**, 3698.

Gentieu, E. P. and Mentall, J. E. 1970, *Science*, **169**, 681.

Habing, H. J. 1968, *B. A. N.*, **19**, 421.

Hentz, R. R. and Rzad, S. J. 1967, *J. Phys. Chem.*, **71**, 4096.

Herzberg, G. 1968, Conf. on Lab. Astrophysics, Lunteren, Netherlands.

Johnson, H. L. 1965, *Ap. J.*, **141**, 923.

Lambrecht, H. and Zimmermann, H. 1955, *Wiss. Zeit. Fr. Schiller Unv. Jena, Math. Naturwiss Reihe*, **3-4**, 217.

Leiga, A. G. and Taylor, H. A. 1965, *J. Chem. Phys.*, **42**, 2107.

Marmo, F. F. 1968, *Quarterly Progress Report, Contract NASw-1726*, GCA Corporation, 33-34.

Matsunaga, F. M. and Watanabe, K. 1967, *J. Chem. Phys.*, **46**, 4457.

McNesby, J. R. and Okabe, H. 1964, *Adv. Photochem.*, **3**, 157.

Metzger, P. H. and Cook, G. R. 1964, *J. Chem. Phy .*, **41**, 642.

Myer, J. A. and Samson, J. A. R. 1970, *J. Chem. Phys.*, **52**, 266.

Nakayama, T. and Watanabe, K. 1963, *J. Chem. Phys.*, **40**, 558.

Payne, W. A. and and Stief, L. J. 1972, *J. Chem. Phys.*, **56**, 3333.

Person, J. C. 1965, *J. Chem. Phys.*, **43**, 2553.

Ramsay, D. A. and Thistlethwaite, P. 1966, *Can. J. Phys.*, **44**, 1381.

Samson, J. A. R. and Myer, J. A. 1969, *Technical Report TR-69-7-N*, GCA Corporation.

Scudder, J. 1971, *Goddard Space Flight Center Document X-692-71-200.*

Shindo, K. and Lipsky, S. 1966, *J. Chem. Phys.*, **45**, 2292.

Sidhu, K. S.; Gsizmadia, I. G.; Strausz, O. P. and Gunning, H. E. 1965, *J. Am. Chem. Soc.*, **88**, 2412.

Stecher, T. P. 1969, *Ap. J.*, **157**, L125.

Stief, L. J.; De Carlo, V. J. and Mataloni, R. J. 1965, *J. Chem. Phys.*, **42**, 3113.

Stief, L. J.; De Carlo, V. J.; and Payne, W. A. 1971, *J. Chem. Phys.*, **54**, 1915.

Stief, L. J.; Donn, B.; Glicker, S.; Gentieu, E. P. and Mentall, J. E. 1972, *Ap. J.*, **171**, 21.

Takita, S.; Mori, Y. and Tanaka, I. 1969, *J. Phys. Chem.*, **73**, 2929,

Thompson, R. F. 1970, *Goddard Space Flight Center Document X-692-70-261.*

Turner, B. E.; Heiles, C. E. and Scharlemann, E. 1970, *Astrophys. Letters*, **5**, 197.

Watanabe, K.; Matsunaga, F. M. and Sakai, H. 1967, *Appl. Opt.*, **6**, 391.

Watanabe, K.; Nakayama, K. T. and Mottl, J. 1962, *J. Quant. Spec. and Rad. Transfer*, **2**, 369.

Watanabe, K. and Sood, S. P. 1965, *Science of Light*, **14**. 36.

Wilson, R. W.; Jefferts, K. B. and Penzias, A. A. 1970, *Ap. J.*, **161**, L43.

Zelikoff, M. and Aschenbrand, L. M. 1965, *J. Chem. Phys.* **24**, 1034.

Reaction Dynamics and the Interstellar Environment

J. C. Polanyi

Department of Chemistry
University of Toronto
Toronto, Canada

I. INTRODUCTION

Traditionally reaction kinetics has been concerned with the study of reaction rates under conditions of thermal equilibrium. However, over the past decade a new branch of reaction kinetics has come into prominence, which deals with reactions and related collision-processes under non-equilibrium conditions (Polanyi 1971). Under non-equilibrium conditions, because of the removal of a part of the thermal averaging characteristic of equilibrium environments, it becomes possible to examine the *effect of specific forms of motion upon the reaction rate.* The "detailed" rate coefficients obtained under non-equilibrium conditions give information concerning the rate of reaction of reagents in specified energy states, reacting to form products in specified energy states. The degree of "detail" in the rate coefficient (see below for a definition of this quantity) depends upon the extent of disequilibrium, and upon the degree of specificity of the instrumentation being used to follow the rate of disappearance of some species and the consequent rate of formation of others.

From a fundamental standpoint the interest that attaches to this type of investigation arises from the fact that the *detailed* rate coefficient gives a very much clearer indication of the molecular motions involved in a reactive encounter, than do the thermally-averaged rate constants. The subject matter of the new field of "reaction dynamics" at its most fundamental, is the study of the forces operating between atoms and molecules in reactive, and potentially-reactive, collisions. The field encompasses the study of detailed rate coefficients, at every achievable level of detail.

The detailed information that is emerging at present from laboratory studies of certain non-equilibrium situations will, ultimately, make it possible to calculate the rates of chemical reactions under *any* condition of non-equilibrium (or equilibrium). Donn (1972) has drawn attention to the non-equilibrium nature of the interstellar medium, and the importance of this fact for reaction rate.

In this brief paper, following a terse statement of the "normal" equilibrium reaction rate laws (section II), I propose to enlarge a little upon the theme of thermal dis-equilibrium in interstellar space (section III), and upon the related topic of detailed rate constants (section IV). Finally (section V) I shall say a few words about the two principal techniques that are currently being used to explore the dynamical details of an increasing range of chemical reactions, in the laboratory, since I believe that these techniques suggest ways in which our understanding of the chemistry of interstellar space may be extended.

II. EQUILIBRIUM REACTION RATE

The rate of a reaction

$$A + B \rightarrow C + D \tag{1}$$

can be expressed as the rate of disappearance of A or B, or the rate of formation of C or D. In each case one can write a rate expression of the general form,

$$\text{Rate} = -\frac{dn_A}{dt} = k\, n_A\, n_B \tag{2}$$

where the n's are the concentrations of the subscripted species, and k is called the "rate constant" or "rate coefficient". It is termed a rate "constant" in view of the fact that it is most commonly invariant with the total concentration of species A and B. It is, however, very sensitive to temperature. The temperature dependence of k can be expressed to about the accuracy to which k can (even today) be measured, by the Arrhenius expression — now ninety years old,

$$k(T) = A \exp\left(- E_a/k_B T\right) \tag{3}$$

where A is the Arrhenius A-factor, E_a is the Arrhenius activation energy, k_B is Boltzmann's constant, and T is the absolute temperature. The expression is an approximate one (a more realistic expression could certainly be concocted by introducing some temperature dependence in A), but it continues to be enormously valuable. The energy-parameter E_a was understood by Arrhenius, and is still understood today, as giving a measure of the height of some sort of barrier-to-reaction.

It is possible to arrive at a clearer understanding of E_a if one assumes that the only significant effect of changing temperature is to alter the relative translational energy with which A and B collide. In truth the changing temperature will also alter the distribution of A and B over their internal energy states; if this has a major effect on the reaction rate then the following simple interpretation of E_a, due originally to Tolman (1920) (also see Menzinger and Wolfgang[1] 1969), is no longer possible (in fact no general interpretation of E_a has yet been given for this case [Karplus et al., 1965, 1966][2]).

Assuming that the effect of changing temperature is to alter the mean translational energy with which the molecules react, $\langle E_T^r \rangle$, the Arrhenius activation energy can be shown to be

$$E_a = \langle E_T^r \rangle - \langle E_T \rangle \tag{4}$$

i.e., it is the difference between the mean translational energy of the reactive encounters and the mean translational energy of all encounters. The second quantity $\langle E_T \rangle = (3/2)k_B T$. This is virtually the interpretation given originally by Arrhenius, now given a formal justification.

Rate constants, k, are tabulated for many reactions, often along with the A and E which permit calculation of k(T). Since chemists have heretofore worked under experimental conditions which closely approximated thermal equilibrium, these rate constants are not as a rule accompanied by any warning. A warning is, however, necessary. ("Shake well, before using", would be *a`propos*). Under conditions of thermal *dis*equilibrium the rate coefficient of equation (2) can be profoundly altered. The fact that k under conditions of thermal equilibrium is as sensitive to T as its presence in an exponential term of equation (3) implies, is in itself an indication that k will be strongly dependent on the maintenance of thermal equilibrium. An examination of the Boltzmann distribution (over translation, rotation or vibration) for two temperatures shows that the mean energy changes slowly, whereas the high energy tail of the distribution changes markedly. It seems likely, therefore, that this high-energy tail plays a disproportionate role in determining the reaction rate. A redistribution of the reagents A and B over their energy states in such a fashion as to remove this high-energy tail from the energy-distribution would reduce the rate of reaction dramatically, while leaving the overall concentrations, n_A and n_B, (molecules/cc, irrespective of energy state) unaltered. The rate coefficient of equation (2) would, therefore, fall to $k^{ne} \approx 0$ (the superscript designates the k as referring to a non-equilibrium situation).

It is important to consider what types of disequilibrium are most likely in the interstellar medium. This question is discussed, in very general terms, below.

III. DISEQUILIBRIUM IN INTERSTELLAR SPACE

Depending on the region of interstellar space, ambient temperatures ranging from 5 °K to 1000 °K represent reasonable estimates (Donn 1970). The high figure applies only in regions where clouds of gas collide, or following the close passage of a hot star. The question I should like to raise in this section has to do with the confidence we can have in applying thermal rate constants tabulated for the appropriate temperature (in the range 5 - 1,000 °K) as an index of the reaction rate to be expected in the corresponding regions of interstellar space. As noted in section II, a minor disequilibrium in the high energy tail of the Boltzmann distribution can give rise to a major discrepancy between k^{ne} and k. There are, obviously, two directions in which the non-equilibrium distribution may deviate from the equilibrium one; the direction of over-population or under-population. Either can result in $k^{ne} \neq k$. We shall consider the likelihood of non-equilibrium in three different degrees of freedom, in turn.

A. Translational Energy

Though calculations have been made which suggest that the distribution of translational energy will be approximately Maxwell-Boltzmann in form (Spitzer 1968), these calculations have not included the possibility that translationally-excited atoms or molecules may react before being thermalised (Wolfgang 1969, Rowland 1972)[3]. Photo-dissociation of CN, CO, or C_2 at $\sim$1000 Å will give rise to "hot" C, N and O atoms with an initial translational energy of $\sim$150 kcal mole^{-1}. At 5 °K the thermal fraction of C atoms with this energy would be $\sim$10$^{-6,500}$.

Since the most abundant species in interstellar space, by a factor of 10^3 - 10^4, are H and H_2, a large number of collisions $C + H_2$, $N + H_2$ and $O + H_2$ will occur (on the average) before any heavy-atom + heavy-atom collisions take place. The *relative* energy of a heavy particle with respect to a light one is much less than is that of the heavy particle in a fixed frame of reference (the motion in the centre-of-mass frame of the colliding pair is motion predominantly of the light particle). The relative energy of the heavy + light pair will nonetheless be $\sim$20 kcal mole^{-1}. It seems that the colliding pair (in which the heavy atom is "hot") will have an initial translational energy well in excess of the activation energy for any reactions of interest to interstellar chemists. Even if the first collision is unsuccessful in bringing about reaction, thermalisation (so that the hot atoms become 1 part in $10^{6,500}$ at 5 °K, or 1 in 10^{33} at 1,000 °K) will require a significant number of collisions (Benson 1960). The reason for this is the mass-disparity between the hot species and the "coolant" (H_2).

Translational disequilibrium deserves special attention where interstellar space is concerned, since — in contradistinction to rotational, vibrational and electronic disequilibrium — it will not be dissipated by radiational processes even in the long times which characterize galactic events. It is therefore worth considering the possible role not only of non-equilibrium with respect to translational speed, but also with respect to velocity, *i.e.*, directed motion.

The thermal rate coefficient of equation (2) could once again be profoundly altered if the system were in equilibrium in all regards except with respect to the random distribution of directions of motion. A disequilibrium of this sort could have the result that $k^{ne} \ll k$. It could not, however, make $k^{ne} \gg k$. If the reagent species A and B, at concentrations n_A and n_B, were moving radially outward from a localised source (such a source might be a star, a gas cloud, or a dust cloud, responsible for the production of A and B) then the typical orientation of velocities would be parallel; *i.e.*, v_A and v_B having the same sign and similar magnitude. The relative velocity $v_{rel} = v_A - v_B$ would be negligible, and the collisions between A and B would be so few and of such small relative translational energy, E_T, that $k^{ne} \approx 0$. By contrast, in a region along a line joining a source of particles A to a source of B, the most likely orientation for the velocities would be opposed, $\therefore v_{rel} = v_A - v_B = 2v_A$. Since in a randomly moving body of gas the most probable value for $\langle v_{rel} \rangle \approx 1.4 \langle v_A \rangle$, this will not constitute a vast enhancement of k^{ne} with respect to k. [This is not to say that it will be a negligible factor, since $E_T(a\ v^2_{rel})$ figures in the exponent of e in equation (3)].

Reference will be made in section V to two specific mechanisms for producing directed translational motion; chemical reaction, and photo-dissociation.

B. Rotational Energy

There are a number of possible sources of highly rotationally-excited molecules in interstellar space, but the two most prominent would appear to be fluorescence and chemical excitation, with fluorescence the more important of the two. If molecule A is transferred temporarily (following the absorption of light) to a bound electronically-excited state, it will have plenty of time to radiate back to its electronic ground state. The upward transition must originate in vibrational and rotational states $v \approx 0$, $J \approx 0$, since the bulk of A is in these states. The downward transition will be to whatever states of v and J the probability of spontaneous emission (governed by the matrix element of the dipole moment) is greatest. There is no preference for $v \approx 0$, $J \approx 0$, and on purely statistical grounds (i.e., owing to the fact that there are very many more states for which $v \neq 0$, $J \neq 0$) we can expect that the outcome will be to populate excited vibrational and rotational states.

Exothermic chemical reactions (reactions that liberate energy, by virtue of the fact that a new chemical bond is formed which is stronger than the old one) have been found, in the cases so far studied (Carrington and Polanyi 1972)[4], to populate highly rotationally-excited states, preferentially. Since exothermic reactions are in general those with low activation barriers and high reaction rates, they are the most likely to be of importance in interstellar space.

Given these sources of rotational excitation, the question arises as to whether the rotational excitation can persist for time periods sufficient to permit one or more potentially-reactive encounters. There is no such thing as a typical time interval between collisions in interstellar space, since there is a $>10^6$ variation in gas density, a $\sim 10^2$ variation in gas temperature, and a $\sim 10^5$ variation in the abundance of various elements (Donn 1970). However, it is noteworthy that even for the most abundant species, $H + H_2$, in the (vast) regions of low density where the concentration is ~ 1 atom per cm^3, the interval between collisions is

~10^3 years. At the other end of the scale, in clouds of what (in terms of the interstellar medium) are at high density and elevated temperature, the collision frequency with hydrogen as one collision partner is as short as ~1 collision per day.

For simplicity it will suffice to distinguish the cases (a) of dipolar molecules: these can be expected to lose all excess rotational energy through radiation before potentially-reactive collisions can occur, and (b) non-polar molecules: these will retain significant rotational excitation for a long enough period that they can be expected to undergo potentially-reactive collisions in regions of higher interstellar density.

It does not follow from this that there will be insignificant rotational disequilibrium where dipolar molecules are concerned. The disequilibrium will, however, be in the opposite sense from that of the non-polar molecules, since it will consist of an under-population of the excited rotational states in the regions of low gas density. If, as is to be expected in these regions, $J \rightarrow 0$, due to radiation between successive collisions, it cannot be expected that a single collision will have the effect of re-establishing the equilibrium concentration of J > 0. The probability for $\triangle J = 1$ at a collision is typically ~0.1 (~0.01 for H_2) (Cottrell and McCoubrey 1961), and a number of *successful* collisions would be required to bring about equilibration with the "heat bath". (Radiation lost in this fashion by dipolar molecules in low density regions will be absorbed by the same species in high density regions, where it will contribute to rotational and perhaps translational heating).

C. Vibrational Excitation

As already noted, fluorescence is a likely cause of the formation of molecules in v > 0. So also is exothermic chemical reaction (Carrington and Polanyi 1972). In addition it is known that electron-impact can give rise to efficient vibrational excitation (Schulz 1964) (the electrons would come, for example, from photo-ionisation processes).

Despite this variety of sources of vibrational excitation, it appears doubtful whether vibrational disequilibrium can play a significant role in promoting chemical reaction in interstellar space, since even homonuclear molecules (H_2, O_2, N_2, C_2, . . .) should have time to radiate by quadrupole emission during the long intervals between encounters. Exceptions could occur in regions of "high" density where collisions are relatively frequent, and radiational lifetimes can be effectively prolonged through entrapment. By "entrapment" I mean to describe the phenomenon whereby quanta of a resonant frequency are transferred from molecule to molecule. The result is indistinguishable from prolonged radiational lifetime. Early measurements of radiational lifetimes yielded values orders-of-magnitude too high because of this entrapment (or "self-absorption") effect. The true radiational lifetime only becomes evident in the limit of low pressure and short path length (Mitchell and Zemansky 1961).

IV. DETAILED RATE CONSTANTS

There is othing in the expression for the equilibrium rate constant k [equation (2)] to indicate why it should be affected by a deviation from thermal equilibrium. This is, of course, simply due to the way in which we have so far written k. In fact species A has available to it internal energy states governed by quantum numbers i, and so on for species B, C and D. We can re-write equation (1) more explicitly as,

$$A_i + B_j \rightarrow C_\ell + D_n \tag{5}$$

Equation (2) can then be made to include an arbitrary distribution of the reactants over states i and j, in place of the equilibrium distribution that was implicit in the original equation,

$$- \frac{dn_A}{dt} = \left(\sum_{\ell n} \sum_{ij} k_{ij}^{\ell n} \, x_{A,i} \, x_{B,j} \right) n_A \, n_B \tag{6}$$

where $k_{ij}^{\ell n}$ is the detailed rate constant for $A_i + B_j \rightarrow C_\ell + D_n$ and $x_{A,i}, x_{B,j}$ are the fractions of n_A and n_B in states i and j. In each term of the sum over i j the appropriate $x_{A,i}, x_{B,j}$ is used. The second summation is there since we are interested, in the present instance, in the rate of disappearance of A irrespective of the states of C and D into which it is converted.

Since A and B must collide in order to react, we can recast the original rate expression [equation (2)] in terms of the product of three terms: the flux of A incident on B, $v_{rel} \, n_A$ (molecules of A crossing a unit area per sec), multiplied by the number of target molecules per unit volume, n_B, multiplied by the effective target area for a reactive encounter, called the reactive cross-section, S:

$$- \frac{dn_A}{dt} = S \, v_{rel} \, n_A \, n_B \tag{7}$$

It follows that the rate constant, from this more physical standpoint, is

$$k = S \, v_{rel} \tag{8}$$

This expression is only correct if S is not itself a function of v_{rel}. In fact we know that S is a function of v_{rel}; it must be zero for relative speeds which correspond to collision energies E_T below some threshold energy E_0, and must tend to zero again at extremely high v_{rel} when molecule A has a negligible time in which to interact with B. Accordingly we write S as $S(v_{rel})$, or $S(E_T)$. There is in fact a set of such collision-energy-dependent reactive cross sections; $S_{ij}^{\ell n}(v_{rel})$, or $S_{ij}^{\ell n}(E_T)$. The rate of disappearance of A_i in collisions with B_j to form C_ℓ and D_n (all subscripted quantum numbers being specified) is

$$- \frac{dn_{A,i}}{dt}\bigg|_{\ell n j} = \left(k_{ij}^{\ell n} \, x_{A,i} \, x_{B,j} \right) n_A \, n_B \qquad (9)$$

$$= \left\{ [S_{ij}^{\ell n}(v_{rel}) \cdot v_{rel} \cdot f(v_{rel})] \, x_{A,i} \, x_{B,j} \right\} n_A \, n_B \qquad (10)$$

where $f(v_{rel})$ is the fraction of the $A_i + B_j$ collisions that have the specified relative speed of approach. The *total* rate of removal of A [see equation (6)] is

$$- \frac{dn_A}{dt} = \left\{ \sum_{\ell,n} \sum_{ij} [S_{ij}^{\ell n}(v_{rel}) \cdot v_{rel} \cdot f(v_{rel})] \, x_{A,i} \, x_{B,j} \right\}$$
$$\times n_A \, n_B \qquad (11)$$

This is the type of rate expression one would have to use, in place of equation (2), if there were disequilibrium over translation, rotation and vibration. If there is a translational Maxwell-Boltzmann distribution then $v_{rel} \cdot f(v_{rel})$ can be expressed in terms of the (translational) temperature, T

$$- \frac{dn_A}{dt} = \text{Const.} \left\{ \sum_{\ell,n} \sum_{ij} x_{A,i} \, x_{B,j} \int_0^\infty E_T \cdot S_{ij}^{\ell n}(E_T) \cdot \exp\left(-\frac{E_T}{k_B T}\right) dE_T \right\}$$
$$\times n_A \, n_B \qquad (12)$$

where the Const. $= (2/k_B T)^{3/2} (1/\pi\mu)^{1/2}$, and μ is the reduced mass of A and B (Light et al., 1969; Polanyi and Schreiber 1973).

To make this still simpler let us consider a case in which we are observing the rate of disappearance of an atomic species A under circumstances where thermal equilibrium applies to translation and to the internal energy states of B. The temperature is so low that (we shall suppose) the only significant internal excitation of B is rotational. Accordingly we drop the quantum number i (A is assumed to be in its electronic ground state) and replace j, designating the molecule B under attack, by the rotational quantum number J. To make matters still simpler we shall consider B to have such widely spaced rotational levels that only levels J and J' are significantly populated. We shall further suppose that the product energy distribution overstates ℓ and n is insensitive to the collision energy (in the fairly small range of collision energies that account for most of the reaction). In this case we can replace $S_J^{\ell n}(E_T)$ by $S_J(E_T)$, where S_J is larger than $S_J^{\ell,n}$ by some constant factor. (In effect, the summation over ℓ,n has been made prior to the integration). Similarly S_J' takes the place of $S_{J'}^{\ell n}$. The rate constant then reduces to the sum of two terms,

$$
- \frac{dn_A}{dt} = \left[x_{B,J} \int_0^\infty E_T \cdot S_J(E_T) \cdot \exp{-\left(\frac{E_T}{k_B T}\right)} dE_T \right.
$$

$$
\left. x_{B,J'} \int_0^\infty E_T \, S_J'(E_T) \cdot \exp{-\left(\frac{E_T}{k_B T}\right)} dE_T \right] n_A \, n_B
$$

(13)

As before, the term in braces is the rate constant (whether it is k_{ne} or k depends on the choice of $x_{B,J}$ and $x_{B,J'}$).

There is experimental evidence which indicates that, to a limited extent, rotational excitation in reagent molecules is "available" to assist in the crossing of even the large activation barriers characteristic of endothermic reaction (Anlauf et al., 1969; Polanyi and Tardy[5] 1969). In addition there are theoretical data from classical trajectory computations on various potential-energy hypersurfaces which also indicate that reagent rotational excitation at first enhances the likelihood of reaction (and later diminishes it) (Karplus et al., 1965, 1966; Anlauf et al., 1969; Polanyi and Tardy 1969; Jaffe and Anderson 1971; Hodgson and Polanyi 1971). These findings are embodied, qualitatively, in

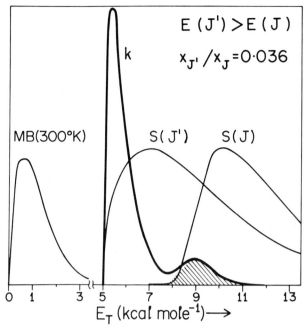

1. The Maxwell Boltzmann distribution of translational energies for $300°K$ is shown at the left (MB). The two curves at the right give postulated reaction cross-sections (for some hypothetical reaction) with the molecular reagent, B, in a lower state of rotation J, or in 2 kcal mole^{-1} higher state, J'. The general form of S(J) and S(J') is correct; the precise form is not yet known for any chemical reaction. The S(J) and S(J') pictured have been used to illustrate the consequences for k of their general form. The total rate constant is given by the area under the curve labelled k [corresponding to the term in braces in equation (13)]. The large unshaded portion of the area under the curve k is the contribution to the rate constant from the 3.6% of B that is in level J'; the small shaded portion is the contribution to the rate constant from the 96.4% of B that is in level J (Polanyi and Schrieber 1973).

Figure 1, which illustrates, graphically, the working of equation (13). Assuming $J' > J$, the threshold energy $(E_0)_{J'}$ at which reaction would become measurable for reagent in rotation state J' has been set lower than the threshold energy $(E_0)_J$ for reagent in level J. For illustrative purposes some rather arbitrary choices have been made for the various constants of equation (13): T = 300 °K, $E(J') - E(J) = 2$ kcal mole^{-1}, $(E_0)_{J'} - (E_0)_J = 2$ kcal mole^{-1}, $x_{B,J'}/x_{BJ} = \exp - ([E(J') - E(J)]/k_B T) = 0.036$. It may not be typical for the reaction threshold to be diminished by the *full amount* of the (2 kcal mole^{-1}) rotational excitation. On the other hand an unrealistically conservative assumption has been made in the choice of 0.036 for the mole-fraction-ratio of the higher rotational level to the lower, since this method of calculation neglects the ratio of degeneracies that significantly favours higher rotational levels at equilibrium. Nonetheless it is evident from the figure that at 300 °K the term in J' is making a much larger contribution to the rate constant than is the term in J; clearly the magnitude of $x_{B,J}'/x_{B,J}$ plays a crucial role in determining the reaction rate, so that k^{ne} can be very different indeed from k. It would be just as easy to show that changes in the form of the distribution over E_T, or the distribution over vibration, v, could profoundly affect the magnitude of the rate constant.

Equation (13) was introduced in order to shed some light on the working of the full rate expression given by equation (11) or (12). Some such simplifications (less drastic than the two-state assumption made here) may be possible in situations of interest to astrochemists, without an intolerable sacrifice in predictive accuracy. Nonetheless we must conclude that, even for what in other contexts might be termed "modest" deviations from thermal equilibrium, in order to obtain useable reaction rates it will be necessary to improve our knowledge of the actual distribution over $x_{A,i}$, $x_{B,j}$ and v_{rel} in interstellar space and also to investigate $S_{ij}^{\ell n}(v_{rel})$ for reactions of astrochemical interest. Fortunately techniques are being developed which should make it possible to measure detailed cross sections, $S_{ij}^{\ell n}(v_{rel})$. These will be mentioned very briefly in the concluding section, below.

V. BEAMS AND CHEMILUMINESCENCE IN INTERSTELLAR SPACE

The two principal techniques currently being used in the laboratory in attempts to obtain information relating to detailed reactive cross sections, are the molecular beam method (Herschbach 1966, 1971; Kinsey 1972) and the infrared chemiluminescence method (Carrington and Polanyi 1972). Since these methods have been described elsewhere, it would not be necessary to refer to them here but for the intriguing fact that each appears to have its natural counterpart in the interstellar medium. If this is so, the laboratory investigations suggest parallel studies that might be undertaken in which the objective would be (in a reversal of the rationale for laboratory experiments) to exploit a prior knowledge of reaction dynamics in order to study the nature of the "apparatus" – the interstellar environment.

In molecular beam studies of reactive scattering each of the reagents, A and B, is localised in space, and one or both of the reagents have a specified direction of motion. The prime measurables are the angular distributions of products C or D with respect to the direction of approach of A, and the translational energies of the products. The fact that reaction products can be, and often are, formed preferentially at certain angles with respect to the reagent directions is an important feature of reaction dynamics that has not previously been mentioned in the course of this brief survey. The measurement of product flux into intervals of solid angle $d\Omega$ gives information regarding differential reactive cross-sections, $\sigma(\Omega)$. The reactive cross-sections used in the equations of the previous section had already been integrated over Ω; $S = \int \sigma(\Omega) \, d\Omega$. Since the total energy available for distribution among the products, E'_{tot}, is often known from thermodynamic considerations, the product translational excitation obtained from molecular beam experiments gives, by difference, information concerning the total internal excitation in C and D. If C is an atomic product and D a molecular one, then one has the probability-distribution over internal energy states $(v + J)$ of particle D.

The infrared chemiluminescence approach is complementary to the molecular beam method, in that it leads to a direct spectroscopic measurement of the *individual* probability distributions over v and J. In essence there is no more to this method than the reaction of A + B under conditions of low pressure, approaching the densities characteristic of molecular beams, with rapid removal of C and D prior to significant radiational, or other, "relaxation". ("Relaxation" connotes a change from the initial v,J distribution to another distribution more-closely resembling equilibrium). A small fraction of the products radiate in the infrared, giving rise to the infrared chemiluminescence spectrum. The relative intensities of vibrational-rotational lines in this spectrum yield, after minor correction for relaxation, the initial distribution of reaction products over states v and J. Since the total energy available for distribution among the products is known (or can be inferred from the observed cut-off in the distribution over v and J at certain $(v + J)_{max} = E'_{tot}$), the initial distribution of translational energies in the reaction products can be obtained for the case that C is atomic and D is molecular. Where data is available for a single reaction from molecular beam experiments and from infrared chemiluminescence experiments, the results are in broad agreement (Anlauf *et al.*, 1970; Maylotte *et al.*, 1972; Anlauf *et al.*, 1972; Grosser and Haberland 1970; McDonald *et al.*, 1972; McDonald and Herschbach 1972; Shafer *et al.*, 1970).

Figures 2 and 3 give some idea of translational energy distributions in the products of some exothermic reactions. Figure 2, from recent molecular beam work on the reaction $F + D_2 \rightarrow DF + D$ (Shafer *et al.*, 1970), actually shows the *angular* distribution (as recorded in the laboratory frame-of-reference). However there is no doubt that the undulations in the laboratory angular distribution of DF are the result of the combined effect of (a) a tendency for the DF formed in different vibrational levels to have differing mean translational energies, and (b) for the DF to be formed at various angles (relative to the attacking F atom) depending on its vibrational quantum number. It is not altogether easy, at the present stage of development of such experiments, to separate the effects of (a) and (b). The results of infrared chemiluminescence experiments, which only

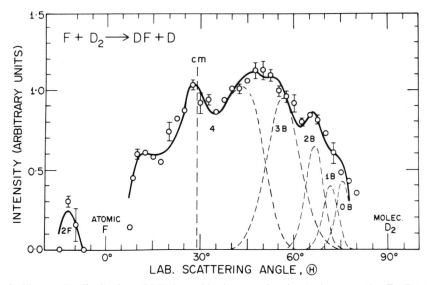

2. The angular distribution of DF, formed in the crossed molecular beam reaction $F + D_2 \rightarrow$ DF + D, as recorded in the laboratory frame of reference. The locations of the reagent beams are at $0°$ and $90°$, and the centre-of-mass is at approximately $30°$. Viewed in the centre-of-mass frame most of the molecular product is scattered "backwards" with respect to the direction of approach of the atomic F (*i.e.*, at $\Theta > 30°$ in the laboratory). This is the only molecular-beam scattering data, up to the present, that has shown "vibrational structure" in the angular distribution. The contributions of the 4 accessible vibrational states of DF can be surmised (broken lines) from the locations and magnitudes of the undulations in the angular distribution ($v = 0$ in the backward direction is labelled "0B", etc.). There is evidence of a small amount of $v = 2$ DF scattered forward (peak "2F"). The various vibrational levels make their distinctive contributions to the laboratory angular distribution because they are associated with markedly different product translational energies; laboratory angular distribution is dependent both on the centre-of-mass angular distribution and the centre-of-mass speed. See Shafer *et al.*, 1970.

yield information concerning (a) (*i.e.*, product translational energy, and not angle), are helpful in this regard. Translational energy distributions have been obtained for a number of reactions by the infrared chemiluminescence approach. Some examples are shown in Figure 3. Since these translational energies were inferred from spectroscopic data concerning distributions over v and over J states, they show a high degree of detail in the translational energy distribution. It is also possible to present the infrared chemiluminescence data in a different type of diagram which shows the translational distribution of newly-formed molecular product in each of the rotation-vibration states to which the reaction

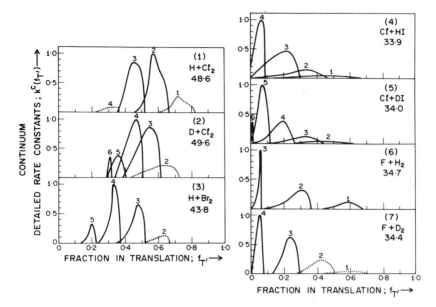

3. Product translational energy-distributions for seven reactions of the type A + BC → AB + C, as obtained from infrared chemiluminescence experiments. The energy available for distribution among the reaction products (kcal mole) is given for each reaction, directly below the reagent designation. The fraction of this energy going into translation is recorded along the ordinate. This fraction is seen to be low for Cl + HI → HCl + I and high for H + Cl$_2$ → HCl + Cl. In some cases the product translational energies corresponding to the various vibrational energy states of the product are markedly different. The reaction F + D$_2$ → DF + F is a case in point (the numbers above the curves indicate the product vibrational quantum state to which that translational energy distribution applies.) This helps to explain why the individual peaks are discernible in the laboratory angular distribution for the same reaction recorded in the previous figure. See Anlauf *et al.*, 1970.

(given the limitations of E'_{tot}) has access. This is shown in Figure 4 for the case of F + D$_2$ → DF + F [this example was chosen for illustration since it is also represented in the two previous figures (for further examples see Maylotte *et al.*, 1972; Anlauf *et al.*, 1972; Polanyi *et al.*, 1972)].

At the level of detail recorded in Figure 4 we are dealing with relative rate constants, $k_{exo}(v', J', T')$, for the formation of the products of exothermic reaction in states of vibrational, rotational and translational energy v', J', T'. It is possible to apply detailed balancing and obtain the corresponding rate constants for the reverse, endothermic, reaction; $k_{endo}(v', J', T')$ (Anlauf *et al.*, 1969; Polanyi and Tardy 1969). In these endothermic rate constants v', J', T' are the vibrational, rotational and translational energies of the reagents. A convenient

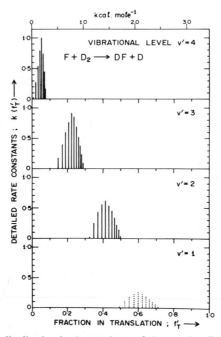

4. Translation energy distribution in the products of the reaction $F + D_2 \rightarrow DF + D$, from infrared chemiluminescence data. The ordinate is the same as that in Figure 3. The level of detail is greater than in the previous figure, since the characteristic translational energies are shown not only for different product vibrational states, but also for the different rotational states of those vibrational states. Since the amounts of products are specified as to product vibrational, rotational and translational energy—states (which we shall symbolise as V', R' and T'), the heights of the vertical lines in this figure give the relative rate constants $k_{exo} (V', J', T')$. (These same rate constants can be pictured on triangular contour plots, in which contours in $V' \text{-} J' \text{-} T'$ space join points of equal detailed rate constant, $k_{exo} (V', J', T')$). See Polanyi and Woodall, 1972.

representation (in lieu of the "stick" diagram type of representation given in Figure 4) ignores the quantisation of v' and J', and simply records contours of equal detailed rate constant in a space of v', J' and T'. Figure 5 gives an example; $HF(v',J') + H \rightarrow F + H_2$. The qualitative features are thought to be typical of endothermic reaction (see Polanyi 1972). These features are (a) a moderate increase followed by a decrease in endothermic reaction rate with increasing reagent translation or rotation (to identify these coordinates refer to the caption of Figure 5), and (b) an orders-or-magnitude increase in endothermic reaction rate with increase in reagent vibration. The marked effect of reagent vibration on reaction probability has recently been demonstrated in a beam experiment (Odiorne *et al.*, 1971).

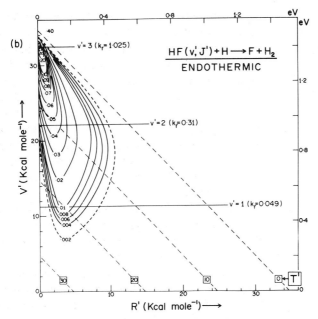

5. An example of a contour plot for *endothermic* reaction. Contours join points of equal detailed rate constant, k_{endo} (V', J', T'), for the endothermic reaction HF (v', J') + H $\rightarrow$ F + H_2. (A very similar plot would apply to DF + D $\rightarrow$ F + D_2, spanning a similar range of vibrational energy V', though encompassing four vibrational quantum numbers, v'). The energies V' (ordinate), R' (abscissa) and T' (increasing toward the $V' = 0$, $R' = 0$ corner of the figure, as recorded by the diagonal grid) refer to the *reagent* vibrational, rotational and translational excitation for endothermic reaction. The contours delineate a mountain with its peak at high V', and low R' and T' (see text). Note that k_{endo} (V', R', T') is given for a constant total reagent energy of $E'_{tot} = 34.7$ kcal mode^{-1}. See Polanyi and Tardy, 1969.

The findings from laboratory studies employing the molecular beam or infrared chemiluminescence techniques, could be of value in characterising reactions occurring in interstellar space. One or more of the following attributes of interstellar reactions may prove to be accessible to experimental study. Each would provide clues as to the nature of the reactions taking place, or, if these reactions are known, they would add to our understanding of the evolution of the interstellar medium.

A. *Angular Distribution of Interstellar Reaction-products*

The direction of product motion, θ, must be measured with respect to the direction of reagent motion. (The customary convention is that a product, C or

D, is described as "forward-scattered" if it is formed along the continuation of the direction of approach of the atomic reagent A — the whole event being viewed by an observer moving with the centre-of-mass of the colliding species A + B). For such a measurement to be meaningful there must, therefore, be a preferred direction of relative motion of A with respect to B. In laboratory crossed-beam experiments the distance from the sources of A and B to the interaction region is typically a few centimetres. The direction of the relative velocity vector $v_{rel} = v_A - v_B$ can be adequately defined either by collimating both reagents into beams, or by making $|v_A| \gg |v_B|$ and collimating A only. It is common practice in molecular beam experiments to use interrupted ("chopped") beams, so that the rate of progress of the reagents (from source to crossing-point to detector) can be measured.

The collision of interstellar clouds may be analogous to the crossing of a pair of chopped beams. The dense interstellar clouds have an approximate diameter of ~1 parsec (3×10^{18} cm). Their density is ~10^8 x that of the intercloud region (in a typical molecular-beam apparatus beam density is ~10^4 x background pressure). The distance travelled by the clouds is characteristically $\geqslant$ 100 pc. This corresponds to an angular uncertainty of $\leqslant 0.6°$ in the direction of motion of the reagent clouds. If the product-cloud location is measured at a comparable distance from the collision-zone, the requirements for a molecular-beam scattering experiment are fulfilled. A resolution of ~$1°$ is sufficient to give important information regarding the angular dependence of the reaction-probability, $\sigma(\theta)$ (cf. Figure 2). There are even cruder questions which are of the first importance; namely whether the product is predominantly forward, backward or sideways scattered. The form of $\sigma(\theta)$ would appear to be an aspect of the dynamics of interstellar reaction that should be included in a model of the evolution of interstellar matter.

Very much the same considerations apply to a second angular-dependent process, namely the photolysis of reagent B to give product fragments C + D. The analogue in the laboratory, in this instance, is the crossed photon + beam experiment (Wilson 1970). If we represent the photons by A, then the condition $|v_A| \gg |v_B|$ is invariably met, and v_{rel} depends on v_A only. The ultraviolet radiation emanates from bright stars, and the direction of v_{rel} is given by the line joining the star to the cloud of B, at the time that C + D were formed. If the time during which B is being photolysed is short compared with the time required for B to move a significant distance (a significant distance being several parsec) then the direction of v_{rel} is unique, and C + D emanate from a crossing zone at the "point" where A intersects B. If the time of photolysis is long, then v_{rel} sweeps out an arc, and C + D emanate from a crossing "line" along which A interacts significantly with B. A cloud of B would in this case give rise to elongated clouds of C and D. The angles corresponding to the vector differences

(v_C - v_{rel}) and (v_D - v_{rel}) would, nonetheless, be well-defined in one plane. These angles would be characteristic of the parent molecule of C + D, and of its electronic state.

B. Energy Distribution of Interstellar Reaction-products

The reactions of importance in the interstellar environment will have low activation energies. Such reactions tend to be exothermic (releasing energy). The reaction-products, as initially-formed, will contain this exothermicity in the form of translation, rotation, vibration and occasionally electronic excitation. Translational excitation is unique in this catalogue (as already noted), in that it cannot be lost by radiation. It will persist over time-spans measured in years, until collisional deactivation takes place. If a means can be found to measure the translational energy distribution precisely, then, with the aid of laboratory studies such as those described in section IV of this paper, the reaction could be unambiguously identified. Unfortunately the measurement of the translational-energy distribution of interstellar matter poses severe problems. Doppler-shapes of emission lines must be measured with high precision. The interpretation must take into account the possibility of asymmetric deviations from the Doppler line-shape, due to preferred directions of motion and non-thermal translational-energy distribution. (Ambiguities will arise; for example emitters moving perpendicularly to the observer will exhibit negligible Doppler broadening at all translational energies).

Even if the initial translational energy of interstellar molecules cannot be measured with sufficient accuracy to provide a "finger-print" of the chemical reaction (cf. Figures 3 and 4) or of the photolytic dissociation process (Wilson 1970), the existence of these energies, which persist over considerable spans of time, cannot be ignored in constructing models of reaction rate (section IV) or of the spatial distribution of matter (cf. section V-A).

The initial rotational and vibrational distributions in the products of chemical reaction, as noted in the previous section, also provide clear evidence as to the nature of the reaction. The difficulty in making use of these clues arises from the fact that radiational relaxation will in the majority of cases be much more rapid than the rate of chemical reaction. The observed "steady-state distribution", which represents a balance between the rates of formation of the product molecules in their various v-J states and their rate of removal by radiation, will only depend weakly on the rates of formation. Nonetheless, accurate measurements of the radiationally-relaxed population distribution will show deviations from a Maxwellian distribution over J-states and v-states, characteristic of the particular chemical reaction (Heaps and Herzberg 1952, Polanyi 1963, Mumma and Donn[6] 1973).

Such measurements could possibly be facilitated by laser action. Thus if level v+2 is populated by reaction and v+1 is not, then there will be a complete population inversion between v+2 and v+1 following radiational decay, since a significant fraction of v+2 radiates directly to level v, without populating v+1 (Heaps and Herzberg 1952; Polanyi 1961, 1965). This could be significant for the steady state distribution of vibrationally-excited H_2. This case appears to be the most favourable for vibrational laser action since H_2 is relatively abundant, the lifetime with respect to spontaneous radiation is low, and the quadrupolar-emission transition probabilities allow $\Delta v = 2$ radiational transitions with significant probability as compared to $\Delta v = 1$ (Karl and Poll 1967; Dalgarno et al., 1969).

The partitioning of reaction-energy between translational excitation in the products and rotational-vibrational energy has been measured in the laboratory both for chemical reactions and also for the products of photolytic dissociations. This energy-partitioning has an interest that goes beyond the characterisation of the reactive or photolytic event. Even if the nature of the interstellar reactions and photodissociations were to be known, the important question would remain as to the manner in which gas-clouds loose their energy. [In this connection one is interested not only in the channelling of exothermicity into product vibration and rotation, but also the distribution among the product degrees of freedom of the energy that was *already present* in the reagents (Cowley et al., 1971). These questions are important to the understanding of energy-loss, since translational energy, so long as it remains translational energy, is "imprisoned" in the interstellar system, whereas rotational, vibrational and electronic energy can be lost through radiation.

It is a pleasure to thank Dr. Bertram Donn who is responsible for my recent initiation into the field of astrochemistry (but blameless for my deficiencies as a student). The experiments performed in this laboratory, referred to in the text, were made possible by grants from the National Research Council of Canada and the United States National Aeronautics and Space Administration (Grant NGR 52-026-028).

FOOTNOTES

1 A lucid discussion of the meaning of E_a.

2 It has been demonstrated for a particular example that the Arrhenius equation remains entirely adequate even if internal excitation is contributing to the reaction rate to a differing extent at each temperature, due to differing (equilibrium) concentrations of molecules in the various internal energy-states.

3 Recent reviews of this type of behavior.

4 A review.

5 In these experiments, involving three different reactive systems, approx. 5 kcal mole^{-1} removed from reagent translation, E_T, and put into reagent rotation, E_R, increased the rate constant of endothermic reaction by about one order-of-magnitude, after which a further 5 kcal transferred from E_T to E_R decreased the rate constant by about one order-of-magnitude (see, for *e.g.,* Figure 5 of this paper).

6 A proposal for the measurement of infrared chemiluminescence arising from vibrationally excited CO formed in the combination reaction C + O.

REFERENCES

Anlauf, K. G.; Maylotte, D. H.; Polanyi, J. C. and Bernstein R. B. 1969, *J. Chem. Phys.,* **51,** 5716.

Anlauf, K. G.; Charters, P. E.; Horne, D. S.; Macdonald, D. H.; Maylotte, D. H.; Polanyi, J. C.; Skrlac, W. J.; Tardy,, D. C. and Woodall, K. B. 1970, *J. Chem. Phys.,* **53,** 4091.

Anlauf, K. G.; Horne, D. S.; Macdonald, R. G.; Polanyi, J. C. and Woodall, K. B. 1972, *J. Chem. Phys.,* **57,** 1561.

Benson, S. W. 1960, *The Foundations of Chemical Kinetics* (New York: McGraw Hill), 394.

Carrington, T. and Polanyi, J. C. 1972, in *Reaction Kinetics, MTP International Review of Science,* ed. J. C. Polanyi (Oxford: Butterworths), Chapter 5.

Cottrell, T. L. and McCoubrey, J. C. 1961, *Molecular Energy Transfer in Gases* (London: Butterworths), 77.

Cowley, L. T.; Horne, D. S. and Polanyi, J. C. 1971, *Chem. Phys. Letters,* **12,** 144.

Dalgarno, A.; Allison, A. C. and Browne, J. C. 1969, *J. Atmos. Sci.,* **26,** 946.

Donn, B. 1970, *Science,* **170,** 1116.

————. 1972, *Ann. N. Y. Acad. Sci.*, **194**, 98.

Grosser, J. and Haberland, H. 1970, *Chem. Phys. Letters*, **7**, 442.

Heaps, H. S. and Herzberg, G. 1952, *Z. Physik*, **133**, 48.

Herschbach, D. R. 1966, in *Advances in Chemical Physics*, ed. J. Ross (New York: John Wiley and Sons), Chapter 9.

————. 1971, in *Proceedings of the Conference on Potential Energy Surfaces in Chemistry*, ed. W. A. Lester, Jr. (San Jose: IBM Research Laboratory Pub. RA18), 44.

Hodgson, B. A. and Polanyi, J. C. 1971, *J. Chem. Phys.*, **55**, 4745.

Jaffe, R. L. and Anderson, J. 1971, *J. Chem. Phys.*, **54**, 2224.

Karl, G. and Poll, J. D. 1967, *J. Chem. Phys.*, **46**, 2944.

Karplus M.; Porter, R. N. and Sharma, R. D. 1965, *J. Chem. Phys.*, **43**, 3259.

————. 1966, *Ibid.*, **45**, 3871.

Kinsey, J. 1972, in *Reaction Kinetics, MTP International Review of Science*, ed. J. C. Polanyi (Oxford: Butterworths).

Light, J. C.; Ross, J. and Shuler, K. E. 1969, in *Kinetic Processes in Gases and Plasmas*, ed. A. R. Hochstim (New York: Academic Press), Chapter 8.

Maylotte, D. H.; Polanyi, J. C.; Woodall K. B. 1972, *J. Chem. Phys.*, **57**, 1547.

McDonald, J. D.; LeBreton, P. R.; Lee, Y. T. and Herschbach, D. R. 1972, *J. Chem. Phys.*, **56**, 769.

McDonald, J. D. and Herschbach, D. R. 1972, *J. Chem. Phys.*, in press.

Menzinger, M. and Wolfgang, R. 1969, *Angew. Chem.*, **8**, 438.

Mitchell, A. C. G. and Zemansky, M. W. 1961, *Resonance Radiation and Excited Atoms* (Cambridge: University Press), 106.

Mumma, M. J. and Donn, B. D. 1973, this volume.

Odiorne, T. J.; Brooks, P. R. and Kasper, J. V. V. 1971, *J. Chem. Phys.*, **55**, -1980.

Polanyi, J. C. 1961, *J. Chem. Phys.*, **34**, 347.

————. 1963, *J. Quant. Spectros. Rad. Trans.*, **3**, 471.

————. 1965, *J. Appl. Optics. Suppl.*, **2**, 109.

————. 1971, *Ibid.*, **10**, 1717.

————. 1972, *Accts. Chem. Res.*, **5**, (April).

Polanyi, J. C. and Tardy, D. C. 1969, *J. Chem. Phys.*, **51**, 5717.

Polanyi, J. C. and Woodall, K. B. 1972, *J. Chem. Phys.*, **56**, 1563.

Polanyi, J. C. and Schreiber, J. L. 1973, in *Physical Chemistry, An Advanced Treatise*, ed. H. Eyring, W. Jost, and D. Henderson (New York: Academic Press), **6**, Chapter 9.

Rowland, S. 1972, in *Reaction Kinetics, MTP International Review of Science*, ed. J. C. Polanyi (Oxford: Butterworths), Chapter 4.

Schulz, G. J. 1964, *Phys. Rev.*, **135**, A988.

Shafer, T. P.; Siska, P. E.; Parson, J. M.; Tully, F. P.; Wong, Y. C. and Lee, Y. T.

1970, *J. Chem. Phys.*, **53**, 3305.

Spitzer, L. 1968, *Diffuse Matter in Space* (New York: Wiley Interscience).

Tolman, R. C. 1920, *J. Amer. Chem. Soc.*, **42**, 2506.

Wilson, K. R. 1970, in *Chemistry of the Excited State*, ed. J. N. Pitts, Jr. (New York: Gordon and Breach), 33.

Wolfgang, R. 1969, *Accts. Chem. Res.*, **2**, 248.

Molecule Formation by Inverse Predissociation

P. S. Julienne*
M. Krauss

National Bureau of Standards
Washington, District of Columbia

*NRC-NBS Post-Doctoral Associate 1969-1971

I. INTRODUCTION

Molecular species have long been known to exist in interstellar gas clouds. One of the earliest theories proposed to account for their existence, was that of radiative association (Kramers and ter Haar 1946, Bates 1951), in which a stable molecule is formed upon the collision of two atoms by the emission of a photon. Radiative association was generally disfavored because of the low rate constants estimated and the belief that it was not applicable to observed molecules such as OH. (Dieter and Goss 1966, Herbig 1968, McNally 1968). However, interest has again been shown in this process due to the recent work of Klemperer and Solomon (1969), who estimated rates for CH and CH+ formation significantly larger than earlier calculations. It is usually assumed that in order to have radiative association there must be an attractive potential energy curve arising from the ground state atoms which has oscillator strength to a stable lower molecular state. A continuum-to-bound transition in such a system produces a stable molecule. In addition to this *direct* process, we wish to show that it is also necessary to consider the *indirect* radiative association by means of resonance states embedded in the continuum. This corresponds to inverse predissociation. We have already shown how this leads to radiative association for OH (Julienne, Krauss, and Donn, 1971).

This paper will present the general theory of inverse predissociation and apply it to specific molecules. The list of molecules which can be formed by radiative association will be lengthened considerably by considering this phenomenon. Typical rate constants for diatoms will lie in the range 10^{-16} to 10^{-20} cm^3 sec^{-1} molecule^{-1}, with the heavy atom molecules such as CN, CO, or NO generally having larger rates than the hydrides. The results are summarized in Table I. It is not our purpose to propose a model of molecule production in the interstellar medium. Rather, our goal is to point out what molecular characteristics

TABLE I

Approximate radiative association rate constants in the
temperature range 50 - 100°K

Range of γ cm^3 sec^{-1}		Inverse Predissociation	Direct
large	$10^{-18} - 10^{-16}$	NO, CN, CO, C$_2$	CH$^+$, CH, CO, C$_2$
intermediate	$10^{-20} - 10^{-18}$	OH, CH	
small	≈ 0	NH, CH$^+$	

determine the radiative association rate and also to make estimates of what rates may be expected for various interstellar molecules. It is the task of those who construct models of interstellar clouds to determine under which conditions inverse predissociation may or may not be significant.

Section II will present the formal theory and a discussion of the molecular properties which determine the inverse predissociation rate. Section III will discuss the details of the theory as applied to specific interstellar molecules.

II. THEORY

We wish to outline here in some detail the general quantum mechanical theory of radiative association, noting the various factors which determine the magnitude of the rate constant. We shall be primarily concerned with diatomic molecules, although much of what is said is also applicable to polyatomic species. If $A_J(E)$ dE is the rate at which the two atoms A and B with total energy E between E and E + dE colliding with total angular momentum J associate to form a molecule AB, then the two-body radiative association rate as a function of temperature, assuming a Maxwell-Boltzmann velocity distribution, is

$$\gamma(T) = \frac{1}{Q_{tr}Q_A Q_B} \sum_J (2J + 1) \int_0^\infty e^{-E/kT} A_J(E) \, dE \qquad (1)$$

where k is the Boltzmann's constant, Q_{tr} is the translational partition function $Q_{tr} = (\mu kT/2\pi\hbar^2)^{3/2}$ and Q_A, Q_B the partition functions for the atoms A and B. Under interstellar conditions almost all atoms are in the lowest fine structure level and Q_A and Q_B simply become the degeneracies of the ground atomic levels. In Equation (1), $A_J(E)$ dE may be interpreted as the Einstein spontaneous emission coefficient for a transition from the vibrational continuum of the molecule AB between energy E and E + dE to all possible stable bound levels of AB.

As mentioned in the introduction, radiative association can occur through two quite distinct mechanisms. The direct radiative recombination of two atoms A and B occur when they collide on an attractive excited potential energy curve of the molecule AB which has a finite transition moment connecting it to a lower stable bound state of the molecule. This is illustrated schematically in Figure (1a) and has been discussed for the interstellar molecules CH and CH+ by Bates (1951) and more recently by Klemperer and Solomon (1969). For this case $A_J(E)$ is a relatively smooth function of the energy E and is proportional to the squared dipole matrix element $|\langle\psi_J(E) | M(R) | \psi_f\rangle|^2$, where $\psi_J(E)$ is the energy normalized continuum wavefunction, M(R) the electronic dipole transition moment, and ψ_f the final bound vibrational wave function; the sum must be taken of the contributions from all possible final states. Although direct radiative association is possible for any incident energy, the rate constant γ for this process tends to be small since the time required to emit a photon is several orders of magnitude longer than the time the two molecules are close to each other in a collision. Typical rates are on the order of 10^{-18} - 10^{-17} cm^3 sec^{-1} per molecule. The rate may be significantly enhanced if there are shape resonances at specific E and J values which permit the atoms A and B to be trapped temporarily behind a barrier, either centrifugal or natural. This effect has been considered for CH by Bain and Bardsley (1971). The detailed temperature dependence of the radiative association rate is strongly dependent on the exact location of such shape resonances, especially at low temperature.

We do not wish to discuss direct radiation in detail in this article. Instead, we wish to consider another process, inverse predissociation, which has not previously been discussed in any systematic way in its relation to interstellar molecule production. Just as the shape resonances of a single excited potential curve can lead to enhanced direct radiative association rates, it is also possible that resonance states may exist which result from excitation of molecular excited states different from the state on which the atoms collide. These Feshbach-type resonance states may lead to an appreciable radiative association rate. This is possible even when potential curves for direct radiative association do not exist, as for OH. This process is illustrated schematically in Figure (1b), where atoms incident either on the ground potential V_1 or some upper potential V_3 can populate bound levels of the excited state V_2, which has an allowed transition to V_1 and subsequently radiates to produce a stable molecule.

These Feshbach-type resonance states can occur because the normal quantum numbers used to describe molecular electronic states, for example, Λ and S for diatomic species, are not exact quantum numbers due either to the inherent incorrectness of the Born-Oppenheimer approximation or to the neglect of the spin-orbit interactions. The exact molecular Hamiltonian matrix is not diagonal in the Λ,S representation but contains small off-diagonal matrix elements connecting different molecular electronic states. When the interaction is between a bound excited level and a continuum state whose asymptotic potential energy lies lower than the bound level, this effect can manifest itself in photoabsorption as a broadening of the excited level due to predissociation into the continuum and in atomic collisions as radiative association through inverse predissociation when the incident kinetic energy is very nearly in exact resonance with the bound level. Thus, in Figure (1b), if there is a strong enough coupling between the states V_2 and V_3 (or V_1), the bound levels of V_2, now thought of as resonance states with a finite width, will predissociate into the continuum of V_3

(or V_1), and atoms A and B colliding along V_3 (or V_1) at the appropriate energies will produce stable ground state AB molecules through radiation from the state V_2.

The bound-continuum interaction and the predissociation and inverse predissociation rates may be conveniently calculated by configuration interaction theory (Fano 1961, Mies 1968, 1969). If ϕ_n and ψ_E represent the respective bound and continuum Born-Oppenheimer electronic-vibrational-rotational product wavefunctions and if $V_n = \langle \phi |H| \psi_E \rangle \neq 0$ is the matrix element between them, then the broadening of the spectral line associated with ϕ_n due to the predissociation is given by the width $\Gamma_{np} = 2\pi |V_n|^2$. The line is also broadened due to the natural radiation width Γ_{nr}. An ensemble of molecules prepared initially in the state ϕ_n decays into the continuum channel at the rate $\Gamma_{np}/\hbar$ and emits photons at the rate $\Gamma_{nr}/\hbar$, the total lifetime of ϕ_n being $\tau = \hbar/(\Gamma_{np} + \Gamma_{nr})$.

The average flux per unit energy $A_{nJ}(E)$ for radiative association through the resonance state ϕ_n with total angular momentum J and line center E_n is given by the standard Breit-Wigner formula (Mies 1969, Mott and Massey 1965).

$$A_{nj}(E) = \frac{1}{h} \; \frac{\Gamma_{np} \; \Gamma_{nr}}{(E - E_n)^2 + 1/4(\Gamma_{np} + \Gamma_{nr})^2} \tag{2}$$

When this expression is substituted in Equation (1) and the integrations carried out with the assumption that the resonance widths are very much less than kT, the final expression for the radiative association rate constant is

$$\gamma(T) = \hbar^2 \left(\frac{2\pi}{\mu kT} \right)^{3/2} \frac{1}{Q_A Q_B} \sum_n (2J_n + 1) \; \frac{\Gamma_{nr} \; \Gamma_{np}}{\Gamma_{nt}} \; e^{-E_n/kT} \tag{3}$$

where Γ_{nt} is the total width $\Gamma_{nr} + \Gamma_{np}$; E_n is the energy of the resonance level n above the lowest atomic fine structure asymptotic energy, *i.e.*, E_n is the kinetic energy of the colliding atoms when each is in its ground state. This expression may be simplified in two limiting cases. When the radiation probability is much larger than the predissociation probability, *i.e.*, the case of weak predissociation where $\Gamma_{np} \ll \Gamma_{nr}$, the ratio $\Gamma_{nr}\Gamma_{np}/\Gamma_{nt}$ may be replaced by Γ_{np} alone and the rate γ is determined by the rate at which the resonance level is formed. This is the situation for OH. On the other hand, when the predissociation is strong, $\Gamma_{np} \gg \Gamma_{nr}$, the ratio $\Gamma_{nr} \Gamma_{np}/\Gamma_{nt}$ may be replaced by Γ_{nr} alone and the rate γ is determined by the rate at which the resonance level can radiate. This is the situation for $C^2\pi$ state of NO.

Equation (3) shows that in order to calculate the temperature dependence of the recombination rate constant, we must know the radiative and predissociation rates of each resonance level and the exact location of the resonance levels relative to the ground state dissociation limit. The rate constant at low temperature is especially sensitive to the near-threshold resonance levels which must lie within ~ kT of the dissociation limit in order for the rate to be appreciable. Since k = 0.696 cm^{-1} deg^{-1}, it is clear that at temperatures on the order of 10 - 20 K the E_n values need to be known to an accuracy of a few wavenumbers. Although the term values of the resonance levels are normally known this accurately from spectroscopic data, the error limits on diatomic molecule dissociation energies typically are much greater than kT (Gaydon 1968). For such molecules it is impossible to calculate the rates accurately at low temperature, and one is restricted to estimating upper and lower bounds to the rate constant.

Calculation of Rate

The situation where the predissociation rate greatly exceeds the radiation rate is usually known experimentally, since no emission will be observed from the upper levels in such a case. Thus, it is only necessary to know Γ_{nr} from experimental absorption oscillator strength measurements in order to calculate the rate γ. However, in the opposite case where $\Gamma_{np} < \Gamma_{nr}$, the predissociation widths can not usually be measured well experimentally, and they must be calculated theoretically in order to determine γ. We wish to consider in some detail the various factors which determine the magnitude of the predissociation widths Γ_{np} for a diatomic molecule. Basically, to calculate these widths requires (1) the electronic potential energy curves for the bound and continuum states, (2) the nondiagonal electronic coupling matrix elements, and (3) the bound and continuum radial wavefunctions. We will consider each of these in turn below.

1. The potential energy curves of the upper radiating state and the final bound state are usually known from spectroscopic data, since the two states must be connected by a relatively strong oscillator strength. At the very least this permits the potentials to be approximated by Morse functions, although Rydberg-Klein-Rees (RKR) potentials are often available. However, spectroscopic data only gives the potentials over a restricted range about R_e and requires extrapolation both to the small R repulsive region and to the large R asymptotic limit. Furthermore, the RKR tabulated potentials require interpolation between the tabulated points. The extrapolation to small R can usually be carried out fairly well with an exponential repulsion term, but the potential in the region which lies to the right of the spectroscopic region but where perturbation expressions are not yet valid is usually unknown. This region is of utmost importance for determining if there is a barrier in the incident

continuum channel, either a natural barrier from the adiabatic potential or a centrifugal barrier in the effective potential. A barrier generally tends to significantly reduce the radiative association rate at low temperatures, although it may result in enhanced rates at higher temperatures, at least for the direct radiative association process. It is this intermediate region that may be treated best at the present by accurate correlated *ab initio* quantum mechanical calculations. In addition, *ab initio* calculations can show the behavior of all the states arising from the atomic ground state asymptote, including the states unobservable spectroscopically, and show whether or not such states may be important with regard to inverse predissociation. Thus, data from experiment and from calculation complement each other. Use has been made of both types of information in estimating radiative association rates of individual molecules in the next section.

2. The electronic coupling may be due either to nonadiabatic or to spin-orbit interactions. The detailed forms of the matrix elements for specific coupling situations have been tabulated extensively by Kovacs (1958a,b; 1969). These have also been discussed by Van Vleck (1951) and Kolos and Wolniewicz (1963). When the total spin S is not zero, it is necessary to distinguish between the two Hund's coupling cases a and b (Herzberg 1951). Case a is valid when the spin-orbit splitting is much larger than the rotational constant, and the electron spin is quantized along the rotating molecular internuclear axis, whereas case b applies when the spin-orbit interaction is much smaller than the rotational splitting. In case a the projection Ω of orbital and spin angular momentum on the axis is coupled with the nuclear rotation angular momentum N to form the total angular momentum J. in case b, the projection Λ of electron momentum alone is coupled with N to form a resultant K, which is then coupled with S to form J. Case b is always used for Σ states, whereas either case a or b or an intermediate coupling between these two limits may occur for states with $\Lambda > 0$. The form of the electronic matrix elements V_n determining the predissociation width depends on which Hund's case is relevant.

In order to provide a concrete example, the squared matrix elements for the interaction between a bound $^2\Sigma^+$ state and a $^2\Pi$ continuum are shown in Table II for both case a and b coupling assumed for the $^2\Pi$ state (Kovacs 1969). Only the non-zero couplings are shown. Each $^2\Sigma^+$ vibrational-rotational doublet, designated by vibrational and rotational quantum numbers v and K, is split slightly into two ρ-doublet components (Herzberg, 1951), the f_1 component with J = K + 1/2 and the f_2 component with J = K - 1/2 (except for K = 0 levels, which are single). Each ρ-doublet component is a separate resonance state and must be included in the sum in Equation (3).

The integrals η and ξ in Table I are the vibrational overlap integrals of the respective nonadiabatic and spin-orbit electronic coupling matrix elements, V_{NA}

and $V_{S\,O}$, *i.e.*,

$$\eta = \langle \phi_{vK}(R) \mid V_{NA}(R) \mid \psi_E(R) \rangle \qquad (4a)$$

$$\xi = \langle \phi_{vK}(R) \mid V_{SO}(R) \mid \psi_E(R) \rangle \qquad (4b)$$

where $\phi_{vK}(R)$ is the $^2\Sigma^+$ bound state vibrational wavefunction and $\psi_E(R)$ is the $^2\Pi$ continuum wavefunction normalized on an energy scale to approach $(hv)^{-1/2}\sin(kR + \delta)$ asymptotically, where v is the velocity, k the wavenumber, and δ the phase shaft. The nonadiabatic electronic matrix element V_{NA} is

$$V_{NA} = \frac{\hbar^2}{2\mu R^2} \; \langle \Sigma \mid L_x \mid \Pi_{-1} \rangle \qquad (5)$$

where L_x is the electronic angular momentum operator perpendicular to the intermolecular axis. It can be noted from Table II that the interference between the spin-orbit and nonadiabatic interactions will lead to unequal predissociation widths for the two ρ-doublet components when both interactions are present.

Similar nonadiabatic and spin-orbit matrix elements occur for interactions between states other than $^2\Sigma$ and $^2\Pi$. Although the nonadiabatic mixing will only be large if the two states have the same spin multiplicity, the spin-orbit interaction can mix states differing in total spin by ± 1, for example $^2\Sigma$ and $^4\Pi$ states. As a general rule, the nonadiabatic coupling will increase approximately in proportion to total angular momentum J, whereas the spin-orbit coupling does not vary very strongly with J. The selection rules for predissociation are given by Herzberg (1951) and in much greater detail by Kovacs (1969). The most rigorous selection rules are that both states must have the same total angular momentum J and parity, *i.e.*, $\Delta J = 0$ and both states must be positive or both negative. The total spin S of the two states can differ only by 0 or ± 1. The electronic angular momentum projections Λ along the intermolecular axis can differ only by 0 or ± 1. In Hund's case *b* the selection rule $\Delta K = 0, \pm 1$ is valid, and in Hund's case *a* the spin angular momentum projections Σ on the axis and also the total angular momentum projections Ω can differ only by 0 or ± 1.

It is often possible to make simple estimates of the magnitude of the electronic couplings V_{NA} and V_{SO}. For example, if the molecular orbitals determining the Σ - Π coupling in Equation (5) are approximately $\eta\ell\pi$ and $\eta\ell\sigma$ orbitals, then $\langle \Sigma \mid L_x \mid \Pi_{-1} \rangle \approx \langle \eta\ell\sigma \mid L_x \mid \eta\ell\pi \rangle = 1/2\sqrt{\ell(\ell + 1)}$. To the extent that one or both of these orbitals does not have this limiting behavior, the matrix element will be smaller and the coupling will be less. Similarly the spin-orbit matrix element V_{SO} can in certain cases be related to the spectroscopically determinable spin-orbit coupling constant of one of the molecular electronic states (Kovacs 1958 a,b, 1969). In any case, the matrix elements V_{NA} and V_{SO} may be evaluated directly using *ab initio* wavefunctions if it is not possible to

TABLE II

Matrix elements determining the partial and total predissociation widths for a bound $^2\Sigma^+$ ρ-doublet predissociated by a $^2\Pi$ state for Hund's cases a and b

	Case a		Case b	
	coupling with	$\Gamma_K/2\pi$	coupling with	$\Gamma_K/2\pi$
$^2\Sigma^+_{K=J-1/2}$ (f_1)	$^2\Pi^+_{1/2}$	$4\eta^2K^2 + \xi^2 - 4\eta\xi K$	$^2\Pi^+_K$	$8\eta^2K(K+1) + \dfrac{K}{2(K+1)}\xi^2 - 4\xi\eta K$
	$^2\Pi^-_{3/2}$	$4\eta^2K(K+2)$	$^2\Pi^-_{K+1}$	$\dfrac{K+2}{2(K+1)}\xi^2$
SUM			$8\eta^2K(K+1) + \xi^2 - 4\xi\eta K$	
$^2\Sigma^+_{K=J+1/2}$ (f_2)	$^2\Pi^-_{1/2}$	$4\eta^2(K+1)^2 + \xi^2 + 4\xi\eta(K+1)$	$^2\Pi^+_K$	$8\eta^2K(K+1) + \dfrac{K+1}{2K}\xi^2 + 4\xi\eta\,(K+1)$
	$^2\Pi^+_{3/2}$	$4\eta^2(K-1)(K+1)$	$^2\Pi^-_{K-1}$	$\dfrac{K-1}{2K}\xi^2$
SUM			$8\eta^2K(K+1) + \xi^2 + 4\xi\eta(K+1)$	

estimate their value in any other way. Specific examples will be discussed later.

3. The final requirement to be able to evaluate the predissociation widths is to obtain the bound and continuum vibrational wavefunctions occurring in Equations (4a and b). These can be calculated by numerically integrating the Schrodinger equation for the appropriate potentials. At the present we obtain the single channel numerical solutions by using Gordon's algorithm (1969). Although it is quite straightforward to calculate the bound vibrational wavefunctions, obtaining the correct continuum wavefunctions is not as simple, since to solve this problem properly requires solving the set of coupled differential equations with the appropriate scattering boundary conditions in a space-fixed coordinate system. Such a procedure is undoubtedly necessary to obtain a qualitative and quantitative understanding of radiative association (from the direct process as well as from inverse predissociation), especially at low temperature where the continuum-continuum interactions and the asymptotic atomic fine structure splittings become an important consideration. However, since the predissociation widths are determined by the overlap of the bound and continuum wavefunctions around the small R region near R_e, the single channel continuum wavefunctions on the individual molecular adiabatic potentials should be sufficiently accurate in this region to determine the magnitude of the predissociation widths to be a good approximation. Although it is not always clear when using the single channel solutions whether to use the Hund's case a or b coupling matrix elements to calculate the widths, when the contribution to the rate of association through a given resonance level is summed over the couplings to all possible continuum channels, the differences introduced by this ambiguity disappear if the integrals η and ξ do not depend on the continuum channel. This is shown in Table II.

An additional effect, mentioned by Bates (1951) in relation to CH and CH+, is that the adiabatic molecular potential for the state along which the atoms must collide to give rise to radiative association may not correlate asymptotically with the lowest atomic fine structure states. Although under laboratory conditions the fine structure levels are populated according to their statistical weights and Boltzmann factors, the atoms under interstellar conditions will normally be predominantly in their lowest fine structure levels. When the radiating state correlates adiabatically with a higher fine structure level, a nonadiabatic collision is required for radiative association to occur. In the absence of a full close coupling calculation for the crossing probability, we can assume that collisions occur along each adiabatic potential according to its statistical weight, although this assumption may not be valid in some cases (Dalgarno 1961).

It is quite important to note that for radiative association to occur does not require an actual crossing of a state arising from the lowest asymptote with a bound radiating state, that is, the Franck-Condon principle may be violated. A crossing leads to a very favorable vibrational overlap in the expression Equation (4) for the predissociation width, but the width may still be non-negligible even in the case of no crossing. This is the situation for OH. In such a case the vibrational overlap is quite sensitive to the relative position of the two potential curves. The OH predissociation discussed in the next section provides a good example. An error of 0.005 Å in locating the distance between the repulsive limbs of the $A^2\Sigma^+$ and $X^2\Pi$ potentials causes an order of magnitude error in the predissociation widths.

III. RATES FOR INTERSTELLAR DIATOMIC MOLECULES

OH

We have already reported a radiative association rate for OH of about 10^{-20} cm^3 sec^{-1} at temperatures as low as around 20 K (Julienne, Krauss, and Donn, 1971). The species OH has no attractive potential capable of producing direct radiative association and the three repulsive $^2\Sigma^-$, $^4\Sigma^-$ and $^4\Pi$ states arising from the ground $O^3P + H^2S$ asymptote cross the $A^2\Sigma^+$ state at energies too high to be responsible for inverse predossociation at low temperature (Michels and Harris 1969). The effect we considered is the weak predissociation of the $A^2\Sigma^+$ levels near the dissociation threshold due to nonadiabatic interaction with the $X^2\Pi$ ground state. We have now refined this earlier calculation by including the $A^2\Sigma^+$ - $X^2\Pi$ spin-orbit interaction and by using the appropriate scattering formalism for evaluating the rate constant, Equation (3). The nonadiabatic and spin-orbit integrals η and ξ have comparable magnitudes but opposite signs. Some of our calculated values for these two interaction terms are shown in Table

TABLE III

Calculated coupling matrix elements for the $A^2\Sigma^+ - X^2\Pi$

interaction in OH[a]

		v = 0	v = 1	v = 2
K=1	η		1.7	2.7
	ξ	---	-1.2	-3.0
K=14	η	0.47	1.2	1.8
	ξ	-0.34	-0.8	-2.0

[a] in units of 10^{-5} $(cm^{-1})^{1/2}$

III. According to the discussion in the previous section on the general $^2\Sigma$ - $^2\Pi$ interaction, the f_1 component of the $A^2\Sigma^+$ ρ-doublets will be predissociated at a larger rate than the f_2 components (see Table II) due to the interference term proportional to $\eta\xi$. This effect may provide at least a partial explanation for the differences in predissociation probabilities observed of the f_1 and f_2 levels (Gaydon and Wolfhardt 1951, Naegeli and Palmer 1967), although at somewhat higher energies than considered here the repulsive $^4\Sigma^-$ state most likely has a role also (Gaydon and Kopp 1971). Some of our calculated widths are shown in Table IV. For purposes of comparison, the radiative lifetime of 8.5 x 10^{-7} sec for $A^2\Sigma^+$ v = 0 (Smith 1970) corresponds to a natural radiative width Γ_r of 6 x 10^{-6} cm^{-1}.

The calculation of these widths is rather difficult due to the small vibrational overlap integrals. The electronic nonadiabatic and spin-orbit matrix elements of Equations (4a,b) are large, and may be obtained in a straight-forward way. The molecular orbital configurations in OH are

$$X^2\Pi_i \; 1\sigma^2 \; 2\sigma^2 \; 3\sigma^2 \; 1\pi^3$$

$$A^2\Sigma^+ \; 1\sigma^2 \; 2\sigma^3 \; 3\sigma \; 1\pi^4$$

The nonadiabatic coupling V_{NA} is determined by the integral $\langle 3\sigma \, |L_x| \, 1\pi \rangle$. Since the 3σ and 1π orbitals are approximately $2p\sigma$ and $2p\pi$ orbitals on the oxygen atom, the calculated Hartree-Fock value 0.66 atomic units is close to the

MOLECULE FORMATION

TABLE IV

Predissociation widths for the $A^2\Sigma^+$ state of OH[a]

	v = 0		v = 1		v = 2	
	f_1	f_2	f_1	f_2	f_1	f_2
K=0	--------		0.01	--	0.08	--
1	--------		0.33	0.18	0.98	0.36
2	--------		0.93	0.68	2.6	1.6
3	--------		1.8	1.5	5.0	3.6
$\vdots$						
13	$\sim 10^{-6}$					
14	2.4	2.3	14	13	33	31
	Experimental widths [b]					
K_{eff}	<6		8 ± 3		34 ± 5	

[a] The widths are given in units of 10^{-7} cm^{-1}. The natural radiative width is 60 in these units.

[b] Smith (1970).

limiting value $\sqrt{2}/2 = 0.707$ (where the 1π orbital is the complex orbital with unit projection on the molecular axis). The experimental value for the matrix element has been determined to be 0.677 and constant for the v = 0, 1, 2, 3, and 4 vibrational levels (Clough, Curran, and Thrush 1970). Similarly the spin-orbit coupling V_{SO} is determined by the integral $\langle 3\sigma\ |L\cdot S|\ 1\pi\rangle$ (Kovacs 1969), and V_{SO} is given in terms of the ground state spin-orbit coupling A as $V_{SO} = A/\sqrt{2}$ = 100 cm^{-1}. A detailed consideration of V_{NA} and V_{SO} shows them to have opposite phases.

Although the bound and continuum vibrational wavefunctions can be calculated to very high accuracy for a given potential, the magnitudes of the η and ξ integrals depend very sensitively on the relative shapes of the $A^2\Sigma^+$ and $X^2\Pi$ potentials. Although RKR potentials exist for both states (Fallon, Tobias, and Vanderslice 1961), the difference between the $^2\Pi_{3/2}$ and $^2\Pi_{1/2}$ ground state potentials has not been treated, and furthermore, the RKR potentials

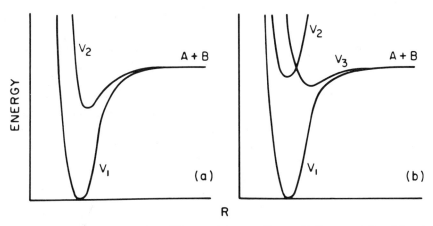

1. Schematic diagram indicating difference between direct radiative association (a) and inverse predissociation (b).

require both numerical extrapolation and interpolation. We constructed various Morse and modified Morse potentials to determine the effect of small changes in the potentials on the calculated predissociation widths. The result of these model studies indicate that our calculated predissociation widths are probably reliable to no more than a factor of two or three. An experimental measurement of predissociation rates in the $A^2\Sigma^+$ state of OH has been made by Smith (1970), who does not resolve rotational lines. Smith's production of excited OH by electron impact on H_2O should result in a rotational distribution peaked at high K values, with an average effective rotational quantum number K_{eff} between 10 and 20 (Carrington 1964). It will be seen from Table III that quite

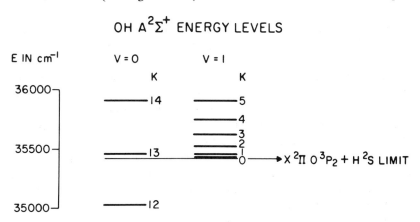

2. Energy levels of the OH $A^2\Sigma^+$ state near the ground state dissociation limit.

reasonable agreement exists between our calculated widths and those measured by Smith.

The OH energy levels of Dieke and Crosswhite (1962) in the region of the dissociation limit are shown in Figure 2, and our calculated two body radiative association rate constant is shown in Figure 3. The contributions to the rate from the v = 1 and v = 0 rotational levels are shown separately, as are the contributions to the v = 1 rate from the f_1 and f_2 p-doublet sublevels. The radiative association through v = 0 occurs only at higher temperatures and through levels with $K \geqslant 14$. Radiative association through the v = 0, K = 13 level near threshold is not possible due to a centrifugal barrier in the ground state effective potential for K = 13. At low temperature the predominant radiative association is through the f_1 sublevels of low K, mainly from v = 1, K = 1. An experimental measurement of the OH radiative association rate at room temperature by Ticktin, Spindler, and Schiff (1967) gave a value of 3×10^{-21} cm^3 sec^{-1}. However, the experiment was done at relatively high pressure and the observed emission was appreciably thermalized, coming mainly from v = 0, K < 13. Even assuming the calculated widths are a factor of two or three too large, there is still an order of magnitude discrepancy between the calculated and experimental rates. This probably results from quenching and other pressure effects in the experimental measurement. One implication of the experiment of Ticktin *et al.* and our calculation is that collisional energy transfer from $A^2\Sigma^+$ v = 1, K = 1 to the nearly degenerate v = 0, K = 13 is highly probable. A similar effect has been noted by Welge, Filseth, and Davenport (1970) for energy transfer between v = 0, K = 20 and v = 1, K = 15, where the energy difference is 27 cm^{-1}.

CH

Bates (1951) calculated a radiative association rate of about 2×10^{-18} cm^3 sec^{-1} at 100 K assuming that a carbon and hydrogen atom collide along the $B^2\Sigma^-$ state arising from the C^3P and H^2S asymptote. Recently Klemperer and Solomon (1969) have reconsidered this process and estimate a rate constant larger than 10^{-17} cm^3 sec^{-1} at low temperature. We wish to consider the rates of CH formation from both the direct and inverse predissociation mechanisms.

The spectroscopic data of Herzberg and Johns (1969) on absorption and emission in CH and CD indicate that there is a barrier in the $B^2\Sigma^-$ state at least 500 cm^{-1} high peaking around 2 Å. A simple model calculation shows even a barrier this small causes a considerable reduction in the radiative association rate at low temperature, due to the decreased amplitude of the continuum function contributing to the Einstein A coefficient in Equation (1). In analogy to the Feshbach resonance states occurring in inverse predissociation, shape resonance states behind the barrier can provide a path for radiative association at low

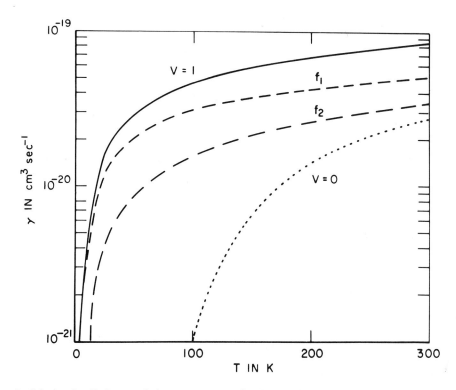

3. Calculated radiative association rate constant for OH as a function of temperature. The contributions of the f_1 and f_2 sublevels for association through $v = 1$ are shown separately by the dashed lines.

temperature if the shape resonance widths are large enough. Bain and Bardsley (1971) have considered the role of such states for the CH $B^2\Sigma^-$ radiative association. The uncertainty in locating the resonances with respect to the ground state dissociation limit restricts them to estimating upper and lower bounds to the rate constant which will most likely be less than 10^{-18} cm^3 sec^{-1} below 20 K.

We have carried out an *ab initio* calculation to verify the deduction of Herzberg and Johns that a barrier of approximately 500 cm^{-1} does exist in the $B^2\Sigma^-$ state of CH. Although a semi-empirical analysis of the correlation energy and an approximation to the Hartree-Fock energy led Liu and Verhaegen (1971) to conclude that there is a barrier of around 900 cm^{-1}, an actual *ab initio* calculation is needed to establish the reality of the barrier. Our calculation, using the techniques pioneered by Das and Wahl (1966) for calculating correlation energies, substantiates the existence of a barrier on the order of 1000 cm^{-1} around 2 Å.

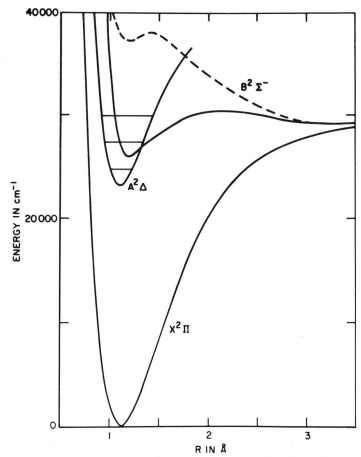

4. Potential energy curves for CH. Morse curves are shown for the $X^2\Pi$ and $A^2\Delta$ states. For the $B^2\Sigma^-$ state the dashed line gives the Hartree-Fock potential and the solid line gives the potential after including the calculated correlation energy.

In addition to the direct radiative association on the $B^2\Sigma^-$ potential, we should investigate whether there are any resonance states which may play a role in CH production. The only candidate state is $A^2\Delta$, which has levels $v = 1$, $K \geqslant 12$ above the dissociation limit (see Figure 4). Since $A^2\Delta$ is $1\sigma^2 2\sigma^2 3\sigma 1\pi^2$ and the ground state is $X^2\Pi$ $1\sigma^2 2\sigma^2 3\sigma^2 1\pi$, there are large nonadiabatic and spin-orbit electronic matrix elements just as in OH. The vibrational overlap between $A^2\Delta$ bound states and the $X^2\Pi$ vibrational continuum states is quite small. Since RKR potentials do not exist for these states of CH, we used the spectroscopic constants to obtain Morse potentials. These lead to an estimated width for $v = 1$, $K = 12$ of 1×10^{-5} cm^{-1}. This number may be considerably in error due to the use of Morse potentials. A width this large would indicate that the higher rotational levels of $A^2\Delta$ vibrational states are predissociated at a rate comparable to the natural radiative decay rate, although the $A^2\Delta$ state has not yet been observed to predissociate (Herzberg and Johns 1969, Hesser and Lutz 1970). Assuming the dissociation limit of Herzberg and Johns (1969) results in an inverse predissociation rate of 1×10^{-17} cm^3 sec^{-1} at 100 K. Due to the uncertainty in the width, this should be understood as an upper limit, with a value an order of magnitude lower being more likely. At low temperature, 20 K and below, the inverse predissociation rate will almost certainly be negligible.

CH^+

The direct radiative association from the upper $^1\Pi$ state to the $X^1\Sigma^+$ ground state has been considered by Bates (1951) and Klemperer and Solomon (1969). There is no barrier in the $^1\Pi$ state and the calculated rate is on the order of 10^{-17} cm^3 sec^{-1} at low temperature. As far as we can tell, there are no resonance states that can contribute through inverse predissociation. The excited $^3\Sigma^-$ state is probably predissociated to some extent by the $^3\Pi$ state arising from the C^+ ^2P + H^2S asymptote, but this $^3\Sigma^-$ state probably lies above the dissociation limit.

NH

There are no states that can contribute to radiative association through either the direct or inverse predissociation mechanisms. The radiative association rate is completely negligible.

NO

Nitric oxide is of some interest since it has a rather large radiative association rate constant but has never been detected in interstellar space, although searches have been made (Turner, Heiles, and Sharlemann 1970). However, this may be related to the abundances of N and O. The excited states of NO have long been studied both spectroscopically and kinetically. The potential energy curves are tabulated by Gilmore (1967). Although there are no states arising from the N ^4S + O ^3P asymptote that give rise to direct radiative association, there are several possibilities for inverse predissociation. The attractive states which arise from this asymptote are the $X^2\Pi$ ground state and a $^4\Pi$ excited state, neither of which is expected to have a barrier. The $A^2\Sigma^+$ state can couple nonadiabatically with the $X^2\Pi$ state, with the Hartree-Fock matrix element $\langle 6\sigma |L_x| 1\pi_{-1} \rangle$ having a value of about 0.2 atomic units. The high rotational levels of v = 3 near the dissociation limit have widths on the order of 10^{-6} cm^{-1}. Another important nonadiabatic coupling is between the $b^4\Sigma^-$ and $a^4\Pi$ states, which should be quite large due to a very favorable vibrational overlap. Almost certainly the predissociation widths of the $b^4\Sigma^-$ levels will exceed their radiation width. The

$a^4\Pi$ state will also be strongly coupled to the $C^2\Pi$ bound levels near threshold due to spin-orbit mixing and in fact predissociates the $C^2\Pi$ levels at a rate 30 times larger than the natural radiative rate (Callear and Pilling 1970).

Inverse predissociation in NO has been studied experimentally by Young and Sharpless (1962). They find emission from $C^2\Pi(v = 0)$ to $A^2\Sigma^+(v = 0)$ and the ground state, from $A^2\Sigma^+(v \leqslant 3)$ to the ground state, and from $b^4\Sigma^-$ to $a^4\Pi$, all of which we would expect if inverse predissociation is occurring. The $C^2\Pi$ emission is the strongest and $C^2\Pi$ to $X^2\Pi$ emission occurs with a rate constant of 2×10^{-17} cm^3 sec^{-1}. The infrared transition $C^2\Pi(v = 0)$ to $A^2\Sigma^+(v = 0)$ occurs at a rate comparable to the ultraviolet $C^2\Pi$ to $X^2\Pi$ emission.

Thus NO has many resonance states contributing to inverse predissociation and probably has an overall rate constant near 10^{-16} cm^3 sec^{-1}. The reduced mass is larger than for hydrides and there is a much higher density of states near threshold. The $C^2\Pi(v = 0)$ rotational sublevels near threshold make the largest contribution to the rate, which will probably remain large to a few degrees Kelvin.

CN

Although we have not carried out a detailed calculation for CN, it is clear that it will have a large radiative association rate, with both direct and inverse predissociation contributions possible. The attractive states $X^2\Sigma^+$, $A^2\Pi$, $^4\Sigma^+$, and $^4\Pi$ arise from the $C^3P + N^4S$ asymptote and the bound states $B^2\Sigma^+$ and $^4\Sigma^-$ lie below the dissociation limit and are crossed by one or more of these curves, (Shaefer and Heil 1971). The electronic $A^2\Pi$ - $B^2\Sigma^+$ nonadiabatic mixing will probably be large with a favorable vibrational overlap also. We would predict a rate between 10^{-16} and 10^{-17} cm^3 sec^{-1} to very low temperature.

CO

The $A^1\Pi$ to $X^1\Sigma^+$ transition, as well as emission within the triplet manifold, can give rise to direct radiative recombination and the $B^1\Sigma^+$, $b^3\Sigma^+$, $C^1\Sigma^+$, and $c^3\Sigma^+$ states near the dissociation limit are known to be predissociated (Krupenie 1966). Again we have not carried out detailed calculations but would expect both direct and inverse predissociation rate constants on the order of 10^{-17} cm^3 sec^{-1}.

C_2

An examination of the potential curves for C_2 (Ballik and Ramsay 1963, Gaydon 1968) shows there are several possibilities for both direct radiative recombination and inverse predissociation. The $A^3\Pi_g$ - $X^3\Pi_u$, $A'\ ^3\Sigma^-_g$ - $X^3\Pi_u$, $b^1\Pi_u$ - $x^1\Sigma^+_g$, and $c^1\Pi_g$ - $b^1\Pi_u$ systems can give rise to direct radiative recombination, if there are no barriers to prevent this. These are moderately strong transitions and a direct radiative association rate on the order of $10^{-17} cm^3 sec^{-1}$ seems quite likely. One strong inverse predissociation possibility is the $B^3\Pi_g$ state, which has an avoided crossing with the $A^3\Pi_g$ state from the atomic asymptote. Another is the $d^1\Sigma^+_u$ state, which has no emission lines above the dissociation limit. An inverse predissociation rate constant on the order of 10^{-17} cm^3 sec^{-1} is expected.

We wish to thank Dr. B. Donn for his interest and encouragement and Dr. D. Neumann for his assistance in carrying out the calculation of the CH barrier. Calculations were carried out at the computing center of the Goddard Space Flight Center, Maryland. This work is supported in part by NASA.

REFERENCES

Bain, R. A. and Bardsley, J. N. 1971 (to be published).

Ballik, E. A. and Ramsay, D. A. 1963, *Ap. J.*, **137**, 61, 84.

Bates, D. R. 1951, *M. N. R. A. S.*, **111**, 303.

Callear, A. B. and Pilling, M. J. 1970, *Trans. Faraday Soc.*, **66**, 1886.

Carrington, T. 1964, *J. Chem. Phys.*, **41**, 2012.

Clough, P. N.; Curran, A. H. and Thrush, B. A. 1970, *Chem. Phys. Letters*, **7**, 86.

Dalgarno, A. 1961, *Proc. Roy. Soc.*, **A262**, 132.

Das, G. and Wahl, A. C. 1966, *J. Chem. Phys.*, **44**, 87.

Dieke, G. H. and Crosswhite, H. M. 1962, *J. Quant. Spectrosc. and Rad. Transf.*, **2**, 97.

Dieter, N. H. and Goss, W. M. 1966, *Rev. Mod. Phys.* **28**, 256.

Fallon, R. J.; Tobias, I. and Vanderslice, J. T. 1961, *J. Chem. Phys.*, **34**, 167.

Fano, U. 1961, *Phys. Rev.*, **124**, 1866.

Gaydon, A. G. 1968, *Dissoication Energies and Spectra of Diatomic Molecules* (London: Chapman and Hall Ltd.).

Gaydon, A. G. and Kopp, I. 1971, *J. Phys. B.*, **4**, 752.

Gaydon, A. G. and Wolfhardt, H. G. 1951, *Proc. Roy. Soc.*, **A208**, 63.

Gilmore, F. R. 1965, *J. Quant. Spectrosc. and Rad. Transf.*, **5**, 369.

Gordon, R. 1969, *J. Chem. Phys.*, **51**, 14.

Herbig, G. H. 1968, *Zs. f. Ap.*, **68**, 243.

Herzberg, G. 1951, *Spectra of Diatomic Molecules* (Princeton: D. van Nostrand Company, Inc).

Herzberg, G. and Johns, J. W. C. 1969, *Ap. J.*, **158**, 399.

Hesser, J. E. and Lutz, B. L. 1970, *Ap. J.*, **159**, 703.

Julienne, P., Krauss, M. and Donn, B. 1971, *Ap. J.*, **170**, 65.

Klemperer, W. and Solomon, P. 1969, in *Highlights in Astronomy* (Dordrecht: Reidel).

Kovacs, I. 1958a, *Can. J. Phys.*, **36**, 309.

————. 1958b, *Can. J. Phys.*, **36**, 329.

————. 1969, *Rotational Structure in the Spectra of Diatomic Molecules* (New York: American Elsevier Publishing Co.).

Kramers, H. A. and ter Haar, D. 1946, *B. A. N.*, **10**, 137.

Krupenie, P. 1966, *The Band Spectrum of Carbon Monoxide*, National Standard Reference Data Series, National Bureau of Standards, 5.

Liu, H. P. D. and Verhaegen, G. 1971, *J. Chem. Phys.*, **53**, 735.

McNally, D. 1968, *Adv. Astr. and Ap.*, **6**, 173.

Mies, F. H. 1968, *Phys. Rev.*, **175**, 164.

————. 1969, *J. Chem. Phys.*, **51**, 787.

Michels, H. H. and Harris, F. E. 1969, *Chem. Phys. Letters*, **3**, 441.

Mott, N. F. and Massey, H. S. W. 1965, *The Theory of Atomic Collisions* (Oxford: Clarendon Press), 408.

Naegeli, D. W. and Palmer, H. B. 1967, *J. Mol. Spectrosc.*, **23**, 44.

Schaefer, H. F. and Heil. T. G. 1971, *J. Chem. Phys.*, **54**, 2573.

Smith, W. H. 1970, *J. Chem. Phys.*, **53**, 792.

Ticktin, S.; Spindler, G. and Schiff, H. I. 1967, *Disc. Faraday Soc.*, **44**, 218.

Turner, B. E.; Heiles, C. E. and Sharleman, E. 1970, *Ap. Letters*, **5**, 197.

van Vleck, J. H. 1951, *Rev. Mod. Phys.*, **23**, 213.

Welge, K. H.; Filseth, S. V. and Davenport, J. 1970, *J. Chem. Phys.*, **53**, 502.

Young, R. A. and Sharpless, R. L. 1962, *Disc. Faraday Soc.*, **33**, 228.

Klemperer: The rate of radiation association of two heavy atoms is slow, even with a radiation association rate constant of 10^{-16} atoms cm^{-3}, compared with radiative association of a hydride, CH^+, followed by chemical exchange.

Solomon: The rate of formation of heavy diatomic molecule [CO, CN] through direct radiative association is three orders of magnitude smaller than formation through the chain

$$C^+ + H \rightarrow CH^+$$
$$CH^+ + O,(N) \rightarrow CO^+, (CN^+) + H$$
$$CO^+,(CN^+) + H \rightarrow CO,(CN) + H$$

or the equivalent chain with CH. The use of the intermediate hydride (CH^+ or CH) increases the rate by the cosmic abundance ration of [H] / [O] .

Molecule Formation on Interstellar Grains

W. D. Watson
E. E. Salpeter

Center for Radiophysics and Space Research
Cornell University
Ithaca, New York

The general requirements for molecule formation on interstellar grains are: (i) a gas particle must stick to a grain long enough for a second gas particle to hit the grain and react with the first particle (10^2 - 10^4 sec) and (ii) the resulting molecule must be ejected from the surface at some time. Both depend upon the nature of the binding to the surface at low temperatures, especially for reactive particles (atoms and radicals) for which there exists little guidance from laboratory experiments. We assume that saturated molecules (*e.g.*, H_2O, CH_4) are physically adsorbed, and that reactive particles are either physically adsorbed or only weakly chemically adsorbed (binding $\leqslant 2$ eV, so that formation of most molecular bonds is exothermic and hydrogen at least can tunnel through typical activation energies). If reactive particles were strongly chemisorbed to the surface, a permanent inert monolayer of chemisorbed hydrogen atoms with some heavy element (C, N, O, *etc.*) atoms would probably cover the surface quickly. On top of this layer the binding (if chemical) would then probably be weak. Atomic hydrogen may be somewhat mobile due to quantum mechanical tunneling (see Hollenbach and Salpeter 1970). If only physical adsorption were possible, the binding would be due to polarizing the bonds of the chemisorbed hydrogen and the adsorption energy should be similar to that for frozen H_2O surfaces (see Hollenbach Salpeter 1970), $\geqslant 0.07$ eV for heavy particles. At expected interstellar grain temperatures (~ 5 - $15\,^\circ$K), this binding is sufficient to hold a gas particle onto the grain until other gas particles stick to the grain and can react with it. In addition, the likelihood of sites on the surface of enhanced physical binding due to heavy atoms of the first monolayer or to surface imperfections (Hollenbach and Salpeter 1971) lends more confidence to the assumption that requirement (i) is satisfied. The physical adsorption energy for atomic hydrogen would be less than for heavy particles and it might not stick long enough for a second particle to hit the grain if there were no enhanced sites. However, we are only interested here in molecules that have a heavy atom, and for these the heavy particles can do the "waiting" for the hydrogen. For this weak binding, hydrogen is highly mobile on the surface (Hollenbach and Salpeter 1970). A heavy atom or radical that sticks to a grain then forms *some* molecule with a probability near unity.

Some of the product molecules are ejected from the surface in the molecule formation process. Steady, thermal evaporation is probably inadequate if requirement (i) is satisfied and we have examined non-thermal ejection due to cosmic rays, X-rays, infrared radiation, as well as to starlight which we find to be the most important under practically all interstellar HI conditions (Watson and Salpeter 1972a). Starlight prevents a physically adsorbed monolayer from forming as long as,

$$\xi = (n\ cm^{-3}/100)\ \exp(2.5\ \tau_v) \leqslant 10^4 \qquad (1)$$

where n is the total hydrogen density (atomic and molecular) and τ_v is optical depth in the visible from the edge of the cloud to the point of interest (this ejection is not rapid enough to influence requirement [i]).

We now present some representative abundances for molecules which are strictly valid only for conditions satisfied by equation (1), typically clouds for which $n \leqslant 10^3$ cm^{-3} and with average observed extinction $\langle \tau_v \rangle \leqslant 5$ (assuming spherical clouds). The interstellar regions that are richest in molecules have an observed extinction greater than $\langle \tau_v \rangle \approx 5$, but our values may still be useful since (a) cloud irregularities can lead to an observed extinction much greater than that appropriate for the reduction of the diffuse radiation field and (b) much of the mass in clouds with $\langle \tau_v \rangle \geqslant 5$ is in the outer parts of the cloud where equation (1) is satisfied.

Molecule formation on grains can be divided into two classes. In one, attachment of hydrogen atoms to heavy particles to form CH, NH, OH, CH$_4$, H$_2$O, etc. is the first step. We do not know what fraction of the heavy atoms that stick to a grain are ejected in unsaturated forms (e.g., OH) and what fractions in saturated form (e.g., H$_2$O), but we expect that neither is negligible (Watson and Salpeter 1972a). For our estimates we assume equal fractions. The result would be about the same if all were ejected in saturated forms since photodissociation would produce the radicals. The CH, NH, OH (or SH, etc.) then undergo exchange reactions in the gas phase to produce CO, CN, CS, etc. in exactly the same manner as would be done if they had been produced by radiative association (Julienne and Krauss 1971, Klemperer and Solomon 1971). The effective rate constant for H-attachment on a grain is $\sim 10^{-17}$ cm^{-3} sec^{-1} (independent of atomic hydrogen abundance n_H for $n_H/n \geqslant 10^{-3}$) which is a factor of 7 smaller than radiative association for formation of CH$^+$ but a factor of 300 greater than has been calculated for OH. However, the actual radiative association rates are dependent on the atomic hydrogen abundance in the gas which begins to be decreased for $\xi \geqslant 1$ (Hollenbach, et al. 1971), so that when $\xi \approx 10$ grain formation certainly dominates in the formation of CH (it can be comparable at $\xi \approx 1$ if the C$^+$ sticking rate is enhanced by a factor of ~ 4 because of a negative charge on the grains). For example, based on formation on grains and destruction by exchange reactions analogous to those for CH (see Watson and Salpeter 1971b) the OH abundance is predicted to be

$$\frac{[OH]}{[H]} \approx \frac{6 \cdot 10^{-7}}{25\,f + 1}$$

where f is the fraction of carbon that is ionized,

$$f = \frac{[C^+]}{[C^+] + [C]} = \frac{[C^+]}{[\langle C \rangle]}$$

For $\xi \leqslant 10^2$, f is $\sim$ 1. This result is in reasonable agreement with observations (*e.g.*, Heiles 1971). Also,

$$\frac{[H_2O]}{[O]} \approx \frac{[H_2CO]}{[CO]} \approx \frac{[NH_3]}{[N]} \approx 10^{-6} \; \xi$$

Exchange reactions produce CN and CO, and photodissociation destroys them; using typical rates we have

$$\frac{[CO]}{[C]} \approx 10 \, \frac{[CN]}{[C]} \approx 10^{-4} \; \xi$$

as long as carbon is ionized. When carbon becomes neutral its ionization edge depletes the radiation field beyond 11.3 eV. Because of the high binding energy of CO, only photons in essentially this energy range can cause dissociation and the CO abundance increases drastically. This seems to be the case only for CO and explains in a natural fashion the high abundance of CO relative to other molecules when $\xi \geqslant 3 \cdot 10^2$ (despite their similar photodissociation cross sections, *e.g.*, Stief 1971).

The second general class of surface reactions is the formation of complex molecules (CH_3OH, *etc.*). The heavy atoms that remain on a grain after reactions are likely to be CH_4, NH_3, H_2O, H_2CO because H-attachments dominate as a result of the larger abundance of H and its greater mobility. Further reactions of these molecules require at least an activation energy and are probably endothermic, so that they cannot occur at the low grain temperatures. However, energy is given to individual particles in this adsorbed layer periodically as a result of absorption of a photon or of chemical reactions. Immediately after a molecular recombination (*e.g.*, $O + H \rightarrow OH^*$) the product has an excess vibrational energy of $\sim$ 2 - 4 ev. If the excited particle can bounce into (saturated) adsorbed particles before the excitation decays (being transmitted to the surface), the excitation energy can be used to stimulate a reaction. The relevant time scales are such that this should be possible in some fraction of the cases, especially when the surface has a large abundance of adsorbed particles (*i.e.*, ξ near 10^4). Unfortunately it is not possible to predict which molecules will be produced by this process, nor what is the efficiency of production (per heavy particle that sticks) which we designate as η. Some examples for dense regions are, assuming a typical photodissociation rate $\sim 10^{-9}$ sec^{-1} for destruction in unshielded regions,

$$OH^* + CH_4 \rightarrow CH_3OH + H$$
$$H_2CO + NH^*_2 \rightarrow HCONH_2 + H$$

$$\frac{[CH_3OH]}{[\langle C \rangle]} \approx 10 \; \frac{[HCONH_2]}{[\langle C \rangle]} \approx 10^{-6} \; \eta \xi$$

Some of these products may remain on the grain to form further generations of yet more complex molecules with a factor of η entering into the abundance for each step. A reasonable estimate for typical values of η seem to be 0.01 - 1.0. Laboratory experiments have shown that complex molecules can be produced from adsorbed layers of simple molecules when they are bombarded with radiation of wavelength 1000 - 4000 Å (Hubbard *et al.*, 1971; Breuer 1971).

Additional aspects of molecule formation on interstellar grains are discussed elsewhere (Watson and Salpeter 1972a,b).

This work was supported by National Science Foundation Grant GP-26068.

REFERENCES

Breuer, H. 1971, this volume.

Heiles C. 1971, *Ann. Rev. Astron. Ap.*, **9**, 293.

Hollenbach, D. J. and Salpeter, E. E. 1970, *J. Chem. Phys.*, **53**, 79.

Hollenbach, D. J. and Salpeter, E. E. 1971, *Ap. J.*, **163**, 155.

Hollenbach, D. J.; Werner, M. and Salpeter, E. E. 1971, *Ap. J.*, **163**, 165.

Hubbard, J. S.; Hardy, J. P. and Horowitz, N. H. 1971, *Proc. Nat. Acad. Sci.*, **68**, 574.

Julienne, P. and Krauss, M. 1971, this volume.

Klemperer, W. and Solomon, P. M. 1971, *Paper presented at the I. A. U. General Assembly* (summer 1970).

Stief, L. 1971, this volume.

Watson, W. D. and Salpeter, E. E. 1972a, *Ap. J.*, **174**, 321.

————. 1972b, *Ibid.*, **175**, 659.

Solomon: The attenuation of radiation at $\lambda < 1100$ Å by H_2 and C is very important in determining the rate of photodestruction of CO. In addition to this, the CO in moderately dense clouds will *shield itself* against photodissociation. A CO column density of 10^{17} cm^{-2} will give an optical depth ~ 1, and many clouds have been observed with $N_{CO} > 10^{19}$ cm^{-2}, indicating no photodestruction.

Photo Chemistry on Solid Surfaces

H. D. Breuer

*Institut für Physikalische Chemie II
der Universität des Saarlandes
Saarbrücken*

The detection of polyatomic molecules in interstellar space raises two questions:

1. by which mechanisms can molecules be formed in such a highly diluted atmosphere?

2. what are the lifetimes in the interstellar radiation field?

Calculations by Stief *et al.* (1972) show that reasonable lifetimes can be expected only in obscured regions. This implies that polyatomic molecules have to be formed in dust clouds where they are observed. From that it can be presumed that dust grains also play a role in the molecule formation. Even if the chemical composition of dust grains is not known it can be understood that the presence of a solid surface is very helpful in molecule formation. By the adsorption of atoms or diatomic molecules, which may be formed by radiative recombination, at the surface of the grains the collision probability is greatly enhanced with respect to the gas phase. The dust particles can absorb reaction energies which in the gas phase would lead to dissociation during the first period of vibration.

Adsorption of reactive gases initially occurs in a weakly bound precursor state. A molecule or atom in this precursor state normally has three courses open to it:

1. it can transfer to the chemisorbed state and thus be held permanently,

2. it can migrate over the surface in search for a vacant site for chemisorption, even at low temperatures,

3. it can desorb.

With interstellar dust grains we have some more possibilities: As we have a mixture of various gases adsorbed at the surface the atoms of molecules can hit other adsorbed particles and form new molecules or radicals which in turn have the same possibilities. We also have to bear in mind that the dust grains are exhibited to the interstellar radiation field. The interaction of photons with adsorbed particles leads to a new kind of photochemistry which is rather different from normal photochemistry in the gas phase. Due to the adsorption bond the spectra of adsorbed atoms and molecules are changed in a way that less energy is necessary to reach an excited state or a dissociation limit. Radicals formed in a photolytical process can be stabilized at the surface and can be involved in reactions with other adsorbed particles.

In our experiments on these photocatalytical reactions we used clean metal surfaces as substrates. The choice of metals as a model surface may be as good or as bad as any other material since very little is known about the chemical composition of interstellar dust grains. Using very clean metals and working under ultra high vacuum conditions has the advantage of minimizing uncontrolled contaminations.

In the first experiments we measured the wavelength dependence of

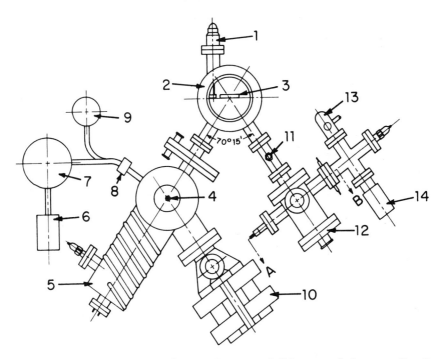

1. Experimental apparatus. *1 to 3:* monochromator; *4:* lightsource; *5, 6:* pumps; *7 to 9:* lightsource accessories; *10:* pump; *11:* valve; *12:* reaction chamber; *13:* gas inlet valve; *14:* pump.

photoreactions for some gases adsorbed on different metal surfaces (Breuer 1969, Breuer and Moesta 1972). The change in the work function was used as an indication of changes in the chemical composition of the adsorption layer. The experimental set up for these measurements is shown in Figure 1. As an example for the changes in the adsorption layer Figure 2 shows the wavelength dependence of the photoreactions of CO adsorbed on a nickel surface. The peaks at 2000 Å and 1400 Å show the occurence of chemical changes in the adsorption layer.

For 2000 Å radiation gaseous CO is completely transparent. In these experiments no identification of the reaction products could be performed. The total reaction cross-sections for both peaks, however, could be estimated from intensity dependence measurements to be in the order of 10^{-16} cm^2.

In order to get some information about the chemical constitution of the reaction products we performed another experiment which is shown in Figure 3. Because of the big energy shift in this experiment we used a Hg-resonance lamp (Pen Ray) (Moesta and Trappen 1970).

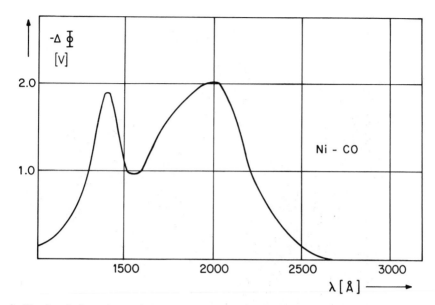

2. Wavelength dependence of photoreactions of CO adsorbed upon a nickel surface.

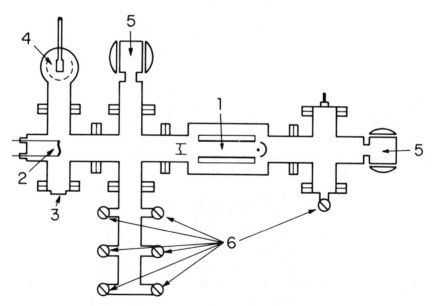

3. Experimental configuration. *1:* quadrupole mass filter; *2:* W-surface; *3:* window; *4:* LHe pump; *5:* ion-getter pumps; *5:* gas inlet.

The products were identified in a very sensitive mass spectrometer. For CO adsorbed on W we found at λ = 2537 Å as photo-products: C_2O, CO_2, C_2O_2, and C_3O_2.

For the reaction mechanism we believe that in the primary step CO molecules are brought to an excited state by the radiation. By interaction with the surface these excited molecules can dissociate and highly reactive carbon and oxygen atoms remain at the surface. These atoms can react with other atoms and undissociated molecules to form the observed products.

If the surface is partially covered with atomic hydrogen or hydrocarbon radicals there should be a large increase in the number of observed photo products. For that reason we irradiated coadsorbed CO and CH_4. CH_4 is known to adsorb dissociatively as H-atoms and CH_3-radicals (Rye and Hansen 1969; Wright *et al.*, 1958). The resulting mass spectrum is shown in Figure 4. In this spectrum as well as in the other spectra shown here only the photo-products are plotted, *i.e.* the "dark" spectra are substracted. An analysis of this spectrum shows fragmentation peaks which are typical for the presence of aldehydes. To make sure of this reaction path we performed the experiments under the same condition with CD_4 (95 per cent) instead of CH_4. The result is shown in Figures 5a and 5b. In 5a H_2CO occurs at m/e = 30 and HCO as a fragmentation product in the ion source of the mass spectrometer at m/e = 29. Using CD_4 we find D_2CO at m/e = 32 and DCO at m/e = 30. The peaks at m/e = 31 and m/e = 29 result from the hydrogen content in the CD_4. Higher aldehydes presumably are formed as well but they cannot be identified only by low resolution mass spectrometric studies.

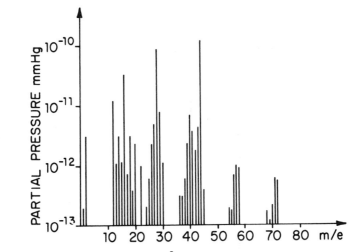

4. Mass spectrum of W - CO + CH_4 at 2537Å

Formaldehyde is known to be an important starting material for the synthesis of different organic compounds. Aldehydes can polymerize under uv irradiation to sugars and sugar-like compounds (Young *et al.*, 1965). In addition aldehydes are reactive with nitrogeneous compounds to form amino acids or their precursors (Miller 1955, Lemmon 1970). The simultaneous adsorption of CO, H_2 and N_2 leads to the formation of the following photo products given in Table I.

Table I
Photoproducts of CO, H_2, and N_2 Adsorption

M/e	Product
25	C_2H
26	CN
27	HCN
29	N_2H
30	NO, N_2H_2
38	C_2N
39	C_2NH
40	C_2H_2N
41	CH_3CN
43	HNCO
44	N_2O
45	$HCONH_2$
46	NO_2
51	HC_3N

For some of the photo products we estimated the cross-section for formation and desorption from the surface. The results are presented in Table II.

Table II
Estimates of Cross-Sections

Compound	$\sigma(\text{cm}^2)$
CN	1.5×10^{-18}
HCN	5×10^{-19}
H_2CO	1×10^{-16}
HC_3N	5×10^{-19}
HNCO	1×10^{-18}
CH_3CN	1.5×10^{-18}
$HCONH_2$	1×10^{-18}

For comparison the cross-section for photo desorption of CO without molecule formation is in the order of 10^{-21} cm^2 (Menzel 1972). Our experiments show that irradiation of adsorbed atoms and molecules leads to a new kind of photo chemistry which is quite effective in producing rather complex organic molecules in a wavelength region in which photo chemistry in the gas phase does not proceed. In accordance with our results are experiments by Hubbard et al., (1972). They found by irradiating CO and water vapor adsorbed on soil or pulverized vycor substratum the formation of CO_2 and of C-organic compounds at wavelengths up to 3000 Å. These experiments and the results of Greenberg (1971) who by irradiating condensed mixtures of gases on a brass plate at 20 °K found rather complex mass spectra, show that the chemical composition of the surface plays a smaller role than the presence of a solid surface itself.

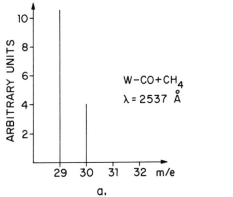 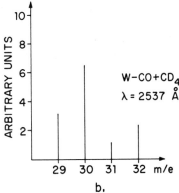

5. Mass spectra.

The importance of light for the formation of molecules on solid surfaces may be emphasized by the negative results in an experiment reported by Yates and Madey (1971). In flash desorption studies of CO and H_2 adsorbed on W without irradiation their search for CH_4, HCO, H_2CO and C_2H_6 was negative. This result was independent of the order of addition of H_2 and CO to the W-surface. The authors, however, observed a gradual decrease in hydrogen desorption activation energy at 100 °K by the interaction of CO with H(ads) at full coverage.

In conclusion, if molecule formation takes place at the surface of interstellar dust grains this "surface photochemistry" certainly cannot be neglected in comparision to other surface reactions such as exchange reactions (Stecher et al., 1966) or Fischer-Tropsch-type reactions.

This work was sponsored by the Deutsche Forschungsgemeinschaft, which gratefully is acknowledged.

REFERENCES

Breuer, H. D. 1969, doctoral dissertation, University of Bonn.

Breuer, H. D. and Moesta, H. 1972, in press.

Greenberg, J. M. 1971, private communication.

Hubbard, J. S.; Hardy, J. P. and Horowitz, N. H. 1972, in press.

Lemmon, R. M. 1970, *Chem. Rev.*, **70**, 95.

Menzel, D. 1972, *Ber Bunsenges. Phys. Chem.*, in press.

Miller, S. L. 1955, *J. Am. Chem. Soc.*, **77**, 2351.

Moesta, H. and Trappen, N. 1970, *Naturwissenschaften*, **57**, 38.

Rye, R. R. and Hansen, R. S. 1969, *J. Chem. Phys.*, **50**, 3585.

Stecher, T. P. and Williams, D. A. 1966, *Ap. J.*, **146**, 88.

Stief, L. J.; Donn, B.; Glicker, S.; Gentieu, E. P. and Mentall, J. E. 1972, *Ap. J.*, **171**, 21.

Wright, P.G.; Ashmore, P. G. and Kemball, C. 1958, *Trans. Farad. Soc.*, **54**, 1692.

Yates, J. T. and Madey, T. E. 1971, *J. Chem. Phys.*, **54**, 4969.

Young, R. N.; Ponnamperuma, C. and McCaw, B. K. 1965, in *Life Science and Space Research*, ed. M. Florkin (Amsterdam: North-Holland).

Gas-Solid Interactions: Laboratory Paradigms
For Reactions on Interstellar Grains

Ralph Klein

National Bureau of Standards
Washington, District of Columbia

I. INTRODUCTION

A number of atomic and molecular species, about twenty, have been positively identified in interstellar clouds. The presence of solid grains of undetermined composition has long been recognized. The problems of chemical dynamics in these clouds, particularly those of molecule formation, are intriguing and have prompted much speculation. Because of the low gas phase concentration in the H I regions, reactions in the gas phase seem to be an unlikely synthetic route. Three body collisions are required, but the probability of such processes are so low that they cannot be invoked. Molecular formation in more favorable regions, that is where pressure and temperature are higher, followed by transport over vast distances also has severe limitations as an acceptable concept. An alternative speculation is that of surface reactions on interstellar grains. Atomic species such as H and O have been observed in the interstellar medium. The temperature is low, below 150 $^\circ$K for the gas and below 20 $^\circ$K for the grains. Hydrocarbons (at least in the form of CH) are present in the gas phase. These are the facts from which appropriately designed laboratory experiments may suggest whether or not it is reasonable to expect that larger molecules may be synthesized on grain surfaces by way of chemical kinetic processes.

Gas-condensed phase reactions have been widely investigated. These include combustion, halogenation, decomposition, and catalytic processes. For the most part, however, high activation energies are involved and relatively high temperatures are required. Reactions that can occur in the cryogenic temperature region, that is below about 150 $^\circ$K at conveniently observable rates, must obviously be characterized by low activation energies, about 2½ kcal/mol or less. In general, only reactions in which at least one of the reactants is an atom, free radical, or ion conform to this requirement. It is convenient to prepare the atomic species in the gas phase, with transport of the atoms to a reactive surface. Within the past decade and a half, considerable progress has been made in the study of gas-condensed phase reactions in the temperature region below 150 $^\circ$K (Klein and Scheer 1958, 1962, 1970; Klein et al., 1960; Scheer and Klein 1961). The techniques developed and the experimental observations made furnish laboratory models for the consideration of a class of reactions, those involving atom-molecule reactions, on interstellar particulates. A brief description of the laboratory reaction system and the characteristics of the pertinent reactions will be given. An assessment of feasibility of these types of chemical interactions on grain surfaces under interstellar space conditions follows.

II. EXPERIMENTAL

The reaction vessel is a simple conical flat bottomed flask with a side tube appendage to accomodate a tungsten or rhenium filament. It is illustrated in Figure 1. In operation, the reactant is condensed on the bottom of the flask, the whole flask immersed in the refrigerant (liquid nitrogen or liquid oxygen are commonly used), and hydrogen or oxygen gas introduced to a pressure of a few hundred microns of mercury. Atomic species are produced by thermal dissociation on the heated filament. They become thermally equilibrated by collision with the vessel walls in the course of diffusion to the surface of the condensed reactant. After sufficient reaction has occurred, the excess gas is removed. The condensed layer is then vaporized and analyzed by gas chromatography. It has been shown that all reactions are completed at the bath temperature with no further reactions on warmup (Klein *et al.*, 1960).

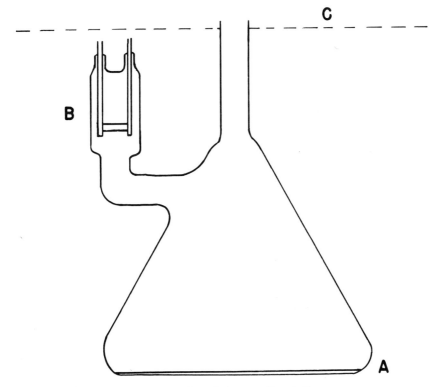

1. Vessel for low temperature gas-condensed phase reaction studies.
   A. Condensed phase; B. Thermal dissociator, for atomic hydrogen or oxygen production; C. Refrigerant level.

Propylene serves as a convenient reactant to illustrate both H atom and O atom addition at low temperatures. Propane is useful as a diluent since it is found to be inert to either atomic hydrogen or oxygen. The chemical reactions are simple and uncomplicated by any carbon-carbon bond scission. The atomic hydrogen addition will be considered first. The initial step is the addition of H to either of the two carbons, terminal and non-terminal, of the double bond. The n-propyl or s-propyl radical results. The addition is primarily terminal with about 99.5 per cent of the alkyl formation being s-propyl at 90 °K. Further reactions, combination and disproportionation, of the alkyl radicals ensue. That is,

$$CH_3\text{-}CH = CH_2 + H \rightarrow CH_3\text{-}\overset{\text{o}}{C}H\text{-}CH_3 \tag{1}$$

$$2 \ CH_3\text{-}\overset{\text{o}}{C}H\text{-}CH_3 \begin{cases} \rightarrow \ \begin{matrix} CH_3-CH-CH-CH_3 \\ \quad \ \ \ \diagdown \quad \ \ \diagdown \\ \quad \ \ \ CH_3 \ \ \ CH_3 \end{matrix} & (2a) \\ \\ \rightarrow \ CH_3-CH=CH_2 + CH_3-CH_2-CH_3 & (2b) \end{cases}$$

III. MECHANISM AND DISCUSSION

The mechanism of interaction in this two-phase system (gas-liquid or gas-solid) needs to be established both for the detailed interpretation of the sequence of chemical reactions and the implication it may have for possible molecular hydrogen formation on interstellar grains. It is certain that atomic hydrogen in the interior of a liquid or even a solid phase above a few degrees absolute is characterized by rapid diffusion. In the two phase system described here, the question that arises is whether the reaction with atomic hydrogen is only on the surface, or if there is a transport mechanism across the surface resulting in a significant concentration of atomic hydrogen in the condensed phase. If H does not cross the interface, alkyl radicals form at the surface only, diffuse into the interior and react by combination and disproportionation according to (2). In either case, allowance must be made for additional reactions, the H atom recombination to molecular hydrogen, and the disproportionation-combination reactions between atomic hydrogen and the alkyl radical.

$$H + H \rightarrow H_2 \tag{3}$$

$$CH_3-CH-CH_3 \quad + \quad H \overset{o}{<}\begin{array}{l} \longrightarrow CH_3CH_2CH_3 \qquad\qquad (4a) \\ \\ \longrightarrow CH_3CH=CH_2 \quad + \quad H_2 \qquad (4b) \end{array}$$

Molecular hydrogen is produced in (3) and (4). Proof that (4) occurs may be obtained by increasing the viscosity of the condensed phase by the addition of 3 methyl pentane. The decrease of the diffusion of the hydrocarbon molecules favors (4) over (2) and hence the ratio of 2,3 dimethyl butane to propane should increase. In fact, the ratio becomes increasingly large with addition of 3 methyl pentane, Table 1, in addition to the overall rate becoming considerably slower. To ascertain whether the H atom addition reaction occurs at the surface or in the interior of the condensed phase requires additional observations. Linearity of

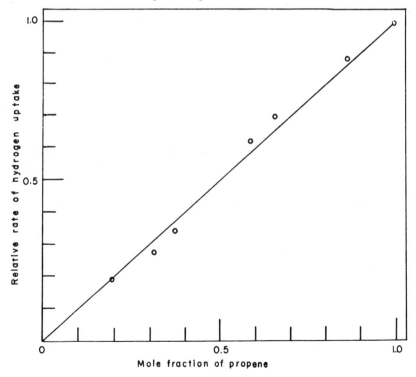

2. Relative reaction rates at 77 °K for hydrogen atom additon to propene as a function of concentration. Propane is the diluent.

rate with dilution by propane is evidenced in Figure 2. The ratio of alkyl dimer to alkane, that is 2,3 dimethyl butane to propane formed from propene is independent of gas phase H atom concentration at least over two orders of magnitude. A hysteresis effect is found in the addition reaction rate which becomes more pronounced as the diffusional processes in the condensed phase are depressed. This hysteresis is evidenced by an increase in rate at the commencement of the H atom addition following an interruption (Klein and Scheer 1962). The weight of evidence in favor of a surface reaction of atomic hydrogen with an olefinic molecule and of the non-penetration of the condensed layer by H leaves little doubt that this mechanism is correct. The detailed arguments are fairly direct and have been given elsewhere (Klein and Scheer 1962). Finally, very convincing experimental support of the model of non-penetration of the surface of condensed hydrocarbons by atomic hydrogen is furnished by calorimetric measurements where pure propane and propane films containing olefins such as propene, were exposed to H atoms (Lomanov *et al.*, 1962). The heat release measured for the former was less than 5 per cent of the latter. If H atom penetration of the condensed phase with concomitant recombination had occurred, the heat release for the pure propane layer should have been comparable to that of the olefin layer since $H + H \rightarrow H_2$ $\Delta H = -103$ kcal and $H + H + C_3H_6 \rightarrow C_3H_8$ $\Delta H = -130$ kcal. A system in which H does not permeate the condensed phase cannot form molecular hydrogen via (3) since the surface concentration of atomic hydrogen is extremely small. H_2 formation may

TABLE 1

PROPANE/2,3-DIMETHYLBUTANE PRODUCT RATIOS RESULTING FROM THE REACTION OF PROPENE WITH ATOMIC HYDROGEN AT 77 °K IN A MIXTURE OF PROPENE AND 3 METHYLPENTANE

Mole Percentage 3-methylpentane	Propane ÷ 2,3-dimethylbutane
0	7
13	12
18	20
22	45

occur by (4). However, the rate is proportional to the surface concentration of alkyl radicals. Other conditions being maintained constant, the relationship between surface radical concentration R_o and the diffusion coefficient, D, is (Klein *et al.,* 1964)

$$D = \frac{(C - C'R_o)^2}{R_o^3}$$ where C and C' are constants.

The rate of production of molecular hydrogen is evidently enhanced by lowering of the temperature or adding substances forming the glassy state in the reaction system, either of these causing a decrease in the diffusion coefficient. The reaction type for molecular hydrogen formation alluded to here is circumscribed because in the process a part of the olefin is converted to the inactive alkane. Experiments with cis-2-butene have shown that the ratio of occurrence of type reaction 4b to 4a at 90 °K is 1.9. Even if the interstellar grains consisted completely of frozen olefins, their capacity to form molecular hydrogen from atomic hydrogen according to equations (4) would be limited.

Oxygen atom addition to condensed olefins at temperatures below 150 °K furnish an interesting contrast to that of H atom addition. The experimental procedures are the same as for the hydrogen-olefin systems. The overall chemistry of the process is given in equations (5) through (7) (Hughes *et al.,* 1966).

The R's represent either H or alkyl groups. If the R groups are all hydrogen, corresponding to the ethylene molecule, reaction (6) and (7) represents the formation of acetaldehyde, CH_3CHO, one of the molecules identified in interstellar clouds. This is not to imply that interstellar CH_3CHO is formed by this route.

The rate of addition at 90 °K is independent of olefin concentration, the O atom supply to the surface being the limiting process (Klein and Scheer 1968). There is rapid transport of oxygen atoms across the gas-condensed phase boundary. An essential difference between the O atom and H atom reactions is that in the latter, stable free radicals are formed. The intermediate from the O atom-olefin combination is very short lived, less than 10^{-9} sec, and aside from the initial addition reaction, no bimolecular processes occur. In the absence of a reactive olefin, the condensed phase being exclusively a condensed alkane for example, a recombination to produce molecular oxygen would proceed readily. At lower temperatures (below 77 °K) the diffusion processes in the condensed phase are depressed and oxygen atom transport across the phase boundary becomes small compared to that at 90 °K. Therefore, even in the oxygen atom-hydrocarbon systems, diffusion plays a significant rate determining role.

Returning to the H atom-olefin system, laboratory experiments have shown that for a layer of pure propylene at 77 °K, approximately 1 out of every 10^4 collisions of hydrogen atoms with the surface is effective in reaction. The activation energy for the reaction has been determined to be 1.3 kcal/mole. An extrapolation to 17 °1 indicates that only 1 out of every 10^{19} collisions would be effective, neglecting the fact that the gas temperature is probably 100 °K. With approximately 10^5 collisions/second, about 10^7 years would be required for one effective collision per square centimeter of surface. The conclusion is that in an H I region, non-catalytic chemical reactions on grains involving atom-molecule, or indeed free radical-molecule type reactions, are most unlikely as a mechanism for molecule production.

The role of the surface of the interstellar grain as a catalyst must be regarded as virtually non-existent. Catalysis is a rather complex process in which adsorption, desorption, and surface migration play a role in surface reactivity. In HI even if the grain temperature were high enough to supply the requisite energy for overcoming the activation barrier, the pressure is too low to maintain the surface concentration required for reasonable rates.

Caution is urged, therefore, in the uncritical attribution to surface reactions as a catch-all for synthetic routes to molecule formation in the interstellar HI regions. Because conditions of concentration and temperature are so unfavorable to gas phase reactions, it is tempting to invoke surface reactions, but it must be emphasized that formidable reservations apply to surface reactivity as a route to molecular syntheses.

This work was supported in part by N. A. S. A.

[1] The upper limit estimate of grain temperature in the HI region.

REFERENCES

Klein, R. and Scheer, M. D. 1958, *J. Am. Chem. Soc.*, **80**, 1007.

Klein, R.; Scheer, M. D. and Waller, J. G. 1960, *J. Phys. Chem.*, **64**, 1247.

Scheer, M. D. and Klein, R. 1961, *J. Phys. Chem.*, **65**, 375.

Klein, R. and Scheer, M. D. 1962, *J. Phys. Chem.*, **66**, 2677.

Klein, R. and Scheer, M. D. 1970, *J. Phys. Chem.*, **74**, 613.

Lomanov, Yu. P.; Ponomarev, A. N. and Talrose, B. L. 1962, *Kinetika i Kataliz,* **3**, 49.

Klein, R.; Scheer, M. D. and Kelley, R. 1964, *J. Phys. Chem.*, **68**, 598.

Hughes, A. N.; Scheer, M. D. and Klein, R. 1966, *J. Phys. Chem.*, **70**, 797.

Klein, R. and Scheer, M. D. 1968, *J. Phys. Chem.*, **72**, 616.

Klemperer: I do not think that it is possible to extrapolate the rate of a chemical reaction far beyond the temperature range of the experiment. It is not obvious that a sharply defined activation energy exists for all chemical reactions. Also, I think that the abundant elements Fe and Ni are quite general catalysts. Radiation-damaged sites are effective catalysts, and these are expected in grain surfaces.

Salpeter: I agree that surface effects are too complicated to predict exactly which molecular reactions go on, but it is much safer to say that *some* molecule is formed when a bare atom or radical hits a grain surface. At least e^{40} vibration periods elapse before an atom evaporates, so the atom has e^{40} chances to initiate a reaction.

Experimental Interstellar Organic Chemistry: Preliminary Findings

B. N. Khare
Carl Sagan

Laboratory for Planetary Studies
Cornell University
Ithaca, New York

In experimental simulations of Jovian atmospheric chemistry, a mixture of the cosmically most abundant fully saturated hydrides – H_2, CH_4, NH_3, H_2O – was subjected to electrical discharge; the primary gas phase products produced were the hydrocarbons, acetylene, ethylene and ethane; the nitriles, hydrogen cyanide, and acetonitrile; and the aldehydes, formaldehyde and acetaldehyde (Sagan and Miller 1960). Computer experiments on quenched thermodynamic equilibrium in similar mixtures of gases produced just this same array of gases as principal products, along with CO (Sagan et $al.,$ 1967; Lippincott et $al.,$ 1967). The discovery of HCN and HCHO in the interstellar medium prompted the suggestion (Sagan 1970, Sagan and Khare 1970) that such experiments may be of relevance to the emerging field of interstellar organic chemistry. The recent announcement of the probable detection of CH_3CN in the interstellar medium (Solomon, Jefferts, Penzias, and Wilson 1971) makes this suggestion more attractive, and has led us to attempt an explicit experimental simulation of interstellar organic chemistry.

An accurate laboratory simulation of average interstellar conditions is impractical. To employ simultaneously nontrivial UV optical depths as well as mean interstellar densities, laboratories of interstellar dimensions are required. It seems clear that to synthesize polyatomic organic molecules three-body reactions must be invoked, reactions which are unlikely in the extreme under average interstellar conditions. Typical three-body reaction rate constants are ~ 10^{-32} cm^6 sec^{-1}, implying reactant densities $> 10^8$ cm^{-3} for reaction rates to be competitive with typical interstellar UV photodissociation rates (Sagan 1971). It follows that interstellar organic chemistry does not occur in gas phase under mean interstellar conditions. At least two alternatives may be suggested ($cf.$ Sagan 1971): in dense clouds with the product molecules perhaps subsequently ejected by radiation and proton pressure after nearby star formation; and in the close vicinity of interstellar grains. Making a virtue of a necessity, we have not hesitated to employ laboratory high vacuums in interstellar simulation experiments.

The experiment is a low-temperature high-vacuum UV irradiation of condensed simple gases known or suspected to be present in the interstellar medium. Because of the discovery of 10μ silicate reststrahlen bands (Stein and Gillett 1969; Hackwell et $al.,$ 1970) and the microwave line identification of SiO (Wilson et $al.,$ 1971) the most appropriate matrix for the condensed gases seems to be a silicate glass. To perform organic chemistry, precursor gases containing H, C, N, and O are required. The molecules chosen were water, ammonia, formaldehyde and ethane. The first three are obvious choices because of their high microwave abundances. Simple hydrocarbons are not accessible to microwave line techniques, although there is every reason to expect their abundant presence in the interstellar medium. Ethane rather than methane was chosen, for experimental convenience and because our experience indicates that in similar experiments it is a useful source of methyl radicals. Ethane is the principal hydrocarbon product from the irradiation of methane under usual laboratory conditions.

The bulk of the interstellar electromagnetic energy density lies longward of 2000 Å; yet most of the identified and suspected interstellar molecules are transparent at these wavelengths. Formaldehyde is an important exception, exhibiting a characteristic absorption band between 2300 and 3400 Å, probably arising from the transition to an anti-bonding π orbital of the carbonyl group of a non-bonding 2p electron of the oxygen atom. The H-CHO bond has an energy of about 3.8 eV. Thus irradiation at, say, 2537 Å will produce "hot" hydrogen atoms, superthermal by about 1.1 eV. (By momentum conservation almost all of the excess energy is acquired by the H atom.) Comparable hot H atoms from the photodecomposition of H_2S are known to initiate chain reactions leading to quite complex organic molecules (Sagan and Khare 1971a, 1971b).

At room temperature the HCHO photolytic products are wavelength-dependent:

$$HCHO + h\nu \rightarrow H + CHO, \text{long UV} \quad \text{(I)}$$
$$HCHO + h\nu \rightarrow H_2 + CO, \text{short UV} \quad \text{(II)}$$

Pathway II is a few times more probable at $\lambda \approx 2600$ Å than is pathway I (McQuigg and Calvert 1969). Although the photolytic pathway may be a sensitive function of the vibrationally excited singlet or triplet state formed, these room temperature results were the only guides available on wavelength choice. A 200-watt Hanovia quartz high pressure mercury vapor lamp, emitting 1.1 watt in the 2537 Å Hg resonance line was employed. The principal Hg line at shorter wavelengths, 1849 Å, has an intensity in the cold finger which we estimate as $< 2 \times 10^{-3}$ that of 2537 Å; the 1942 Å line is $< 3 \times 10^{-4}$ the intensity of 2537 Å at the reaction site.

The experimental apparatus is displayed schematically in Figure 1. The gases are frozen at a conveniently low temperature — here 77 K — on a UV-transparent spectrosil quartz cold finger in a high-vacuum system. Successive layers of formaldehyde, water, ethane, and ammonia were deposited on the cold finger. Initial gases of high purity were used — as, *e.g.,* Matheson research grade ethane, containing a maximum of 60 ppm CH_4. The amounts of material introduced were H_2O, 10.6 millimoles, ethane, 30.7 millimoles and ammonia, 30.8 millimoles. Paraformaldehyde was heated to give off 76.7 millimoles of formaldehyde. It was first condensed at liquid nitrogen temperatures in a separate trap and later evaporated to condense on the cold finger for photolysis. Water, C_2H_6, and NH_3 were then successively condensed on the cold finger. HCHO and NH_3 were separated to inhibit the formation of

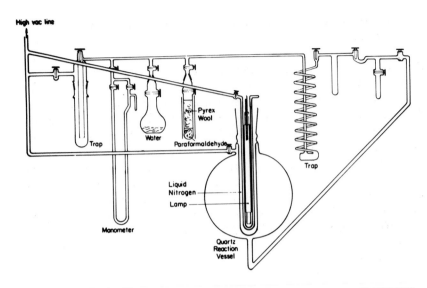

SYSTEMATIC SKETCH OF THE PHOTOLYSIS APPARATUS FOR LOW TEMPERATURE

1. Schematic diagram of the experimental apparatus. Formaldehyde, water, ethane, and
   ammonia from condensed phases, are introduced into the liquid-N_2 cooled cold finger
   within the reaction vessel. Frosts deposited on the quartz finger are UV-irradiated in
   cylindrical geometry from within, under high vacuum.

hexamethylenetetramine — a molecule which is nevertheless of interest in
interstellar organic chemistry and which is referred to below. Access of vacuum
system Hg to the reaction vessel was prevented by a spiral trap maintained at -78
° C. After deposition and pumping at a pressure ≤ 10⁻⁵ torr with a high
performance mercury diffusion pump, the layered frosts were irradiated for
three hours. During these three hours of irradiation no pumping was carried out.

At the end of the experiment roughly 30 cm³ of non-condensible gases
produced during the photolysis were detected manometrically, and analyzed
with a Perkin-Elmer Model 621 double-beam spectrophotometer, revealing the
presence of CH_4 and CO. The noncondensible gases were also examined with an
AEI MS 902 double-focussing mass spectrometer, which confirmed the presence
of methane and carbon monoxide, and also revealed the production of H_2.
Experimentally matching the strengths of the infrared bands, we find
approximately 96 per cent of the noncondensible gas to be CO, and
approximately 3.9 per cent to be CH_4. Assuming no other condensible gases to
be present, we deduce about 0.1 per cent H_2. By summing the appropriate mass
peaks of the mass spectrum, we find, independently, about 97 per cent CO, 3
per cent CH_4, and 0.2 per cent H_2, in good agreement with the infrared results.

The production of CO and H_2 are expected from Pathway II of the formaldehyde photolysis. The $[H_2]/[CH_4]$ ratio $\sim$ 5 per cent is much larger than the ratio $\sim$ 0.1 per cent expected from cracking CH_4. Because the system was pumped before irradiation, the H_2 and CH_4 found here cannot be contaminants in the initial constituents. We attribute the great departure from $[H_2]/[CO]$ $\sim$ 1, expected from Pathway II, and the synthesis of CH_4 to reaction of hydrogen to form other constituents.

The detection of CH_4 is an important result. Its abundance was more than two orders of magnitude above background. It can only have been produced by a chain reaction initiated through Pathway I, probably by superthermal H atoms. There are several possible pathways to methane, including collisional decomposition of ethane, followed by the recombination of hydrogen atoms and methyl radicals, or the formation and subsequent photolysis of acetaldehyde. The presence of CH_4 demonstrates the mobility of radicals in such irradiated frosts at 77 K, and suggests the formation of a variety of other compounds which remain condensed out at 77 K.

Accordingly the photolytic products were gradually warmed to -37 C. The temperatures were monitored by Cu-constantan thermocouple in thermal contact with the deposited frosts. The radiation products were then divided into fractions. The first fraction was transferred again to a 77 K trap and samples successively stored in five receptacles at -196 C, -160 C, -126 C, -111 C and -78 C. Because of diffusion limitations in the gas handling apparatus this represents only a zeroth order fractional distillation. No samples in the first fraction had access to liquid water at any time. The remaining fraction of irradiated products was gradually warmed to room temperature, where a small quantity of liquid water condensed out in the access tube at the bottom of the reaction vessel (Figure 1). Even for this fraction we believe it unlikely that gas phase chemistry was significantly affected by liquid water.

The various samples were examined with a Hewlett-Packard Model 5750 gas chromatograph, and with a Perkin-Elmer Model 270 mass spectrometer/gas chromatograph. With the latter instrument $\sim$ 1000 individual mass spectra of gas chromatographic elution peaks of the various samples were obtained and examined. Typical gas chromatograms, with peaks identified by gas chromatographic techniques alone, are represented in Figure 2. Typical GC programs are given in the figure legend.

The principal molecules identified, with the corresponding analytic technique(s), are exhibited in Table 1. Each of the compounds determined mass spectrometrically corresponds to a gas chromatographic peak on the GC/MS, but here GC is a separation technique, not a detection technique. Table 1 lists gas chromatography as a detection technique only when compounds are independently recognizable by their retention time characteristics. A number of

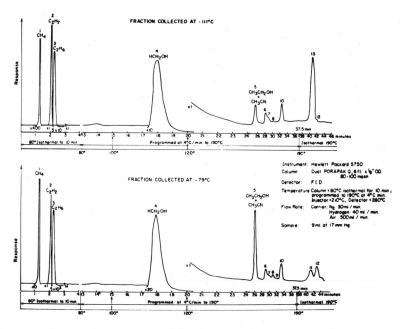

2. Representative gas chromatograms of irradiation products raised to -37 C and then collected at -79 C and -111 C. Several of the numbered peaks were subsequently identified by GC/MS.

results, such as a larger CH_3OH/C_2H_5OH ratio at -111 C than at -78 C are consistent with the relative vapor pressures of the compounds. We attribute the larger yield of organic molecules in the fraction which was allowed to come to room temperature to the low vapor pressures of these compounds at -37 C, preventing their substantial transfer to other low temperature baths. Control samples revealed none of the compounds listed in Table 1.

The production of methanol, acetonitrile, and possibly formic acid are of interest, because these molecules have been reported in the interstellar medium by microwave line radioastronomy techniques. Although other reaction pathways are possible, the production of methanol and possibly formic acid in this experiment suggests a surface-catalyzed (or gas phase) ultra-violet-induced Cannizzaro reaction, $2\ HCHO + H_2O \rightarrow CH_3OH + HCOOH$. Methyl formate is probably produced by a Tischenko reaction, $2\ HCHO \rightarrow CH_3OOCH$. A set of unidentified peaks exists in the gas chromatograms (Figure 2). Some but not all of these correspond to compounds identified in GC/MS. The unidentified compounds having larger retention times are all of higher molecular weight than those identified. A variety of unidentified molecules also exist in the GC/MS

TABLE 1

SYNTHESIZED MOLECULES AT $-196^\circ C$ UV IRRADIATION OF HCHO, H_2O, C_2H_6, AND NH_3

Molecule	Detection Technique	Molecule	Detection Technique
Fraction collected at $-111^\circ C$		**Fraction collected at room temperature**	
methanol, CH_3OH	GC,MS	methanol, CH_3OH	GC,MS,IR
ethanol, C_2H_5OH	GC	ethanol, C_2H_5OH	GC,MS
acetone, CH_3COCH_3	MS	acetone, CH_3COCH_3	MS
acetylene, C_2H_2	GC	acetonitrile, CH_3CN	GC,MS
acetonitrile, CH_3CN ?	GC	acetaldehyde, CH_3CHO	MS
Molecule X	MS	methyl formate, CH_3OOCH	MS
Molecule Y	MS	formic acid, HCOOH ?	MS
Fraction collected at $-78^\circ C$		Molecule X	MS
methanol, CH_3OH	GC,MS	Molecule Y	MS
ethanol, C_2H_5OH	GC,MS		
acetone, CH_3COCH_3	MS		
acetylene, C_2H_2	GC		
acetonitrile, CH_3CN ?	GC	GC = gas chromatography	
Molecule X	MS	MS = combined gas chromatography/mass spectrometry	
Molecule Y	MS	IR = infrared spectroscopy	

mass spectra. Only two of the most prominent of these are given in Table 1. Molecule X has cracking pattern peaks at m/e = 55, 57, 70, and 71. Molecule Y has peaks at m/e = 73, 83, and 85. Work on the identification of these gas phase unknowns of high vapor pressure and high molecular weight is continuing.

The suggestion (Sagan 1970, Sagan and Khare 1970) that acetonitrile might, on the basis of laboratory experiments, be present in the interstellar medium seems to be supported by the most recent observational evidence (Solomon *et al.*, 1971). From the present experiments, we suggest searches for acetaldehyde, acetone, ethanol, and methyl formate. The presence of acetonitrile in these experiments strongly implies the production of HCN; and acetylene suggests the presence of ethylene, although, because of masking, such molecules have not been detected directly. The synthesis of amino acids in related experiments (Sagan and Khare 1971a), probably formed by a Strecker synthesis, suggests that higher nitriles and, in particular, the nitriles of the simpler amino acids should be suitable targets for further microwave line searches. Because of their very low vapor pressures, the absorption frequencies of the amino acids themselves are difficult to determine.

One implication of this work is that some of the more complex organic molecules discovered in the interstellar medium may be genetically related — as, *e.g.*, by the Cannizzaro and Tischenko reactions. This implies that the cross-correlation of molecular species in the same cloud may eventually provide a powerful method for verifying proposed reaction schemes. The fact that molecules such as acetone and the aldehydes are very UV-labile, while compounds such as CO are very UV-stable, must of course be taken into account in such correlation attempts.

In the course of experiments preparatory to the one reported here (Sagan and Khare 1970) the ease of formation of hexamethylenetetramine (HMTA) was brought dramatically to our attention. This compound forms stoichiometrically from formaldehyde and ammonia, $6\ HCHO + 4\ NH_3 \rightarrow C_6H_{12}N_4 + 6\ H_2O$. HMTA is a colorless solid with a melting point $>$ 500 K. Its high melting point is associated with a symmetry and rigidity of its cage structure. It was the first organic compound to have its structure determined by X-ray diffractometry. We calculate that in dense clouds HMTA can form stoichiometrically by collision in times short compared with the formaldehyde photodissociation time, allowing for the ultraviolet extinction within such a cloud. Molecules of this sort have a marked absorption feature in the vicinity of 2000 Å; and because of their stability commend themselves as candidate constituents of the interstellar grains.

We are pursuing the photochemistry of HMTA and related compounds, in the expectation that some of the molecules recently detected in the interstellar medium may be photolytic fragments of larger organic molecules rather than the interaction products of smaller molecules (Sagan 1971). Work is also being pursued in this laboratory on experiments like those reported here, but in which the product analysis is done exclusively at low temperatures — *e.g.*, by time-dependent infrared spectroscopy of mixed frosts deposited and irradiated on CsI windows.

A remarkable similarity exists between the mass spectrometric cracking patterns of interstellar molecules identified at microwave frequencies, and the radicals in cometary tails identified at optical frequencies (Sagan 1971). In the expectation that comets are composed at least in part of cosmically abundant ices (Whipple 1963) and have been UV-irradiated during their history, we propose that experiments of the sort reported here are also of relevance to cometary chemistry.

We are grateful to Jeremy Hribar, Francois Raulin, Dennis Ward, Mohan Khare and Lawrence Wasserman for technical assistance, and are particularly indebted to Leon H. Hinman for glassblowing of high excellence. This research was supported in part by NASA Grant NGR 33-010-101.

REFERENCES

Hackwell, J. A.; Genrz, R. D. and Woolf, N. J. 1970, *Nature*, **227**, 822.

Lippincott, E. R.; Eck, R.; Dayhoff, M. O. and Sagan, C. 1967, *Ap. J.*, **147**, 753.

McQuigg, R. and Calvert, J. 1969, *J. Am. Chem. Soc.*, **91**, 1590.

Sagan, C. 1970, in "Highlights of Astronomy," *Trans. I.A.U.*, **XIV B** (Dordrecht: D. Reidel).

————. 1971, Cornell CRSR Rept. No. 453, also submitted for publication.

Sagan, C. and Miller, S. L. 1960, *A. J.*, **65**, 499.

Sagan, C.; Dayhoff, M. O.; Lippincott, E. R. and Eck, R. 1967, *Nature*, **213**, 273.

Sagan, C. and Khare B. N. 1970, *Bull. A.A.S.*, **2**, 340.

————. 1971a, *Science*, **173**, 417.

————. 1971b, *Ap. J.*, **168**, 563.

Solomon, P.; Jefferts, K. B.; Penzias, A. and Wilson, R. W. 1971, *Ap. J., (Letters)*, **168**, L107.

Stein, W. A., and Gillett, F. C. 1969, *Ap. J., (Letters)*, **155**, L197.

Whipple, F. L. 1963. In *The Moon, Meteorites, and Comets*, B. M. Middlehurst and G. P. Kuiper, eds. (Chicago: University of Chicago Press).

Wilson, R. W.; Penzias, A. A.; Jefferts, K. B.; Kutner, M. and Thaddeus, P. 1971, *Ap. J., (Letters)*, **167**, L97.

Molecule Formation in Normal Clouds

Per A. Aannestad*

Astronomy Department
University of California, Berkeley
Berkeley, California

*Now at Goddard Institute for Space Studies, NASA, New York, N. Y.

I want to report briefly on a part of a study of molecule formation in normal interstellar clouds, where "normal" means densities n < 100 cm^{-3}.

This morning we discussed both gas phase reactions and catalysis on grain surfaces. As we heard from Dr. Klemperer, gas phase reactions may account for the molecules CH, CH$^+$, CN, and CO in spite of the high UV photodestruction rates in these almost unshielded regions. However, in the case of OH, which is found also in such clouds [cf. observations by Davies (1971) where, typically, n ~ 30 cm^{-3}] there is no known gas phase reaction that will give an abundance of OH in agreement with observations. For the type of clouds considered here, the radiative association mechanism proposed by Julienne and Krauss fails by at least two orders of magnitude.

As was also done this morning by Salpeter and Watson, one may thus explore the possibility of catalytic formation of OH on a grain surface, with subsequent escape from the surface. Because the details of the surface chemistry are not known (as stressed by Dr. Klein), we have to assume a probability for this process and have arbitrarily assumed 0.5. This means that half of the oxygen atoms hitting the grain leave as OH, building up the molecular density in the gas phase while the other half remain on the grain, and may react further to form an H$_2$O mantle. We have neglected any destruction of this mantle, either by ultraviolet, cosmic-ray or x-ray radiation. However, as pointed out by Watson, this may be questionable.

We then have a picture of the trace elements freezing out on the grains while molecules are being formed both by gas phase reactions and catalytic surface reactions. In our study we used the gas phase reactions proposed by Solomon and Klemperer (1971) with the addition of the reactions OH + C$^+$ → CO + H$^+$, NH + C$^+$ → CN + H$^+$ with the usual rate constant of 10^{-9} cm^3 s^{-1}. For the photodestruction rates of OH and NH we took 5 x 10^{-12} s^{-1} and 1 x 10^{-11} s^{-1}, respectively. The catalytic reactions were O, C$^+$, N + grain → OH, CH, NH + grain. The initial grain radius was 0.05 μ and the density of the molecular mantle was 1.5 g cm^{-3}.

Since the depletion of trace elements decreases the cooling rate in a cloud, the cloud will heat up and expand at a constant pressure as discussed by Dr. Field on Monday. If one selects a cloud of a given mass one can follow the evolution of its molecular abundances as well as its temperature and density as the depletion proceeds. It is convenient to discuss the results as a function of the depletion parameter $\xi(C^+)$ as the time dependence of this parameter may be greatly changed by the intervention of cloud collisions.

The results for a cloud of 500 M$_\odot$ and an intercloud pressure of 1800 cm^{-3} °K are shown in Figure 10 on page 38. The assumed value of the intercloud pressure is consistent with recent temperature determinations of the intercloud medium. We see that the density n goes from ~ 100 cm^{-3} to ~ 10 cm^{-3} and the temperature T from $\sim 20°$ K to $\sim 200°$ K as C^+ is depleted by about two orders of magnitude. The time scale for the variation in temperature and density is a few 10^7 years. The visual extinction A_v through the cloud rapidly increases from 0.1 mag to 0.5 mag as the grains grow mantles and their cross sections increase. However, for $\xi(C^+) < 0.3$, A_v decreases slowly because of the expansion of the cloud and the consequent decrease of N (which is proportional to the inverse square of the cloud radius).

The column densities of CH^+, CH, and CN are highest at the start, where n is large, because their formation rate is proportional to n^2. OH, on the other hand, because it forms at a rate proportional to nv times the cross-sectional area of the grains per cm^3, would be expected to have a column density proportional to $nT^{1/2} A_v$. This reaches a peak later because the increase in $T^{1/2} A_v$ more than compensates the decrease in n.

We have compared these predictions for a 500 solar mass cloud with recent observational data given by Frisch (1972) and Davies (1971). Frisch has reanalyzed the Adams' optical data on CH^+, CH, and CN, using the most recent f-values. One sees that both the CH^+ and CN observations can be satisfied with $\xi > 0.3$, although the theory fails by a factor of 2 to account for the CH observations.

Davies (1971) has discussed the properties of clouds in which OH is found in absorption. Again, we find fair agreement between the mean of the observations with a theoretical 500 solar mass cloud having $\xi < 0.1$, except that OH fails by a factor of somewhat more than 3. Considering the uncertainties in the rate coefficients and the spread in the observational data, the agreement is acceptable.

The following point emerges from a study of Figure 10. Roughly speaking, we see that CH clouds are "young" – they have relatively large ξ, while OH clouds are "old", having small ξ. The theoretical reason for this is that OH (as a product of grain catalysis) persists over a much larger period of cloud evolution, while CH^+, CH, and CN (as products of gas-phase reaction) can be found only in "young" clouds where depletion has not yet occurred, so that the gas is cool, and the intercloud pressure is able to compress the cloud to favorably high densities. This may explain why OH has not been seen in radio observations of regions where CH^+, CH, and CN have been seen optically.

It is obvious that such a model has many assumptions and uncertainties in it, but the calculations may give a hint of the correlations between cloud properties which may soon become much more meaningful.

REFERENCES

Davies, R. 1971, *Proc. of 17th Liege Symposium,* in press.
Solomon, P. M. and Klemperer, W. 1972, *Ap. J.,* **178**, 389.
Frisch, P. *Ap. J.,* **173**, 301.

Session 6

Biological Implications

Interstellar Molecules: Significance for Prebiotic Chemistry

Cyril Ponnamperuma

Laboratory of Chemical Evolution
Department of Chemistry
University of Maryland
College Park, Maryland

I. INTRODUCTION

The discovery of organic molecules in the interstellar medium has a profound impact on the hypothesis of chemical evolution. In this paper an attempt will be made to outline the recent experimental work in the field of prebiotic chemistry and demonstrate how these investigations may have acquired new meaning and significance in a cosmic perspective.

The objective of chemical evolution is to retrace the path by which life appeared on this earth. The cardinal premise implies that life is an inevitable consequence of the evolution of matter. If this sequence of events took place upon the earth it can be argued that similar processes may have occurred elsewhere in the universe. The earth is thus considered to be the model laboratory in which these events took place.

The idea that a long chemical evolution was a necessary preamble to the appearance of life on earth, is implicit in the writings of Oparin (1924), Haldane (1928), and Bernal (1949). Long before the origin of life was discussed in scientific terms by Oparin, Charles Darwin had, in 1861, with profound intuition, written to his friend Hooker: "If we could conceive in some warm little pond, with all sorts of ammonia and phosphoric salts, light, heat, electricity, etc. present that a protein compound was chemically formed ready to undergo still more complex changes." (De Beer 1959). It is Darwin's warm little pond that the student of Chemical Evolution is endeavoring to recreate in his modern laboratory.

In the experimental work performed over the last few years, emphasis has been placed on both analysis and synthesis. In the analytical approach an attempt is made to see the record in the rocks and sediments of the earth going back to the earliest times. Further information could be obtained from lunar samples and meteorites about events during the very early stages of the formation of the solar system.

The Geologic clock, after Schopt (1967), (Figure 1) represents the age of the earth in a twelve hour diagram. Although at one time it was generally believed that nothing of significance for life existed beyond the base of the Cambrian at 600 million years ago, the painstaking work of the micropaleontologists, Barghoorn, Schopt (1972), and Tyler, has provided evidence for microfossils in sediments of precambrian age. The Bitter Spring formation of Australia, 10^9 years old, the Greenflint chart of Ontario, about 2×10^9 years old, and the Fig tree chart and Onverwacht shales from South Africa, over 3×10^9 years in age, have been assiduously studied for the presence of microfossils. Highly organized and well defined structures have been found in these sediments of great antiquity.

The most interesting era, however, for the prebiotic chemist is that period

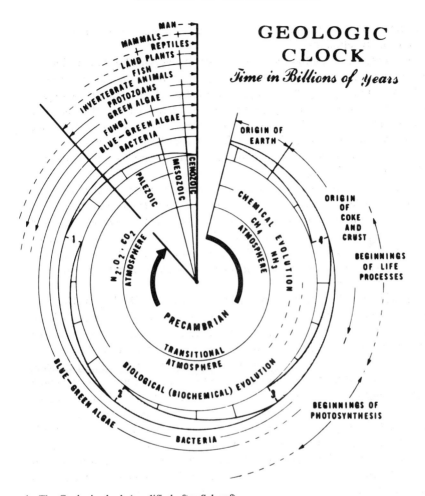

1. The Geologic clock (modified after Schopf).

which extends between the appearance of the first life and the formation of the earth. Unfortunately there appears to be little hope of finding any rocks older than 3.5×10^9 on the earth. This leads the prebiotic chemists to rely on meteorites and lunar samples for evidence of the early organic chemistry of our planetary system.

II. SYNTHESIS OF MICROMOLECULES

In the synthetic approach we try to recreate Darwin's "warm little pond" or Haldane's "hot dilute soup" and search for these molecules which have a significance for life. Although much work has been done in other laboratories on the synthesis of prebiotic molecules, in this paper I shall concentrate on the work done in my laboratory.

In our experiments we have adopted the simple working hypothesis that the molecules which are important now were important at the time of the origin of life. We are investigating the abiogenic synthesis of the nucleic acids and proteins. We synthesize the "primordial soup" described by Haldane, and proceed to analyze it. In the course of these experiments, we have established that some of the micromolecules of biological significance can be synthesized and that, under the same conditions, they can be condensed or polymerized to give rise to macromolecules.

In experiments starting with methane, ammonia, and water, an electron beam was used to simulate potassium-40 on the primitive earth (Ponnamperuma et $al.$, 1963). The gas mixture was irradiated with electrons from a linear accelerator in the Lawrence Radiation Laboratory at Berkeley. In a 45-minute period, the total energy absorbed was about 7×10^{10} ergs per gram. The results of this investigation clearly established that adenine was a product of the radiation of methane, ammonia, and water.

The production of adenine is enhanced by the absence of hydrogen. This is to be expected since methyl carbon has to be oxidized in order to appear finally in the purines. In any event, the high concentration of organic matter on the prebiotic earth probably arose when most of the hydrogen had escaped.

Formaldehyde was formed by the action of electric discharges, or ionizing radiation, or a mixture of primitive gases. Experiments in which formaldehyde was used as starting material have shown that sugars are formed (Ponnamperuma and Mariner 1963, Gabel and Ponnamperuma 1967). A preliminary separation indicates that the highest yield by far is in the pentoses and hexoses. There is evidence that the biological sugars, ribose and deoxyribose, can be formed by this means. Similarly, hydrogen cyanide, which is readily formed in simulation experiments, gives rise to adenine and guanine by photochemical reactions (Ponnamperuma 1963).

This brief summary of previous work indicates that micromolecules can, indeed, be formed in simulation experiments. The results from other laboratories have confirmed our findings, and have extended the list of compounds so formed.

Let us now turn to the question of condensation reactions giving rise to polypeptides and oligonucleotides.

III. SYNTHESIS OF MACROMOLECULES

Dehydration-condensation reactions are generally involved in the formation of more complex molecules. On the primitive earth, this type of condensation could have taken place in water, namely in the ocean, or, in the relative absence of water, on the shore of the ocean or the dried-up bed of a lagoon. In our simulation experiments we have reconstructed both models:

1. The condensation reaction taking place in the presence of water, and
2. The condensation reaction taking place in the relative absence of of water or in a hypohydrous condition.

The condensation reaction taking place in the presence of water will be illustrated by our synthesis of peptides, while the condensation reaction in the relative absence of water will be demonstrated by our synthesis of oligonucleotides.

IV. SYNTHESIS OF MACROMOLECULES UNDER AQUEOUS CONDITIONS

We have two examples of the synthesis of peptides in aqueous solution:

1. The photochemical synthesis of dipeptides and tripeptides from glycine and leucine (Ponnamperuma and Peterson 1965).
2. The synthesis of a polymer during the electric discharge through methane, ammonia, and water (Ponnamperuma *et al.,* 1969).

When an aqueous solution of glycine and leucine was exposed to ultra-violet light in the presence of cyanamide, the dipeptides glycyl-glycine, glycyl-leucine, leucyl-glycine, leucyl-leucine, and the tripeptides glycyl-glycyl-glycine and leucyl-glycyl-glycine were formed.

We have recently found that in our experiments with electrical discharges on the earth's primitive atmosphere, most of the amino acids synthesized appear to be present in the condensed form. The amino acids appear to have been linked together as soon as they were synthesized. It was only on hydrolysis that free amino acids were detected.

V. SYNTHESIS OF MACROMOLECULES
UNDER HYPOHYDROUS CONDITIONS

The condensation reactions taking place in the relative absence of water, or in the hypohydrous condition, are exemplified in the case of the oligonucleotide synthesis (Ponnamperuma and Mack 1965). Assuming that nucleosides could be formed in aqueous solution, we investigated the possibility of their phosphorylation under hypohydrous conditions. Such a situation could have prevailed on the primitive earth when organic material may have been deposited on the beds of lagoons and pools which may have fringed the early coastline. An intitimate contact between nucleosides and phosphates could have been brought about by a process of evaporation under the action of solar heat.

When the nucleosides adenosine, guanosine, cytidine, uridine, and thymidine

ADENINE AND GUANINE FROM HCN TETRAMER

2. Synthesis of the two purines adenine and guanine from hydrogen cyanide.

were heated with sodium dihydrogen orthophosphate, NaH_2PO_4, the nucleoside monophosphates were formed. The heating was conducted at about 160° C. At this temperature, the highest yield of monophosphate obtained was about 18 per cent. This reaction could take place at much lower temperatures. At 50° C, a small yield was obtained, but this required at least three days' heating as compared to two hours at 160° C.

The conditions of these experiments may be considered to be genuinely prebiotic. The temperatures used are within reasonable limits. Although the reaction is favored by the absence of water, it is not completely inhibited by water.

The results presented so far show that micromolecules of biological significance can be synthesized in the laboratory under conditions which may have prevailed on the prebiotic earth. These results also show that under the same conditions, condensation reactions of the micromolecules could have taken place to give rise to polymers which may have been the forerunners of the nucleic acids and proteins of today.

VI. CHEMICAL MECHANISMS INVOLVED IN ABIOTIC SYNTHESIS EXPERIMENTS

Recent work on the elucidation of the mechanisms involved in these reactions points to relatively simple chemical pathways for the origin of the micromolecules. Hydrogen cyanide is formed in copious yield from methane, ammonia, and water. It is the pathway for the purines adenine and guanine, as shown in Figure 2 (Sanchez, Ferris, and Orgel 1966). Formaldehyde is an intermediate in the reaction of primitive atmospheres. It is the simplest of the sugars. Condensation reactions with formaldehyde give rise to sugars of biological importance. This is a reaction known to organic chemists since 1861 (Figure 3) (Butlerow 1861). The cyanide and the aldehyde together, by the Strecker synthesis, give rise to amino acids, as shown in Figure 4.

A brief survey of the mechanisms involved in these reactions suggests that all the ingredients necessary for the production of nucleic acids and proteins can be generated from methane, ammonia, and water.

The mechanisms involved in the condensation reactions are somewhat obscure. In the case of the polypeptide, it is reasonable to assume that the hydrogen cyanide tetramer may be involved since the solution contains 18 per cent cyanide. The effective use of the tetramer in condensation reactions has been recently established (Chang, Flores, and Ponnamperuma 1969). In the polynucleotide synthesis, it seems likely that the dehydration is mediated by the linear polyphosphates. The formation of linear phosphates from orthophosphates has been shown to take place under laboratory conditions.

$$2 \ CH_2O \longrightarrow CH_2OH \cdot CHO$$

$$CH_2OH \cdot CHO + CH_2O \longrightarrow CH_2OH \cdot CO \cdot CH_2OH$$

$$CH_2OH \cdot CHO + CH_2OH \cdot CO \cdot CH_2OH \longrightarrow CH_2OH \cdot CHOH \cdot CHOH \cdot CO \cdot CH_2OH$$

$$2 \ CH_2OH \cdot CHO \longrightarrow CH_2OH \cdot CHOH \cdot CHOH \cdot CHO$$
$$OR \ CH_2OH \cdot CO \cdot CHOH \cdot CH_2OH$$

$$2 \ C_4 \longrightarrow \left[C_8 \right] ? \longrightarrow C_3 + C_5$$

3. Mechanism for the formation of sugars from formaldehyde.

VII. JUPITER SIMULATION STUDIES

The type of synthesis which may be taking place on the planet Jupiter has also been one object of our studies (Woeller and Ponnamperuma 1969). These investigations are of great interest because of the present atmosphere of Jupiter approximates that of the early solar nebula. The recent models of the Jovian atmosphere based on the calculations of Peebles (1964) and Gallet (1963) suggest the presence of liquid water and higher temperatures beneath the outer layer of clouds of ammonid crystals.

The simulation studies have shown that diaminonitriles, which are precursors of amino acids, can be synthesized. Some form of chemical evolution may be taking place on Jupiter (Chadha *et al.,* 1970). Our experiments also lead us to believe that the red color on the planet may be due to a ruby-red organic polymer formed when a mixture of methane and ammonia is exposed to electric discharges.

The formation of aminonitriles during the electric discharge of a mixture of methane and ammonia has been further substantiated by recent work. The acid hydrolysis of the reddish-brown material led to the formation of the amino and imino acids, demonstrating the significance of the corresponding nitriles which are formed under the conditions of electric discharge. The polymeric products have molecular weights in the 2,000 - 3,000 range (Noda and Ponnamperuma 1970). It is possible that polymerization results from the formation of an amidine linkage by the interaction of cyano and amino groups.

$$RCHO + NH_3 + HCN \rightleftharpoons RCH(NH_2)CN + H_2O$$

$$RCH(NH_2)CN + 2H_2O \rightarrow RCH(NH_2)COOH + NH_3$$

$$RCHO + HCN \rightleftharpoons RCH(OH)CN$$

$$RCH(OH)CN + 2H_2O \rightarrow RCH(OH)COOH + NH_3$$

4. The Strecker Synthesis of amino acids from aldehydes, hydrogen cyanide and ammonia.

VIII. METEORITES - PREVIOUS STUDIES

Meteorites have been analyzed for organic compounds for over a century. Berzelius (1834) examined the Alias, Wohler and Hornes (1859), the Kaba, and Berthelot (1868) the Orgueil, and reported the presence of substances of organic origin. These investigations have continued over the years, and there is general agreement about the presence of polymeric organic matter in carbonaceous chondrites. However, the inherent potentiality for contamination resulting from the ubiquitous distribution of biomolecules on earth leads us to believe that many of the results reporting the presence of organic compounds in meteorites are inconclusive. The results of our recent investigation seem to resolve some of these ambiguities and provide evidence for amino acids and hydrocarbons of possible extraterrestrial origin.

IX. THE MURCHISON METEORITE - THE FALL

The Murchison meteorite fell on September 28, 1969, near Murchison, Victoria, Australia (latitude $36°36'$, longitude $145°12'$). The parent object broke up during flight and scattered many fragments over an area of about 5 square miles. Most fractured surfaces on the individual pieces have fusion crusts.

For this study we selected those stones which had the fewest cracks, the least exterior contamination, and which generally appeared to have a massive character. The samples we examined contained 2.0 weight per cent carbon and 0.16 weight per cent nitrogen. Chemically, the Murchison meteorite is a type II carbonaceous chondrite (C-2).

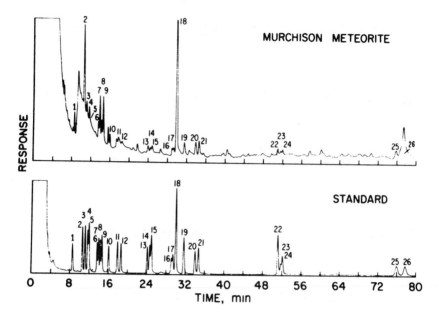

5. Gas chromatogram of amino acids in the Murchison meteorite.

X. THE MURCHISON METEORITE - AMINO ACIDS

Ten grams of an interior piece of the meteorite were extracted, hydrolyzed, and examined for the presence of amino acids by the different techniques developed for the analysis of lunar samples.

Conventional ion exchange chromatography revealed a number of peaks having retention times similar to common amino acids. The presence of the amino acids suggested by ion exchange chromatography and their enantiomeric distribution were established by gas chromatography of the N-trifluoroacetyl-D-2-butyl esters of the amino acids. Both enantiomers of amino acids with asymmetric centers are present. The percentage of D and L amino acid enantiomers present was calculated from the gas chromatograms. Within the range of experimental error, they appear to be present in equal quantity.

To obtain an unambiguous identification, the compounds eluted from the gas chromatogram were introduced through a membrane separator into the mass spectrometer. The mass spectra obtained were compared with the spectra of known standards. In this manner, the identity of the amino acids glycine, alanine, valine, proline, glutamic acid, and aspartic acid were confirmed (Kvenvolden *et al.*, 1970).

In addition, twelve non-protein amino acids appear to be present in the Murchison meteorite. The identity of eight of them has been conclusively established: N-methylglycine, β-alanine, 2-methylalanine, α-amino-n-butyric acid, β-amino-n-butyric acid, γ-amino-n-butyric acid, isovaline, and pipecolic acid. Tentative evidence is available for the presence of N-methylalanine, N-ethylglycine, β-aminoisobutyric acid, and norvaline (Kvenvolden, Lawless, and Ponnamperuma 1971) (Figure 5).

XI. THE MURCHISON METEORITE - HYDROCARBONS

The analysis of the aliphatic hydrocarbons indicates that they are largely saturated alkanes. There is a marked resemblance between the gas chromatographic traces of the aliphatic hydrocarbons synthesized by the action of electrical discharges on methane and those from the meteorite. The mass spectral data reveal the same dominant homologous series in both samples. The similarity between the aliphatic hydrocarbons from the meteorite and the spark discharge material, on these three counts, may be suggestive of a possible abiogenic origin for the hydrocarbons in the Murchison meteorite.

XII. THE MURCHISON METEORITE - ISOTOPE FRACTIONATION

The ^{13}C values relative to the Peedee Belemnite standard were determined for various fractions of the meteorite using the techniques of Smith and Kaplan (1970). The following values were obtained:

Total Carbon	^{13}C PDB -7.27, -7.06
Carbonate	+45.37
Material soluble in organic solvents	+4.43, +4.76, +5.93
Residual organic matter (after solvent extraction and demineralization with HF)	-10.64

By comparison with the results obtained by Smith and Kaplan in their recent study of seven carbonaceous chondrites, these figures indicate that, except for the carbonate, the Murchison meteorite is enriched with ^{13}C in every fraction. Since carbon-hydrogen compounds on the earth are reported to have a value ranging from -15 to -35 per mil (Degens 1965), and carbon from petroleum and kerogen from Precambrian rocks to have values of -25 to -35 per mil, the heavy $\delta^{13}C$ value of the carbon in the extractable organic matter suggests minimal terrestrial contamination.

These results have provided evidence for the indigenous nature of amino acids and hydrocarbons in the Murchison meteorite. The $\delta^{13}C$ values of 4.43 to 5.93 for the extractable organic material of the meteorite fall into a range widely different from terrestrial organic matter. The presence of the amino acids was established by the use of ion exchange chromatography, gas chromatography, and gas chromatography combined with mass spectrometry. The presence of almost equal amounts of amino acids with optically active centers minimizes the possibility of terrestrial contamination and suggests a possible extraterrestrial origin. The presence of twelve amino acids which are not generally found in biological systems is indicative of a possible abiogenic synthesis. The gas chromatographic distribution pattern and the mass spectrometric fragmentation pattern of the aliphatic hydrocarbons are similar to the patterns produced by hydrocarbons synthesized abiotically in the laboratory. These similarities support the contention that the organic molecules identified here are abiotic and possibly extraterrestial in origin.

If the amino acids are indigenous to the meteorite and therefore extraterrestrial, the question immediately arises as to their origin. Two possibilities present themselves: either these amino acids were present at some period of time in the meteorite in one stereoisomeric form and were then racemized in the course of time, or the two forms were present in nearly equal amounts. The possibility that an extraterrestrial biota was responsible for D or L amino acids initially cannot be discounted entirely. However, by analogy with terrestrial life, it would be difficult to explain the presence of several non-protein amino acids. The production of either the D or the L form by an abiogenic process and subsequent racemization may also be considered unlikely, since no well-defined evidence is available for abiotic processes which would produce one form rather than the other, in spite of extensive experimentation. We are therefore led to believe that the amino acids in the Murchison meteorite were most likely produced in both forms by an abiotic process. Of the twelve non-protein amino acids positively and tentatively identified by us, five have been identified in laboratory experiments simulating primitive earth or planetary conditions (Miller 1955). The presence of all the isomers of the amino acids of two and three carbon atoms suggests a random synthesis. In addition, the previously reported random distribution of isomers of aliphatic hydrocarbons and the heavy $\delta^{13}C$ isotopic values of the extractable carbon point to an abiogenic origin of the organic matter in this meteorite. We conclude that our analysis of the Murchison meteorite provides a new basis for the study of chemical evolution and the search for extraterrestrial life.

The finding of several organic molecules such as hydrogen cyanide, formaldehyde, and cyanoacetylene in the interstellar medium, brings us immediately to recognize their role in prebiotic synthesis. In the formation of amino acids both cyanides and aldehydes play an important role. The purines can be traced to rearrangements of hydrogen cyanide. Pyrimidines have been synthesized from cyanoacetylene and urea. The sugars are a direct condensation product of formaldehyde.

The molecules that have been found in the interstellar medium may themselves make no direct contribution to the origin of molecules leading to life. However, they may represent for us conceptually a satisfying pathway suggesting the sequence from atom to small molecules leading to the polymers necessary for life. The pathways of chemical evolution appear to be commonplace in the cosmos.

REFERENCES

Bernal, J. D. 1949, *Proc. Phys. Soc.*, **62A**, 537-558.

Berthelot, M. 1868, *Compt. Rend.*, **67**, 849.

Berzelius, J. J. 1834, *Ann. Phys. Chem.*, **33**, 113.

Butlerow, A. 1861, *Ann.*, **120**, 296.

Chadha, M. S.; Lawless, J. G.; Flores, J. J. and Ponnamperuma, C. 1971, in *Molecular Evolution 1*, ed. R. Buvet and C. Ponnamperuma (Amsterdam: North-Holland), pp. 143-151.

Chang, S.; Flores, J. and Ponnamperuma, C. 1969, *Proc. Nat. Acad. Sci.*, **64**, 1011-1015.

De Beer, G. 1959, *Notes and Records - R. Soc. London*.

Degens, E. T. 1965, *Geochemistry of Sediments: A Brief Survey* (Englewood-Cliffs: Prentice Hall).

Gabel, N. W. and Ponnamperuma, C. 1967, *Nature*, **216**, 453-455.

Gallet, R. M. 1963, *Physics Today*, **16**, 19.

Haldane, J. B. S. 1928, *Rationalist Annual*, **148**, 3-10.

Kvenvolden, K.; Lawless, J. G. and Ponnamperuma, C. 1971, *Proc. Nat. Acad. Sci.*, **68**, 486-490.

Miller, S. L. 1955, *J. Amer. Chem. Soc.*, **77**, 2351-2361.

Noda, H. and Ponnamperuma, C. 1971, in *Molecular Evolution 1*, ed. R. Buvet and C. Ponnamperuma (Amsterdam: North-Holland), pp. 236-244.

Oparin, A. I. 1924, *Proischogdenie Zhizni*.

Peebles, P. J. E. 1964, *Ap. J.*, **140**, 328.

Ponnamperuma, C. 1963, *Abstract - 21st Congress, International Union of Pure and Applied Chemistry, London*, pp. 288-289.

Ponnamperuma, C. and Mack, R. 1965, *Science*, **148**, 1221-1233.

Ponnamperuma, C. and Mariner, R. 1963, *Rad. Res.*, **19**, 183.

Ponnamperuma, C. and Peterson, E. 1965, *Science*, **147**, 1572-1574.

Ponnamperuma, C. *et al.*, 1963, *Proc. Nat. Acad. Sci. U. S.*, **49**, 737-740.

Ponnamperuma, C. *et al.*, 1969, *Advan. Chem. Ser.*, **80**, 280-288.

Proceedings of the Conference on the Organic Analysis and Carbon Chemistry of Lunar Samples: Their Significance for Exobiology, Space Life Sciences, in press.

Sanchez, R. A.; Ferris, J. P. and Orgel, L. E. 1966, *Science*, **153**, 72-72.

Schopf, J. W. 1967, 1967 McGraw-Hill Yearbook of Sci. & Tech., p. 46.

————. 1972, in *Exobiology*, ed. C. Ponnamperuma (Amsterdam, London: North-Holland), pp. 16-61.

Smith, J. W. and Kaplan, I. R. 1970, *Science*, **167**, 1367-1370.

Woeller, F. W. and Ponnamperuma, C. 1969, *Icar*, *10*, 386-392.

Wohler, M. F. and Hornes, M. 1859, *Sitzber. Akad. Wiss. Wien, Math.-Naturw. Kl.*, **34**, 7.

Interstellar Molecules: Formation in Solar Nebulae

Edward Anders

Department of Chemistry
University of Chicago
Chicago, Illinois

I. INTRODUCTION

Carbon is largely missing from the inner solar system. Even carbonaceous chondrites (the meteorites richest in C) contain only 6 x 10^{-2} their cosmic complement of C. The Earth and Venus contain even less, about 10^{-4}. This carbon, along with other volatiles (H, O, N, noble gases) seems to have been lost to interstellar space when the solar nebula was dissipated. At least 3 x 10^{-3} M$_\odot$ of volatiles, complementary to the inner planets, were thus expelled from our own solar system. If this is typical of other planetary systems, and if planetary systems are as common as most authors believe, then a large part of interstellar matter must once have passed through solar nebulae.

Herbig (1970) has suggested that solar nebulae might be the principal source of interstellar grains and molecules. Such nebulae provide a high-density environment ($\sim 10^{15}$ molecules/cm^3) in which matter can be rapidly transformed to grains and molecules and returned to interstellar space.

Two lines of evidence support Herbig's hypothesis. First, a variety of organic compounds have been observed in model experiments simulating processes in the solar nebula (Studier, Hayatsu, Anders 1968, 1972; Hayatsu, Studier, Oda, Fuse, Anders 1968; Hayatsu, Studier, Anders 1971 Hayatsu, Studier, Matsuoka, Anders 1972 Yoshino, Hayatsu, Anders 1971). Second the interstellar gas is depleted in those elements (Ca, Ti, possibly Al; Herbig 1970) which are the first to condense from a cosmic gas in the pressure range of the solar nebula, 10^{-2} to 10^{-6} atm (Larimer 1967, Grossman 1972).

We can test Herbig's hypothesis in three stages. First, we determine the physical and chemical conditions in the early solar system. Second, we see what happens to carbon under these conditions; what organic compounds are likely to form. Third, we compare the distribution of these organic compounds with the observed distribution of interstellar molecules. The closer the match, especially for larger molecules, the better the chance that the right process has been found.

The first two stages are nearly completed. Results will be summarized in Sections II to IV. The third stage has barely begun; a brief status report will be given in Section V. A few remarks about biological implications are given in Section VI.

II. CONDITIONS IN THE EARLY SOLAR SYSTEM

It has become apparent during the past two decades that chondrites are relatively unaltered condensates from the solar nebula (Wood 1962, 1968). Their chemical and isotopic composition thus provides clues to conditions in the inner parts of the nebula. Following early attempts by Urey (1952*a,b*, 1954), a detailed and self-consistent picture has emerged. I have covered the subject in two recent reviews (Anders 1971, 1972) and will therefore give only a brief summary of those findings that pertain to interstellar molecules.

1. The inner solar nebula seems to have been hot enough at one time to vaporize all solids, *i.e.* > 2000° K. Temperatures probably declined further outward, but even so it is doubtful that any interstellar organic material survived.

2. Dust grains (10^{-5} - 10^{-6} cm) condensed on cooling from 2000 to ~ 700° K. Little or no accretion took place in this range. This suggests that the nebula

Table 1

ACCRETION TEMPERATURES OF PLANETS AND METEORITE PARENT BODIES

Source	Temperature (°K) Based on:		Low-T Fraction
	Tl[*]	O^{18}/O^{16}[†]	(%)
Earth (Oceanic Basalts)	458	450–470	(12)
Earth (Cont. Basalts)	455	450–470	11
Moon	502	455	2
Eucrites	460	475	0.8
Nakhlites	478	460	38
Shergottites	471	455	28
L-Chondrites	460–560	450–470	24
Cl Chondrites	358±12[‡]	360±5	≥90
		360±15	

[*] Laul et al. (1971); Keays, Ganapathy, Anders (1971); Anders, Ganapathy, Keays, Laul, Morgan (1971). The Tl thermometer is pressure dependent; these values apply to a total nebular pressure of 10^{-4} atm. Other volatile metals (Bi, In, Pb) give very similar temperatures. The two values for the Earth are independent estimates, based on oceanic and continental basalts.

[†] Onuma, Clayton, Mayeda (1971). The O^{18}/O^{16} thermometer is independent of pressure above P = 10^{-6} atm. The two values for Cl chondrites refer to oxygen-isotope fractionations H_2O-carbonate and H_2O-silicate.

[‡] Lancet (1971). This value refers to C^{13}/C^{12} fractionation between carbonate and organic carbon.

cooled rapidly through this range. Nonetheless, some partial separations of gas and dust seem to have occurred, leading to regional variations by up to a factor of 2 in the proportions of refractories (Mg, Ca, Al), siderophiles (Fe, Ni), and two major volatiles, C and O.

3. The bulk of the accretion in the inner solar system seems to have taken place upon cooling below 700° K (Table 1). Accretion temperatures, determined by 3-6 chemical or isotopic thermometers, are remarkably concordant for each body. The Earth, Moon, and most meteorites, all presumably formed between 1 and 3 AU, give values between 450 and 500° K; carbonaceous chondrites, presumably formed at greater distances, give lower temperatures, $350 - 400^\circ$ K.

4. More or less concurrently with accretion, a variable fraction of the dust was remelted and outgassed by brief, presumably local, reheating events (collisions or electric discharges; Whipple 1966, 1972; Cameron 1966). Estimated amounts of unremelted material are given in the last column of Table 1.

5. Pressures in the regions of ordinary chondrites and carbonaceous chondrites were $10^{-4} - 10^{-6}$ atm and $10^{-5} - 10^{-6}$ atm, respectively.

III. PRODUCTION OF ORGANIC COMPOUNDS IN THE SOLAR NEBULA

A. Chemical State of Carbon

On cooling of the nebula, carbon remains in the form of CO down to at least 700° K (Figure 1). If equilibrium is reached, it transforms to CH_4 by the reaction

$$CO + 3 H_2 \rightleftharpoons CH_4 + H_2O \qquad (1)$$

B. Free-Radical Reactions

CH_4 is a dead end, being thermodynamically more stable in a cool cosmic gas than any organic compound. Thus, if reaction (1) has gone to completion, organic compounds can be made only by endoergic porcesses, requiring energy input from an external source. This is the fundamental idea behind the classical Miller-Urey reaction in which energy is supplied as UV, β- or γ-radiation or electric discharges (Urey 1952c; Miller 1953, 1957; see Lemmon 1970 for a recent review). These reactions are dominated by free radicals.

C. Catalytic and Other Spontaneous Reactions

Reaction (1) is exceedingly slow in the absence of a catalyst. And when a catalyst is present, the reaction tends to stop at intermediate stages of

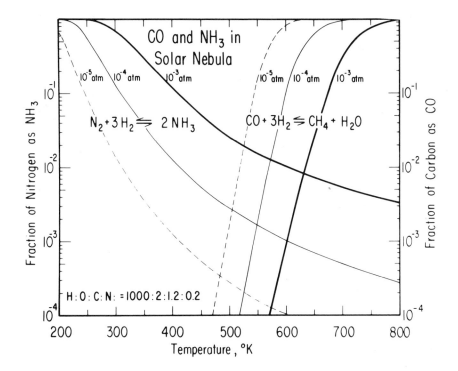

1. If equilibrium is maintained on cooling, carbon will be converted largely to CH_4 (solid lines), before formation of more complex compounds becomes thermodynamically possible (dashed lines). However, hydrogenation of CO to CH_4 is very slow, and thus CO can probably survive to $360 - 380°$ K where carbonaceous chondrites formed.

hydrogenation, giving metastable products of H/C $\approx$ 2 rather than stable methane of H/C = 4. In fact, this process (the Fischer-Tropsch reaction) is used industrially for the production of gasoline:

$$8\,CO + 17\,H_2 \rightarrow C_8H_{18} + 8\,H_2O \qquad (2)$$

Some major constituents of the nebular dust (Fe, Ni, Co, Fe_3O_4, clay minerals) are good catalysts for this reaction, and thus there is reason to expect that the hydrogenation of CO proceeded to metastable organic compounds rather than stable CH_4 (Urey 1953; Oro 1965; Studier, et al. 1968, 1972).

What might be regarded as a third category of reactions circumvents the thermodynamic problem by heating the gas to a high temperature where various energy-rich molecules (e.g. CO, C_2H_2) form spontaneously. Suitable energy sources are heat, shock waves (Bar-Nun, Bar-Nun, Bauer, Sagan 1970), or plasma discharges (Eck, Lippincott, Dayhoff, Pratt 1966; Griffiths, Schuhmann, Lippincott 1970). On cooling, these molecules transform in part to stable CH_4, in part to metastable species of greater complexity.

The classification of these reactions has caused some confusion. A useful criterion is whether significant amounts of ions and radicals survive down to temperatures where ordinary chemical reactions among neutral molecules have ceased. If they do survive, then the synthesis should be classified as Miller-Urey. But if the radicals recombine while reactions among neutral species are still rapid, the product distribution will be governed mainly by these processes, and the reaction must be classified as "spontaneous". Several syntheses of aromatic hydrocarbons by arc source or rf plasma discharge are cases in point (Ponnamperuma and Woeller 1964; Eck et al. 1966). In spite of the violent nature of the energy source, they gave compound distributions corresponding to (metastable) chemical equilibrium (Eck et al. 1966).

IV. EVIDENCE FROM METEORITES

Let us compare the compounds in meteorites (specifically, C1 and C2 carbonaceous chondrites) with those made by the two processes under consideration: Miller-Urey and Fischer-Tropsch-type (= FTT) reactions. [Our treatment will be very brief; more details and references are found in the reviews of Hayes (1967) and Vdovykin (1967) and Studier et al. 1968, 1972; Hayatsu et al. 1971, 1972; Yoshino et al. 1971.]

A. Aliphatic Hydrocarbons

The same few compounds dominate in meteorites and FTT syntheses (Studier et al. 1968, 1972; Gelpi, Han, Nooner, Oro 1970; Gelpi and Oro 1970). Normal (straight-chain) hydrocarbons are most prominent, followed by five slightly branched ones, with a CH_3-group replacing an H atom on the following carbons: 2, 3, 2+3, 3+4, and 4+5. Figure 2 shows a gas chromatogram[1] of meteoritic and FTT hydrocarbons.

At C_{16}, some 10^4 hydrocarbon isomers are possible; the fact that the same 5 dominate in meteoritic and FTT samples suggests that the meteoritic compounds were made by the FTT reaction or a process of the same, extraordinary selectivity. The Miller-Urey reaction, on the other hand, shows no such selectivity. Gas chromatograms of Miller-Urey hydrocarbons show no structure. Apparently all 10^4 possible isomers are made in comparable yield, as expected for random recombination of free radicals.

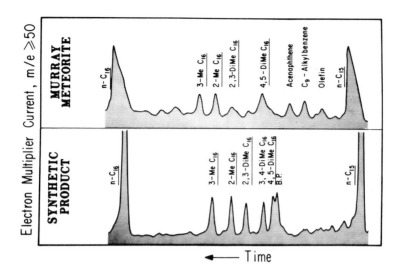

2. Gas chromatogram of hydrocarbons in the range C_{15} to C_{16} (from Studier *et al.*, 1972). The synthetic product was made by Fischer-Tropsch reaction.

Only 6 of the $\sim\!10^4$ isomeric hydrocarbons with 16 C atoms are present in appreciable abundance; 5 of these (underlined) are common to both samples. Acenaphthene, not detectable in this particular Fischer-Tropsch sample, has been seen in other prepared at higher temperatures (Studier *et al.*, 1972).

B. Aromatic Hydrocarbons

Meteorites contain a variety of polynuclear aromatic hydrocarbons, of up to 6 fused benzene rings. Essentially the same compounds form when CO, CH_4, or aliphatic hydrocarbons are heated to 700 - 1200° K for periods of days to seconds, with small amounts of hydrogen. The compound distribution resembles that for metastable equilibrium (Dayhoff, Lippincott, Eck 1964), regardless of whether the heat source was a resistance heater (Studier *et al.* 1968, 1972), a carbon arc (Ponnamperuma and Woeller 1964), or a plasma arc (Eck *et al.* 1966; Griffiths *et al.* 1970).

C Purines, Pyrimidines, and Other Nitrogen Bases

Several of the bases of DNA and RNA (adenine, guanine, *etc.*) have been found in meteorites, along with structurally similar nitrogen compounds (melamine, ammeline, *etc.*) of no biological significance. All these compounds and several more are produced, in yields of 0.1 to 0.5 per cent, by an FTT reaction in the presence of NH_3 (Hayatsu *et al.* 1968, 1972).

D. Amino Acids

Kvenvolden *et al.* (1970, 1971) have shown by an elegant, contamination-proof technique that the Murchison meteorite contains at least 17 amino acids, or compounds hydrolyzable thereto. A somewhat similar distribution of amino acids is produced in an FTT synthesis involving a brief initial heating to 500 - 700° C (Yoshino *et al.* 1971; Hayatsu *et al.* 1971). Such heating events, regenerating CO, might be expected in the solar nebula during collisions, chondrule formation, or passage of shock waves.

E. Porphyrins

Hodgson and Baker (1969) have detected pigments resembling porphyrins in several carbonaceous chondrites. Similar compounds were seen in FTT syntheses (Hayatsu *et al.* 1972).

F. Carbon-Isotope Fractionation

Meteorites show a very large difference in C^{12}/C^{13} ratio between carbonate and organic carbon: 60 to 80 per cent (Clayton 1963; Briggs 1963; Smith and Kaplan 1970). This trend remained unexplained for a number of years, because coexisting carbonate and organic matter on Earth shows a much smaller difference, typically 25 to 30 per cent. Lancet and Anders (1970) have shown, however, that the Fischer-Tropsch reaction gives a fractionation of the right sign and magnitude (60 per cent at 400° K, increasing at lower temperatures), owing to a kinetic isotope effect. A Miller-Urey reaction gave a fractionation of -0.4 ± 0.2 per cent under similar conditions (Lancet 1972).

G. Macromolecular Material

Most of the organic matter in meteorites consists of an ill-defined, insoluble solid, resembling humic acids in soil. It has an aromatic skeleton to which -COOH and -OH groups are attached. No attempts have been made thus far to produce or identify such material in FTT or Miller-Urey syntheses.

H. Conclusions

It appears that FTT reactions account reasonably well for all well-established features of organic matter in meteorites, except for one that has not yet been systematically investigated: the macromolecular material. The only alternative proposed, Miller-Urey synthesis (Urey and Lewis 1966; Kvenvolden *et al.* 1970), completely fails to account for the aliphatic and aromatic hydrocarbons and the carbon-isotope fractionations, though it is at least partially successful in the remaining areas.

A few details have yet to be resolved. All experiments on FTT syntheses have been conducted at CO pressures 5 to 8 orders of magnitude higher than those in the solar nebula. It has not yet been demonstrated that the reaction will proceed at reasonable rates at these low pressures, though there is reason to believe it will (Brecher and Arrhenius 1971, Lancet 1972). Thermodynamically the FTT reaction is certainly feasible. At the formation conditions of Cl chondrites ($358°$ K and $\sim 10^{-5}$ atm; see Anders 1972 for details and references), about 30 per cent the CO can transform metastably to $C_{20}H_{42}$. It is not clear, though, whether enough NH_3 will be present under these conditions. Figure 1 shows that NH_3 becomes less abundant at lower pressures. Only about 4×10^{-3} the total N will be present as NH_3 at $358°$ K and 10^{-5} atm if equilibrium is reached. Perhaps this was sufficient for the synthesis of nitrogen compounds; if not, it is conceivable that some NH_3 was brought in from denser parts of the nebula, or synthesized by shock waves.

V. INTERSTELLAR MOLECULES

Having established that FTT reactions were dominant in the solar nebula, we can check to see how well they account for the interstellar molecules identified to date (Table 2). Ten of the thirteen molecules in this table are indeed produced in FTT syntheses (Hayatsu, Studier, and Anders; unpublished work). Two others, NH_3 and COS, can be produced under similar conditions. Thus the only significant omission is formamide, a reactive molecule that may transform to other products at the relatively high pressures used in the laboratory experiments.

Before any final conclusion can be drawn, the light molecules in FTT syntheses will have to be studied systematically. If Herbig's hypothesis is valid, there ought to be rather close qualitative and even quantitative agreement between the FTT and interstellar distributions, with respect to both positive and negative identifications. Some allowance will have to be made for photochemical reactions during dissipation of the nebula (Sagan 1971). These would affect mainly gaseous molecules, less so those sequestered by grains.

As of the moment, the outlook seems rather favorable for the FTT reaction. The most primitive carbon retained in our solar system bears the imprint of the FTT reaction, and hence the lost C, too, probably had been involved in this process. There also exists circumstantial evidence linking interstellar molecules

Table 2

INTERSTELLAR MOLECULES OF ≥3 ATOMS

Molecules	Observed in FTT Synthesis (= F) or Meteorites (= M)
H_2O	F,M
HCHO	F
CH_3CHO	F
CH_3OH	F
HCOOH	F
HCN	F
HC≡CCN	
HNCO	F
$CH_3C≡CH$	F
HN_3	
CH_3CN	F
NH_2CHO	
COS	M

to protostars. Interstellar molecules tend to occur in regions of very high density, $> 10^6$ H_2/cm^3, *e.g.* the infrared nebula in Orion (Penzias, Solomon, Wilson, Jefferts 1971; Barrett, Schwartz, Waters, 1971). Star formation proceeds at a rapid rate in such clouds, and thus solar nebulae must have formed and dissipated. The lifetime of our solar nebula seems to have been rather short: $<$ 10^6 years, or less than the age of a typical cloud, *e.g.* the Orion Nebula (see

Cameron and Pine 1973, Anders 1972, for details and references). If these values are typical, even a young cloud should contain appreciable amounts of carbon cycled through solar nebulae. Abundances of interstellar molecules relative to CO are at least 2 orders of magnitude lower than yields in FTT syntheses (Penzias *et al.* 1971; Thaddeus, Wilson, Kutner, Penzias, Jefferts 1971). Unless this low abundance is ascribed entirely to their short photolytic lifetimes, even a small degree of star formation and CO processing would suffice to account for the interstellar molecules.

VI. BIOLOGICAL IMPLICATIONS

It is doubtful that interstellar molecules as such have any biological implications. If they are to play a role in the evolution of life on a planet, they must be delivered intact to the planet. Free molecules are out of the question, because their lifetimes against photodecomposition are much shorter than transit times to a planetary system: $< 10^2$ yr (Steif, Donn, Glicker, Gentieu, Mentall 1971) versus $> 10^7$ yr. Molecules embedded in grains have a better chance of survival, as they will suffer only moderate radiation damage from cosmic rays in transit. This is confirmed by the rather good state of preservation of organic compounds in meteorites, with cumulative radiation doses of $\sim 1 \times 10^9$ rads.

However, it seems likely that any such grains drifting into a planetary system from without will be completely swamped by similar grains from within. Let us consider our own planetary system as an example.

The Earth has received an average meteorite influx of 10^{11} g/yr during the past $\sim 10^5$ yr, as shown by measurements of Ir and Os in Pacific and Indian Ocean sediments (Barker and Anders 1968). Similar measurements on lunar soil have confirmed this figure, and have shown, moreover, that most of this material consists of micrometeorites of primitive composition, similar to that of carbonaceous chondrites (Ganapathy, Keays, Laul, Anders 1970). Presumably this represents cometary debris. If its carbon content is 3 per cent, the average influx rate of carbon on the Earth is 3×10^9 g/yr. Such a flux, maintained for 4.6×10^9 yr, would contribute about 3×10^{-4} the Earth's carbon. This is not a very impressive amount, yet it is large compared to the expected interstellar contribution. For a source at interstellar distances, the solid angle subtended by the Earth is so small, that the ratio of incident solar- to extra-solar-system material is $> 10^6$ even on the most optimistic assumptions (*cf.* Anders 1961; Heide 1963, p. 127). Especially if Herbig is right and interstellar grains are expatriates of growing solar systems, the return of these expatriates will contribute nothing new to the chemistry of the system.

I am grateful to M. H. Studier and R. Hayatsu for persevering in an unorthodox line of research. Figure 1 was computed by M. S. Lancet. This work was supported in part by NASA Grant NGL 14-001-010 and NGR 14-001-203.

[1]In this technique, a He-stream is used to flush the sample through a heated, 100-m long stainless steel capillary, coated with a non-volatile oil. Compounds emerge in the order of their volatilities, the least volatile ones spending the most time on the walls and therefore migrating most slowly. The effluent is continuously monitored on a mass spectrometer, to identify each compound. Deuterium rather than light hydrogen is used in these syntheses, to preclude contamination.

REFERENCES

Anders, E. 1961, *Science,* **133**, 115-116.

————. 1971 *Ann. Rev. Astr. Ap.,* **9**, 1-34.

————. 1972 in *Nobel Symposium 21 "From Plasma to Planet",* ed. A. Elvius (Stockholm; Almquist and Wiksell), pp133-152.

Anders, E.; Ganapathy, R.; Keays, R. R.; Laul, J. C. and Morgan, J. W. 1971, *Proc. Second Lunar Sci. Conf.,* 2, *Geochim. Cosmochim. Acta,* Suppl. 2, 1021-1036.

Bar-Nun, A.; Bar-Nun, N.; Bauer, S. H. and Sagan, C. 1970, *Science,* **168**, 470-473.

Barker, J. L., Jr. and Anders, E. 1968, *Geochim Cosmochim. Acta,* **32**, 627-645.

Barrett, A. H.; Schwartz, P. R. and Waters, J. W. 1971, *Ap. J.,* **168**, L101-L106.

Brecher, A. and Arrhenius, G. 1971, *Nature Phys. Sci.* **230**, 107-109.

Briggs, M. H. 1963, *Nature,* **197**, 1290.

Cameron, A. G. W. 1966, *Earth Planet. Sci. Lett.,* **1**, 93-96.

Cameron, A. G. W., and Pine, M. R. 1973, *Icarus,* in press.

Clayton, R. N. 1963, *Science,* **140**, 192-193.

Dayhoff, M. O. Lippincott, E. R. and Eck, R. V. 1964, *Science,* **146**, 1461-1464.

Eck, R. V.; Lippincott, E. R.; Dayhoff, M. O. and Pratt, Y. T. 1966, *Science,* **153**, 628-633.

Ganapathy, R.; Keays, R. R.; Laul, J. C. and Anders, E. 1970, *Proc. Apollo 11 Lunar Sci. Conf.,* 2, *Geochim. Cosmochim. Acta,* Suppl. 1, 1117-1142.

Gelpi, E. and Oró, J. 1970, *Geochim. Cosmochim. Acta,* **34**, 981-994.

Gelpi, E.; Han, J.; Nooner, D. W. and Oró, J. 1970, *Geochim. Cosmochim. Acta,* **34**, 965-979.

Griffiths, P. R.; Schuhmann, P. J. and Lippincott, E. R. 1970, *J. Phys. Chem.,* **74**, 2916-2920.

Grossman, L. 1972, *Geochim. Cosmochim. Acta,* **36**, 597-619.

Hayatsu, R.; Studier, M. H.; Oda, A.; Fuse, K. and Anders, E. 1968, *Geochim. Cosmochim. Acta,* **32**, 175-190.

Hayatsu, R.; Studier, M. H. and Anders, E. 1971 *Geochim. Cosmochim. Acta*, **35**, 939-951.

Hayatsu, R.; Studier, M. H.; Matsuoka, S. and Anders, E. 1972 *Geochim. Cosmochim. Acta*, **36**, 555-571.

Hayes, J. M. 1967, *Geochim. Cosmochim. Acta*, **31**, 1395-1440.

Heide, F. 1963, *Meteorites* (Chicago: University of Chicago Press), p. 127.

Herbig, G. H. 1970, presented at American Astronomical Society meeting, Boulder, Colorado. See also *Mém. Soc. Roy. Sci. Liege*, Tome XIX, 13-26 (1970).

Hodgson, G. W. and Baker, B. L. 1969, *Geochim. Cosmochim. Acta*, **33**, 943-958.

Keays, R. R.; Ganapathy, R. and Anders, E. 1971, *Geochim. Cosmochim. Acta*, **35**, 337-363.

Kvenvolden, K. A.; Lawless, J.; Pering, K.; Peterson, E.; Flores, J.; Ponnamperuma, C.; Kaplan, I. R. and Moore, C. 1970, *Nature*, **228**, 923-926.

Kvenvolden, K. A.; Lawless, J. G. and Ponnamperuma, C. 1971, *Proc. Nat. Acad. Sci.*, **68**, 486-490.

Lancet, M. S. 1972, Carbon-Isotope Fractionations in the Fischer-Tropsch Reaction and Noble-Gas Solubilities in Magnetite: Implications for the Origin of Organic Matter and Primordial Gases in Meteorites. Ph.D. Thesis, University of Chicago.

Lancet, M. S. and Anders, E. 1970, *Science*, **170**, 980-982.

Larimer, J. W. 1967, *Geochim. Cosmochim. Acta*, **31**, 1215-1238.

Laul, J. C.; Keays, R. R.; Ganapathy, R.; Anders, E. and Morgan, J. W. 1972, *Geochim. Cosmochim. Acta*, **36** 329-345.

Lemmon, R. M. 1970, *Chem. Rev.*, **70**, 95-109.

Miller, S. L. 1953, *Science*, **117**, 528-529.

————. 1957, *Ann. N. Y. Acad. Sci.*, **69**, 260-274.

Onuma, N.; Clayton, R. N. and Mayeda, T. K. 1972, *Geochim. Cosmochim. Acta*, **36**, 169-188.

Oró, J. 1965, in *Origins of Prebiological Systems and of Their Molecular Matrices*, ed. S. Fox (New York: Academic Press), pp. 137-171.

Penzias, A. A.; Solomon, P. M.; Wilson, R. W. and Jefferts, K. B. 1971, *Ap. J.*, **168**, L53-L58.

Ponnamperuma, C. and Woeller, F. 1964, *Nature*, **203**, 272-274.

Sagan, C. 1971, this volume.

Smith, J. W. and Kaplan, I. R. 1970, *Science*, **167**, 1367-1370.

Stief, L. J.; Donn, B.; Glicker, S.; Gentieu, E. P. and Mentall, J. E. 1971, *Ap. J.*, **171**, 21.

Studier, M. H.; Hayatsu, R. and Anders, E. 1968, *Geochim. Cosmochim. Acta*, **32**, 151-174.

————. 1972, *Geochim. Cosmochim. Acta,* **36,** 189-215.

Thaddeus, P.; Wilson, R. W.; Kutner, M.; Penzias, A. A. and Jefferts, K. B. 1971, *Ap. J.,* **168,** L59-L65.

Urey, H. C. 1952a, *The Planets* (New Haven: Yale University Press).

————. 1952b, *Geochim. Cosmochim. Acta,* **2,** 269-282.

————. 1952c, *Proc. Nat. Acad. Sci.,* **38,** 351-363.

————. 1953, in *XIIIth Intl. Congress of Pure and Applied Chemistry: Plenary Lectures,* pp. 188-214

————. 1954 *Ap. J. Suppl.* **1,** No. 6, 147-173

Urey, H. C. and Lewis, J. S., Jr. 1966, *Science,* **152,** 102-104.

Vdovykin, G. P. 1967, *Carbon Matter of Meteorites (Organic Compounds, Diamonds, Graphite)* (Moscow: Nauka Publishing Office).

Whipple, F. L. 1966, *Science,* **153,** 54-56.

Whipple, F. L. 1972, in *Nobel Symposium 21 "From Plasma to Planet",* ed. A. Elvius (Stockholm: Almqvist and Wiksell), pp. 211-230.

Wood, J. A. 1962, *Nature,* **194,** 127-130.

————. 1968, *Meteorites and the Origin of Planets* (New York: McGraw-Hill).

Yoshino, D.; Hayatsu, R. and Anders, E. 1971, *Geochim. Cosmochim. Acta,* **35,** in press.

Greenberg: With regard to your comment on the lack of penetration of our solar system by interstellar particles, I should like to mention that the present amount of zodiacal light could be significantly due to particles in a moderate density cloud through which we are presently passing.

Salpeter: There are now upper limits on the UV radiation incident upon us from directions almost normal to the galactic plane. This gives upper limits to the scattering of starlight by particles above us and indicates that we cannot be in a particularly dense interstellar cloud.

Woolf: In the past year I have been partly to blame for arguing that organic materials are produced in stellar envelopes. Here Dr. Anders suggests that they are produced in pre-planetary system nebulae. Other speakers point out that they can be produced in interstellar space.

It seems that the similarity of the processes that can occur in these different regions should be seen as an analogy, and not that the material all moves from one region to another. The only material in interstellar space for which we have definite evidence of production elsewhere are the minerals that make up the grains.

Buhl: I would like to suggest that the collision of a large comet with the earth may have introduced a large amount of organic material into the earth's oceans. This would provide a spectacular increase in the organic material at a particular place on the earth which may have provided the event triggering the origin of life.

Thermodynamics and the Origin of Life

Philip Morrison

Physics Department
Massachusetts Institute of Technology
Cambridge, Massachusetts

I. INTRODUCTION

The task defined for me by the committee for the conference is one which could inspire many results, but so far has not. I have to restrict myself to a pedagogical account, in which I want to speak fairly dogmatically. I believe what I say; I think I can persuade you in a longer time that most of it is true; I hope that there will be some dissidents who will raise a point or two at the end. But this dogma has the support of what I think the most substantial accomplishment of Physics since Archimedes, indeed, since Newton's time: namely, the laws of thermodynamics. The laws of thermodynamics are, of course, so general and so powerful mainly because they subtly restrict themselves to the broadest statements. Therefore, though you can never contradict them, they do not help you much; that's the first principal of thermodynamics, the minus-first law. You had better not violate them, that's all! We physicists know very well the second law, say, is only as good as it has to be and not perfect, since $N \ll \infty$; but it will work every time well enough, provided you have good enough statistics; that's like all physics, and therefore it's fair enough to say that it is not a separate domain.

The lesson I'm going to preach to is this one: all that we're hearing today, even the origin of life, is a beautiful exemplification of the true laws of thermodynamics, in a perfectly straightforward fashion. How they come in will vary in detail from model to model, and a little bit of that is what I want to talk about. That is the frame of mind in which I'm beginning.

II. THERMODYNAMICS AND TIME

First, I want to claim that thermodynamics is a study of systems idealized to such a degree they hardly ever really exist. It deals with equilibrium, or if you are fancier, it deals with linear departures from equilibrium, which you can treat pretty well too. It defines as its most sacred concept the idea of temperature equilibrium. But nothing is ever in equilibrium, or ever has a temperature; that's the way of the world. Of course if that were not true there would be no observers, no observations, no interests, all would be made of a vapor of ^{56}Fe, or something like that, and the world would be quite uninteresting. The great trick is not said in the text books of thermodynamics: the world is sufficiently hierarchical, or at least our science is sufficiently hierarchical. You're always prepared to make dominant approximations. In the gas system on the chemist's bench he has pure hydrogen gas. He says he has thermal equilibrium as soon as the thermometer stops moving, with good lagging and the hot water running for a few minutes. He doesn't worry about the fact that his sample is truly undergoin nuclear change, and converting very slowly into deuterium. He holds that it never happens at all, and he's quite wise to do so!

On the other hand, as you recognize, he's not exactly one hundred per cent right. In a very long time he would be quite wrong. If he simply raises the temperature of the system, he will quickly lose control of this particular variable. Then he's got to start over again and say a new kind of nuclear equilibrium is approaching. He leaves it up to other people to try to handle that equilibrium. The point is that thermodynamics pretends there's no such thing as time; what it really amounts to is that thermodynamics is complementary to time. If anything changes with time, you can measure it in time and forget thermodynamics; I think that's an honest statement. Of course, as always, time is the most important variable. The key is that you can find a hierarchy of relaxation times, or of frequency bands, so that you can say I'm going to ignore everything that hasn't happened yet, and I'm also going to forget how anything happened in the past, if it happened so long ago that it's no longer responsible for detectable changes. Those are the two sides of every thermodynamic approximation. The judgement lies in picking out that right time scale. In my view, the world can be described poetically but not inaccurately as just a play of these limiting times in different regions.

III. THE IMPORTANCE OF FREE ENERGY

Second, you may deal with systems which are not necessarily closed systems, most interesting for astronomy. Think of an analogy such as a plane, an automobile, even the driver of the automobile, and the earth. These things approximate for reasonable times, not closed thermodynamic systems but steady-state systems. Everybody would recognize that energy input is needed. But evidently that is not really fair, because in the first place, if energy were needed alone, it would be reversible by definition. There's no reason why energy doesn't go both ways as all the paradoxes lead you to say. Yet all the plane, the automobile, the driver and the earth achieve is to take energy in and put energy out, averaged over a reasonable time, in the same amount; they do not store it (except in active stages of growth, *etc.*). The point is there is something else which they're really feeding on.

What they feed on of course, is not energy, but is what is called free energy, F. In a very general form can be written in this fashion $F = E - TS$ where the system is defined for any thermodynamic state. No system can spontaneously increase that quantity, that's of course the famous principal. It always decreases, or maybe remains constant; when it doesn't decrease, and there's any event you can call a change then you say that is equilibrium. (Experts will know that there are other variables I might use but they don't change the essential feature.) These two terms are invented to take care of the two laws of thermodynamics the first of which is in one term, E, and the second of which is the other term,

-TS. The order measure is in the negative term, and the energy content in the positive term. The balance of these two is the driving force of all change. (In this case the usual variables, temperature and volume, have been chosen; it's not necessary to stick to that.) Now in the steady-state it is true that energy goes in and out. But it balances. In a way that is not very interesting, conservation of energy guarantees that you don't learn more from it at all.

But irreversibility is characteristic of all these systems. The gas definitely gets burnt and sent out the exhaust, the man eats (he does not uneat, *etc*). What is used is free energy. The notion that energy is what is supplied by the sun to the earth, for example, is quite unfair. By the same token, draw an envelope around the earth, and the sun puts in its energy and the earth puts out just about the same energy in the steady state; there's not much difference.

It isn't just the energy. It is, in fact, a steady state. If you made the polar diagram of the radiation you would see at $6000°$ K a half-degree needle beam from the direction of the sun. Then many tiny little outgoing beams, very, very, many of them, all over the earth. An optical receiver would see all over the earth, lots of little light filaments all sending up little beams, distributed in all directions, but very tiny compared to the sun, in total under a part in ten thousand. The presence of those little beams is some kind of a sign of the availability of the free energy which this energy coming in represents. If I draw the rest of the radiation then I have to drop down to say $300°$ K, room temperature. Of course I have a nearly isotropic curve, the day side a little warmer than the night side; it's this change between this sharp needle with photons of $1/2\ \mu$ and this near-sphere with photons of $20\ \mu$ that is the sign of the machine of the earth and of all that's on it. The energy flow is not different; the photon number of course *is* changed. The free energy of the photons can be calculated. They go out in a very different bag; that difference of free energy is what makes it work. This is characteristic for all the mechanisms which one finds which make things work.

Interstellar clouds must work on free energy content; they must derive irreversibility from a flow of free energy from somewhere. It can be stored in many ways; typically you think of stars, just as the earth depends on a star. One remark which can be made at this point is also very general. Since by definition, so to speak, the flow is irreversible, it destroys the equilibrium conditions which demand detailed balancing for every microscopic process. It biases somehow against detailed balancing; if that happens, as soon as you have quasi-diodes between generalized states describing what the subsystems are part of, as soon as I somehow differentiate the directions of change, and start pumping (or unpumping) in familiar fashion, I will be able to describe a cyclical process. Therefore, the irreversible flow of free energy generally tends to generate cyclic procedures, as part of the way of tapping the free-energy current which makes the whole thing run. If you look at the examples we've seen that's characteristic in almost all proposals. Very few of the models we've seen fail to show it.

IV. LIVING SYSTEMS

From my point of view, living systems, all life, fall under exactly the same rules. I would not take back any statements that I've made so far one bit for the case of living systems. I don't think there's any difference for them in a qualitative thermodynamical account. But the thermodynamical difference, which is a profound difference, is quantitative in nature. What it says I cannot yet formulate with precision, but it is about this — in life, in all life, a fair fraction (not something enormously small compared to one, but maybe as small as 1 per cent or 10 per cent) of the free-energy current can be found by looking at the internal structural information content of the system. Once a system has achieved that degree of structure, we seem to recognize it is living. Systems which have much less structure than that but still some, are easy to demonstrate; indeed, they are all weak analogs to life.

To discriminate them from life is not a trivial matter, but to formulate the discrimination is easy. Consider the growth of a crystal, or the burning of a flame (which of course can reproduce itself just as the crystal does), or the ingenious artifacts of Penrose and other people, who made small toys which put themselves together in one precise form; once you throw a particular "seed" into the mix, the beings multiply. Self-duplicating! But on the quantitative test they fail by 20 orders of magnitude, because all the free energy of moving that heavy one-centimeter object around is represented by the storage of one bit only of structural information, essentially one unit of kT: A fits into B, but not the other way around. That's where the analogy to life fails; the toy is too simple.

But the molecular polymers built by assembly out of monomers of a hundred or a thousand at most atoms — we would call them large molecules, but the biologist calls them small — are the substructure of all life. Only a few hundred of those molecular monomer species, but all arranged in a fantastically complicated integrated sentence, to pick up a figure of speech that everybody knows, "spell out" living forms. It's the presence of such sequences, where the energy difference, step by step, is negligible, but the free energy store is large because the complex order is fixed, which is the characteristic property of life: S is low. There's not much energy difference whether you put in one base pair or another base pair in making the DNA; in fact if there were much difference, then the whole thing would work badly. (There is a little difference; one can find "nonsense" DNA which is just repetitive, just like a teletype running, giving you AB, AB, AB; it doesn't really spell anything, but it shows some signs of the channel through which it's made, or of the free-energy differences where they bond together.)

I believe, moreover, you need to define life by structure. It is not physics, in my view, to define it wholly by function. Such functional definitions are sensible enough: you see a thing and you say it reproduces, it's irritable, *etc.*; that's an operationally significant scheme. But it is not a way that fits in to thermodynamics; to fit in to thermodynamics you have to be able to look at structure, to say what are the positions in space and time, and to define a living system in some way in terms of that pattern. I believe that must be possible. The best argument, of course, is that if you are very careful on a small scale you can freeze everything in life (except the electron motions, which of course you can't freeze at all), but you can cool the organism down to milli-degrees K, and reverse it. So it's not any dynamic structure, at least on the level of the cells; I believe that's probably generally true, certainly up to the level of the central nervous system, where I agree the electrical interactions might just possibly have something intrinsically non-static.

V. CONCLUSIONS

We thus have a rule: no elaboration of any kind of complicated systems which store structure inside the system, like molecules out of star-light upon atoms, *unless you have a flow of free energy*. Once you have any such flow, then you can expect cyclical processes which store order within. That flow can be maintained by any sort of gradients possible in the concentrations of any energy-containing systems. A star certainly radiates to a black universe. But inside a star, in the tremendous flow of heat and light, under the surface, I cannot use that flow at all to make any kind of ordered system, until I contrive for myself a little black box (a little empty box, covered with reflecting metal so that inside is a vacuum, free of photons; then I could very well pump energy through a small port into that little system). But until you can have a structure of that sort you won't be able to do it; the presence of energy alone is not enough; there must be free energy, which of course gradients can bring if the system has finite extent.

I have a few final remarks to make. One, there is no sharp single test, it seems to me, for living systems or for quasi-living systems or pre-biological systems; the test must be a detailed, complex, and quantitative one. Nobody will convince me that finding any precursor molecules, even optical activity, even complicated morphology, none of these things in themselves is an adequate statement that we have here something which is *necessarily* living. The full statement must involve some real tests of the density of complexity, which we always associate with the living systems we have. That doesn't mean that there aren't plenty of interesting partial cases; it is, by the way, very attractive that we find in pre-biotic materials non-optically active mixtures of most amino acids, which suggest to us that the special historical event which selected one channel out of two identically stable channels — a trick which can only come about through history — was not done there, though it is done in all living systems on the face of the earth. Even if it turns out that we had only the one, I would blame that on some initial asymmetry of the external circumstances. I wouldn't be quite convinced that we were dealing necessarily with the historical work of living forms. There is too little order in 1 binary choice!

Two, this afternoon's talks led very much to these remarks, because it seems to me that we are presented with three models: i, the large interstellar cloud; ii, the condensing circumstellar regime, as in the early solar system; or iii, a merely planetary atmospheric-gas and water system, each fed with lots of photon or particle free energy. All three appear to be capable of molecular build-up; all three are examples of the same general thermodynamic principal. Are they only analogies for each other, or are they genetically related? Is the earth surface a descendent of the interstellar cloud? Are we made of molecules generated in some early HII region? I'm somewhat skeptical. Rather, we are seeing in every case what happens when a current of free energy is directed on simple atoms, atoms with the complex potentiality for cohesive rearrangements which our human presence demonstrates they have. That same general process has begun independently at three distinct scales in the Galaxy.

Third, let me remind you: when I land at an airport I always have an unpleasant feeling for 10 or 20 seconds that I am there in configuration space all right, but I'm not yet there in momentum space: I hope the brakes will work! That's the same situation for all the phenomena we talk about. It is much easier to make some material structure where you want it, with the velocity typical of its entire surround, than it is to get it there across a big gap, either of space or of velocity; those two are equally important. That's why it's not easy to export these delightful products from one place to another, unless you find really a very clever way of enveloping them, or decelerating them, or otherwise something like landing at the airport. It is quite plain that being at the same point is not enough; to be home, you've got to be going at the same velocity as home. Otherwise when you interact with home, the result may be irreversible.

Finally, I think we need to calculate quantity. We're not now able to do that. The free-energy flow per square centimeter per second is what determines the rates of chemical reactions. The rates vary steeply for complicated ones, and mildly for very much simpler ones. The quantitative results are going to discriminate for us which step led to the biological systems we know about.

Of course, nothing I say should be construed to mean that there might not be quite other very complex biological systems, very different, never historically related to us, though with an origin of the same kind as our own; analogies to us, as perhaps the interstellar clouds and the chondrites are two distinct analogies to our own earthborn small-molecule legacy. That is a very important problem which we have alluded to both seriously and facetiously during this meeting; I hope that the radio astronomers, who have I believe the best means of inquiry, will not forget that possibility, though I don't yet ask them to put it high on the list of where to point the dish on Monday!

REFERENCES

Morowitz, H. J. 1968, *Energy Flow in Biology* (New York: Academic Press).
Morrison, P. 1964, *Rev. Mod. Phys.*, **36**, 517.
Snell, F. M., editor, 1967 on, *Progress in Theoretical Biology*, **1** and later volumes (New York: Academic Press).

Organic Chemistry and Biology
of the
Interstellar Medium

Carl Sagan

Laboratory for Planetary Studies
Cornell University
Ithaca, New York

I. INTRODUCTION

With the discovery of the microwave lines of formaldehyde and of hydrogen cyanide in the interstellar medium (Snyder *et al.,* 1969; Snyder and Buhl 1971) it became clear that there is an emerging field of interstellar organic chemistry (Sagan 1970, Sagan and Khare 1970). Aldehydes and nitriles are the precursors of amino acids in the Strecker synthesis and HCN is the precursor of the purine adenine in experiments in prebiological organic chemistry (Calvin 1969; Kenyon and Steinman 1969; Horowitz *et al.,* 1970). With the more recent discoveries of cyanoacetylene, an important condensing agent in prebiological organic chemistry, formamide, methanol, formic acid (perhaps), acetaldehyde and acetonitrile (Turner 1971; Zuckerman *et al.,* 1971; Ball *et al.,* 1971; Solomon *et al.,* 1971; Ball *et al.,* 1971), the existence of an interstellar organic chemistry has become secure.

The interstellar medium is composed primarily of hydrogen and helium, with C, N, O, and Ne present in a total relative abundance by number $\sim 10^{-3}$. At 10^{-4} to 10^{-5} are Mg, Si, S, and Fe. Molecular abundances should reflect these atomic abundances. Between clouds of interstellar material the density is 0.1 cm^{-3},. The most common dark clouds have densities $\sim 10^3$ cm^{-3}; recent work on the repopulation by collision of excited states of CH_3OH, CS and H_2O, which have short natural lifetimes, imply densities in some clouds of 10^6 to 10^8 cm^{-3} or larger (Barrett 1971, Solomon 1971, Litvak 1970). In contrast to optically large clouds (Spitzer 1968) with densities ~ 20 H atoms cm^{-3}, masses of 20,000 solar masses, and radii ~ 20 pc, the much denser clouds determined by radio techniques have hydrogen in the molecular rather than the atomic form. Much denser clouds yet must exist because stars form from gravitational instability in dense clouds. Since the mean density of late spectral type main sequence stars is ≥ 1 gm cm^{-3}, there must be occasional clouds with densities $\geq 10^{15}$ times the highest densities detected directly. Typical kinetic temperatures in an ordinary cloud are $\sim 100^\circ$ K; in a dark cloud, $\sim 10^\circ$ K; and in intercloud regions $\sim 10^3$ to $10^{4\circ}$ K.

II. GRAIN SIZES AND COMPOSITION

Dark clouds are dark because of interstellar grains, inferred from the patchy extinction and polarization of starlight traversing the interstellar medium. A typical dark cloud has an extinction at visual wavelengths of 4 or 5 magnitudes. From rocket and satellite observations the extinction in the ultraviolet is now known (Houziaux and Butler 1970): clouds with an extinction of a few mag at 5000 Å exhibit extinctions of 7 or 8 mag at 2000 Å, and 10 to 15 mag short of 1200 Å. The extinction increases from long to short wavelengths through the visible (interstellar reddening), has a local maximum at about 2000 Å, reaches a minimum near 1600 Å, and then rapidly increases to shorter wavelengths. The shape of this curve has been used to infer grain composition. Mixtures of silicates and graphite are a currently fashionable (*e.g.*, Gilra 1971), but hardly unique, explanation of the extinction curve. Independent evidence for silicates exists from the 10μ reststrahlen bands (Stein and Gillet 1969; Jackwell *et al.*, 1970). The failure to find a strong infrared absorption near 3.1μ excludes > 15 per cent water frost on the grains (Knacke *et al.*, 1969; Hunter and Donn 1971).

The grains are usually said to have an effective radius, $a \approx 10^{-5}$ cm. (interstellar polarization implies nonspherical grains.) But this is an artifact of the wavelength, $\hat{\lambda}$, at which the grains are observed. For approximately spherical particles, the dimensionless extinction function $Q(a) \propto a^4$ in the Rayleigh domain (small particles, $a < \hat{\lambda}$), and $Q(a)$ is independent of a in the geometric domain (large particles, $a > \hat{\lambda}$), where $\Lambda \equiv \lambda/2\pi$ (van de Hulst 1957). The cross-section for extinction is $\pi a^2 Q(a)$. The effective particle size observed will then be

$$\langle a \rangle = \int_{0}^{\hat{\lambda}} a^3 Q(a)\, f(a)\, da \Big/ \int_{0}^{\hat{\lambda}} a^2 Q(a)\, f(a)\, da$$

for the Rayleigh domain, and an exactly similar expression, but with the limits of integration from $\hat{\lambda}$ to ∞, for the geometric domain; $f(a)$ is the size distribution function for the grains, which we here take as $\propto a^{-\nu}$, where ν is a dimensionless constant. The Rayleigh $Q(a)$ has a λ^{-4} wavelength dependence, which drops out in the formulation of $\langle a \rangle$. For both the Rayleigh and the geometric domains we find $\langle a \rangle \approx \hat{\lambda}$, provided $\nu \neq 3$, $\nu \neq 7$ to avoid singularities. Most material size distributions, from rock flour to asteroids to interstellar clouds, are fit well by such an $f(a)$ (except for the very smallest particles), and exhibit $7 > \nu > 3$. Thus observers in visible light ($\lambda \approx 5500$ Å) will deduce $\langle a \rangle \approx 10^{-5}$ cm for any of a wide variety of real particle size distributions. Just the same conclusion follows for other, physically quite different, effective particle sizes — *e.g.*, $a = \int a\, f(a)\, da \Big/ \int f(a)\, da$, with the same limits of integration as before. Thus, in the absence of other arguments, we conclude that there may be substantial numbers of grains $\gg$ or $\ll 10^{-5}$ cm.

The ultimate limits on the total mass in grains are placed dynamically. The UV observation of increasing extinction to shorter λ is expected from $Q(a)$ if there are more small particles than large ones, as our choice of $f(a)$ implies. At some shorter λ there should be a leveling off of extinction due to the expected truncation of $f(a)$ for small particle sizes. If we are to understand the generation of large grains from small ones and the degradation of large grains to small ones it will be important to search for this extinction turnover.

III. RELIABILITY OF MOLECULAR IDENTIFICATIONS

Most of the interstellar organic molecules found by microwave radioastronomy have been found in the large radio source Sagittarius B2 towards the galactic center, and in such other regions as W51 and the infrared source in the Orion nebula. The number of lines being discovered is so great at the present time that serious reservations on their identification will soon be appropriate. Identifications have often been made on the basis of a single microwave line. In the two intervals in the mm wave region between 85 and 90 GHz and between 110 and 120 GHz there are, at the time of writing, at least 14 lines identified, *i.e.,* $\sim$ 1 line/GHz. Several more unidentified lines exist. The total widths of such lines in Sgr B2, *e.g.,* are $\sim$ 15 MHz. Thus if we search for a single line of a molecule of interest there is a probability $>$ a few per cent that it will randomly coincide with a line already discovered. This calculation assumes that the rest frequencies of the interstellar lines have been correctly estimated; if not, the probability of a misassignment increases still further. As long as it was believed that only a few molecules existed with detectable transitions in these frequency ranges, identification on the basis of single lines appeared to be acceptable. Indeed few people seem to have anticipated the great variety of moderately complex and detectable interstellar molecules. But at the present time, when the number of lines is increasing very rapidly, identification on the basis of a single line must be viewed with increasing caution. Discovery of several lines, and particularly appropriate hyperfine splitting and isotopic lines, are required for a secure identification. An interesting possibility for the future is the confirmation of many of these molecular identifications by orbiting ultraviolet spectrometers examining stars at the peripheries of dense molecular clouds. Molecules which are indetectable at microwave frequencies may be accessible in the UV, and, while the UV detectivities from orbit are not competitive with microwave detectivities, they are still respectable.

IV. ORGANIC SYNTHESIS AT HIGH DENSITIES

According to Townes a typical dark cloud with $H_2 \approx 10$ cm^{-3} might have abundances of CO $\approx 10^{-1}$ to 10^{-2} cm^{-3}; HCN $\approx 10^{-4}$ cm^{-3}; NH$_3 \approx 10^{-3}$ cm^{-3}; HCHO $\approx 10^{-4}$ to 10^{-5} cm^{-3}; and water in as yet undetermined abundances. This immediately suggests that much of the O, N, and C atoms in the clouds exist in molecules and in grains. In such a cloud one pc across there is approximately one solar mass ($M_\odot$) of these simple molecules. With a usual number density ratio of grains to atoms $\sim 10^{-12}$, there is also ~ 1 $M_\odot$ of grains in such a cloud.

The array of interstellar molecules known to exist in late 1971 bears a tantalizing relation to two other molecular arrays: those detected by optical spectroscopy in cometary comas and tails, and those identified in experiments in prebiological organic chemistry, starting from such simple precursors as methane, ammonia, and water (cf. Tables I and II). The radicals detected in cometary spectra look very much like the dissociation fragments of the identified interstellar molecules. The gas phase products synthesized in experiments in prebiological organic chemistry bear a very close relation to the interstellar molecules. For example, apart from CO and hydrocarbons, the most abundant gas phase products found in sparkings of simulated primitive atmospheres (Sagan and Miller 1960) and in quenched thermodynamic equilibrium computer simulations of such experiments (Sagan *et al.,* 1967; Lippincott *et al.,* 1967) were HCN, CH_3CN, HCHO, and CH_3CHO. All of these molecules have now been reported in the interstellar medium (Snyder *et al.,* 1969; Snyder and Buhl 1971; Turner 1971; Zuckerman *et al.,* 1971; Ball *et al.,* 1971; Solomon *et al.,* 1971; Ball *et al.,* 1971). All are produced by a variety of energy sources.

But it is very difficult to understand the synthesis of such molecules by two-body reactions in the extraordinarily diffuse interstellar medium. Three-body reactions seem to be required, at least to make the more complex molecules. Typical three-body reaction rate constants are known both from

TABLE I

COMPARISON OF INTERSTELLAR AND COMETARY COMPOUNDS

Microwave Identifications of Interstellar Molecules	Optical Identifications of Cometary Radicals
OH, H_2O	O, OH, OH^+
CO	CO, CO^+, CO_2^+
NH_3	NH_2, NH, N_2^+
CN, HCN	CN
$HCHO$, CH_3CN, HC_2CN, HC_2CH_3, many other organics	CH, CH^+, C_2, C_3

experiment and from first principles (*e.g.*, Glasstone *et al.*, 1941) to be $\sim 10^{-32}$ cm^6 sec^{-1}. If the reacting constituents were present at densities of 1 cm^{-3} there would be $\sim$ 1 three-body reaction km^{-3} in the lifetime of the galaxy. On the other hand, at a reactant density of 10^{11} cm^{-3}, there will be $\sim$ one three-body reaction per molecule in 300 years, a respectable rate. But regions with this density of minor constituents are not typical interstellar regions, and are probably collapsing clouds well on the way to star formation if they have sufficient total mass. Also, such high-density regions are UV-opaque; while in low-density regions the photodissociation timescale, calculated below, is $\ll$ the three-body reaction timescale.

V. SYNTHESIS IN CONDENSING CLOUDS

A cloud of such composition, collapsing at low temperature, will produce complex molecular species, particularly as the stars which form out of the cloud reach the high luminosity or Hayashi phase of their pre-main sequence stellar evolution. Shock synthesis of organic compounds is known to be of remarkably high efficiency (Bar-Nun *et al.*, 1970); such synthesis should occur within a large condensing cloud after a bright O or B star inside the cloud turns on. While temperatures in the early solar nebula of our own solar system are expected to be mainly $\leqslant 300°$ K, pressures $\geqslant 10^{-6}$ bars are calculated (*e.g.*, Jastrow and Cameron 1963, Urey 1966), and the sun at one point in its evolution had a luminosity $\sim 10^3$ times its present value (Stein and Cameron 1966, Ezer and Cameron 1967). These conditions sufficiently resemble UV irradiation experiments in prebiological organic chemistry that we can expect the same molecules to have been produced in the early solar nebula (See Table II). In addition to the simple gas phase products of the sort identified to date by microwave radioastronomy, much more complex molecules, including amino acids, sugars and nucleotide bases have been synthesized. Further, a characteristic intractable brownish polymer forms in large quantities in all such experiments, and in the absence as well as in the presence of liquid water (Sagan and Khare 1971). There is every reason to expect that collapsed prestellar clouds will contain such molecules, as well as the comparable Fisher-Tropsch products which resemble carbonaceous chondritic organics (Studier *et al.*, 1968). Organic chemistry in such a preplanetary cloud occurs at low temperatures, in contrast to the chemistry in the extended envelopes of red giants where the temperatures are sufficiently high to pyrolize most organic compounds of interest. As the local star goes through its Hayashi phase, the vast bulk of the surrounding material must be ejected – probably by electromagnetic and corpuscular radiation pressure – back into interstellar space. During ejection the material will be irradiated, and its chemistry will alter. If the particle size distribution

TABLE II

SIMPLE ORGANIC COMPOUNDS OF H, C, N, AND O
IN THE INTERSTELLAR MEDIUM AND IN EXPERIMENTS ON
PREBIOLOGICAL ORGANIC CHEMISTRY

	Interstellar Microwave Identification	Laboratory Identification
CO	X	X
C_2H_2	No accessible lines	X
C_2H_4	No accessible lines	X
C_2H_6	No accessible lines	X
HCHO	X	X
CH_3CHO	X	X
HCN, HNC	X	X
CH_3CN	X	X
HNCO, HCNO	X	
CH_3OH	X	X
C_2H_5OH		X
HC_2CN	X	X
HC_2CH_3	X	
HC_2NH_2		
HC_2CHO		
NH_2CHO	X	X
CH_3NH_2		
HCOOH	X	X
CH_3COOH		X
.		
.		
.		
Higher nitriles & polynitriles		X
Sugars		X
Amino Acids		X
Nucleotide bases		X
Porphyrins		X
Polycyclic aromatics		X
"Intractable polymers"		X

function of interstellar grains turns out to be relatively sharply peaked near a $\sim$ 10^{-5} cm, this may provide additional evidence for the origin of the grains in stellar ejection events: the ratio of radiation pressure to gravity is also peaked when a $\approx \lambda_{max}$, where λ_{max} is the Wien peak of the stellar radiation field. The radiation-stable fraction — *e.g.*, grains of intractable polymer — should preferentially survive the outward-bound journey.

Herbig (1970,1971) has inquired whether such dust ejection during the pre-main sequence evolution of stars of middle to late spectral type might be the principal source of the general population of interstellar grains. The mass of dust in the Galaxy is $\sim 10^{-2}$ the mass of gas, which is, in turn, $\sim 10^{-1}$ the mass of the Galaxy. Thus if $\sim 10^{-3}$ of the galactic mass is in dust, $\sim 10^{-3}$ of the mass of each star must be ejected as grains in this hypothesis. Models of the early solar nebula place its mass between 10^{-2} and 1 $M_\odot$ (Opik 1963), essentially all of which must have been dissipated into space. Thus between 10^{-3} and 10^{-5} $M_\odot$ of C, N, and O must have been ejected back into space — which, if the grains are primarily compounds of these atoms, is barely enough to explain the entire mass of interstellar grains, if we assume that the grain half-life is comparable to the age of the Galaxy and that there is not a significant mass of grains with a $\gg$ or $\ll \lambda$. Hence, at least an important fraction of interstellar grain material may be stable organics ejected during star formation. This hypothesis is to be distinguished from hypotheses of grain origin within stellar atmospheres — *e.g.*, of red giants.

VI. SYNTHESIS IN COMETS

The apparent close similarity between the chemistry of comets and the chemistry of the interstellar medium (Table I) must be explained. One possibility is that comets are aggregations from the contemporary interstellar medium. The principal version of this suggestion, made by Lyttleton, has been attacked on dynamical grounds (Opik 1963). If comets had a different isotopic composition — *e.g.*, $^{12}C/^{13}C$ — than the present solar system this would be consistent with an accretional rather than a primordial cometary origin. No such difference is known. It appears more likely that comets are remnants of the solar nebula, deep frozen for 5×10^9 years and bearing possible genetic relations to some Apollo asteroids (Opik 1963) and the Type I carbonaceous chondrites. They may be material formed in the inner solar system early and ejected to the distant comet cloud (Oort 1963), or preplanetary condensates from the earliest history of the solar system — bearing the same relation to the sun as globular clusters do to the centers of galaxies. Consequently there is every reason to expect a complex and cosmogonically important organic chemistry within cometary heads; a spaceborne gas chromatograph/mass spectrometer flying through a comet would be an experiment of some importance.

Can comets make a significant contribution to interstellar organic chemistry? The total mass of comets mostly resident at 3×10^4 to 3×10^5 astronomical units is 10^{-1} to 10^{-2} earth masses (Oort 1963). With present stellar perturbations of their orbits, the e-folding time for loss of such comets to the interstellar medium is $\sim$ the age of the solar system (Oort 1963). Close stellar encounters, which are infrequent, will dominate the perturbations, and considerably more cometary dissipation may have occurred over geological time than is extrapolated from the present. But it seems unlikely that $>$ one terrestrial mass of comets has been lost over geological time. Therefore comets contribute $\leqslant$ 10^{-5} $M_{\odot}$ to the interstellar medium. If this is typical of all stars, it is a factor of at least 100 too small to explain *all* the interstellar grains. Under the same assumptions just invoked, the mean time between close approaches to the Sun of such interstellar comets is easily calculated to be $\sim 10^3$ years, consistent with the absence of reliable reports of comets in hyperbolic orbits.

VII. DEGRADATIONAL ORIGIN OF SIMPLE ORGANICS

Thus by the dissipation of organics during star formation, and by subsequent loss of comets to the interstellar medium, organic material produced in the early history of solar systems must be making contributions to interstellar organic chemistry. Objects of cometary mass and even grains $\sim 10^{-5}$ cm in dimensions will dissipate in the interstellar radiation environment quite slowly. I therefore suggest that one source of the simple organic molecules discovered by microwave radioastronomy are organic grains (and larger aggregations of organic material), released from the grain by grain-grain and grain-molecule collisions, shock waves, and heat from nearby stars. Single photon events are unlikely to be very effective in grain degradation; from calculations below, it follows that a typical grain will absorb $\sim 10^3$ molecules of ices in the time for it to absorb a single ultraviolet photon. The simple interstellar organics found thus far can then be looked upon as the degradation products of more complex organic compounds, seen briefly until they are themselves dissipated. The microwave line astronomers may be merely skimming the surface of the problem, detecting molecular fragments spalled off large organic molecules which pervade the interstellar medium. This view is the antithesis of the more usual one in which the molecules responsible for the interstellar microwave lines are imagined to be synthesized within the interstellar medium from yet simpler precursors.

VIII. SYNTHESIS ON GRAINS

An alternative locale for the synthesis of moderately large interstellar molecules is on grain surfaces (Salpeter 1971). Because of grain surface defects, there are unbound valence electrons, and high sticking probabilities; and grains should be the locales of growing free radicals and simple non-fully saturated organic molecules. The bond energies connecting these radicals with the grain are comparable to the energies of typical long-λ UV photons in the interstellar radiation field. Collisions on the surface and in grain interstices provide a local environment for three-body reactions. The combination of adjacent radicals attached to the grain also occurs by quantum mechanical barrier tunneling (Salpeter 1971). This is relatively efficient for H and relatively inefficient for larger radicals. Possibly, synthesis on or near grains is not very efficient for larger organic molecules. A typical interstellar atom has a 50 per cent probability of having passed through one pre-stellar nebula during the lifetime of the galaxy. On the other hand an atom has had 10^3 grain collisions during the same period. It is entirely possible that the production efficiency in the nebular scenario is $>$ 10^3 times that on the interstellar grains. (It is even possible that liquid water was available in the solar nebula.) Whether grain or pre-stellar cloud chemistry dominates remains highly uncertain; it seems clear that both will make important contributions to interstellar organic chemistry.

In a recent set of experiments relevant both to interstellar grain and to prestellar cloud scenarios, Khare and Sagan (1971) have UV-irradiated at 77° K condensates of simple interstellar molecules (H_2O, NH_3, HCHO, C_2H_6) frozen on a quartz finger. The results are not expected to depend sensitively on the energy source employed. Some products are synthesized at 77° K, demonstrating low-temperature radical mobility in such systems. When the irradiation products were heated (as, *e.g.*, in the dissipation of a grain, or of a prestellar cloud), a range of more complex organic molecules were detected — some of which (*e.g.*, HCN, CH_3CN, CH_3CHO, CH_3OH) have been identified in the interstellar medium, others of which have not yet been identified. Reaction mechanisms of the Tischenko and Cannizarro type are suggested for certain of these syntheses.

IX. ULTRAVIOLET NATURAL SELECTION
OF INTERSTELLAR MOLECULES

In free interstellar space the radiation field is dominated at short wavelengths by bright O and B stars, and has an energy density $\sim 10^{-13}$ ergs cm^{-3} in the photolytic UV, peaked shortward of 2000 Å. With a typical photon energy ~ 10 eV this is $\sim 10^{-2}$ photons cm^{-3}, and corresponds to a flux $F \sim 3 \times 10^8$ photons cm^{-2} sec^{-1}. The mean lifetime, t_r, of the molecule is $\sim (\sigma F)^{-1}$, where σ is the mean photodissociation cross section over the wavelength region in which the mean flux is specified. For most of the molecules in question, $\sigma \approx 10^{-18}$ cm^2 (*e.g.*, Calvert and Pitts 1966), and $t_r \approx 100$ yrs. Some of the identified molecules, such as HCHO and CH_3CHO and some plausible candidates, such as CH_3COCH_3, H_2S, $(CH_3S)_2$, and $(C_2H_5S)_2$ have nontrivial absorption features near 3000 Å; because more photons are available to photodissociate such molecules, their lifetimes should be correspondingly shorter. Carbon monoxide, on the other hand, is not dissociated until 1110 Å (*e.g.,* Calvert and Pitts 1966) where σ and F are relatively low. Accordingly t_r for CO is ~ ten or more times larger than for aldehydes. Solomon (1971) finds, for Heiles Cloud 2, a sharp decline in the HCHO abundance at the cloud boundary, while CO can be found a considerable distance outside the cloud. This is in accordance with our rough calculations and emphasizes the importance of the interstellar radiation field in determining which molecules will be present. Similarly, despite the comparable cosmic abundances of nitrogen and oxygen the radio line ratio [CN]/[CO] $\approx$ 10^{-3} to 10^{-4}; this is consistent with the bond energies and σs of the cyanyl radical compared with CO.

Within clouds the UV flux may be substantially attenuated, and the lifetime of the molecules correspondingly increased. For an extinction of $A\lambda$ mag, the attenuation will be $10^{-0.4}$ $^{A\lambda}$. Lifetime of typical dense clouds are ~ 10^7 yrs. Hence for t_r of the contained molecules to be ~ the lifetime against dissipation of the cloud they reside in, the UV flux must be down by a factor ~ 10^4, corresponding to ~ 10 mag extinction in the UV, which is not excessive. Accordingly, the association of complex interstellar molecules with clouds is to be expected: they are produced there, and they are preserved there. But because the clouds have themselves lifetimes short compared to the age of the galaxy, the contained molecules must either have been made in a period $\leqslant$ the lifetime of the cloud, or transported to the cloud from elsewhere by a means which protects the molecules from UV. Star formation and ejection of radiation-stable organic grains provide a natural solution to this problem.

There remains a suspicion by some radioastronomers that ultraviolet-labile molecules are being seen at greater distances from dense clouds than would be expected on the basis of their t_r. It is difficult to be certain about such a matter because we do not know about possible very small dense clouds. But if grains are composed of organic compounds, their UV degradation readily provides small organic molecules far from dense clouds.

These examples of molecular differential stability suggest a kind of UV natural selection occurring in the interstellar medium. Those molecules which are relatively stable to interstellar UV may eventually achieve an abundance much higher than their relatively small production rates might suggest. Since t_r $\leqslant$ 10^3 yrs for small molecules, a very radiation-stable molecule can achieve significant abundance even if its production rate is down by a factor of 10^6 or 10^7. It is therefore most important to discover which jmolecules have such properties.

There have been no systematic studies of the prolonged irradiation of cosmic gases, but preliminary results (*e.g.,* Sagan and Khare 1971) indicate that there is a significant trend toward high molecular weight material — the intractable polymers, produced in relative yields $\gg 10^{-6}$. It is not difficult to understand such a result (Sagan 1957). First, large molecules are, all other factors being equal, more stable than small molecules: if a single bond is broken in a diatomic molecules, its constituent atoms fly apart and restructuring the molecule is unlikely. If the same bond is broken in a polyatomic molecule the steric forces provided by adjacent atoms will tend to heal the break. Secondly, the π electrons in heterocyclic organic molecules are able to absorb UV as an excitation rather than as a dissociation event much more readily than simpler molecules can. For these reasons high molecular weight cyclic compounds may be an important constituent of the interstellar grains and gas. Such molecules are found in the carbonaceous chondrites, which are probably the best samples of the Sun's prestellar nebula available to us; and which are possibly (Baldwin and Schaeffer 1971) the most abundant variety of meteoroids in interplanetary space.

X. INTERSTELLAR PORPHYRINS?

Johnson (1967a, 1967b, 1971) has in fact claimed to identify one such molecule from the heretofore unidentified diffuse optical interstellar lines. He was led to consider porphyrins as candidate interstellar molecules in part because of the possible agreement of the porphyrin Soret band with the most prominent unidentified interstellar diffuse line at 4430 Å. More recently, in an attempt to explain all 15 or 20 unidentified such lines by a single molecule, Johnson has called attention to the molecule he calls χ, $(MgC_{46}H_{30}N_6)$, a Mg - chelated tetrabenzoporphyrin with the magnesium in sixfold coordination, two of the bonds attached to pyridine molecules normal to the ring plane. The lines appear appropriately when χ is imbedded in an appropriate paraffin matrix at 77° K. As unlikely as such a phytol-free super-chlorophyll might seem as a pervasive constituent of interstellar space, it would have to be given serious consideration if, as Johnson claims, the 15 strongest lines of χ correspond, within a few Å, to the 15 strongest unidentified interstellar diffuse lines. But a close examination of the most recent report (Johnson 1971) shows that many laboratory lines of χ, which appear in significant strengths, do not correspond to any known interstellar lines. For example, between 6072 Å and 6670 Å there are 6 unidentified interstellar lines. In two runs reported by Johnson (1971) with different paraffin matrices, χ yields a close correspondence ($\pm$ a few Å) to the unidentified lines for 2 out of 8 lines on one matrix, and for 1 or 2 out of 13 lines for another. Consequently there is roughly a 15 per cent probability that one of 10 lines of a laboratory candidate will match the unidentified interstellar lines purely by chance; and a 2 per cent probability that two lines in 10 will correspond. (I am assuming that the line spacings are uncorrelated.) It is only necessary to examine a few dozen molecules which have this many lines in this wavelength region to achieve such coincidences.

There is an additional difficulty with Johnson's suggestion. If the molecule is imbedded in a different paraffin matrix; or if the molecule is chelated with, say, Fe; or if the pyridines are removed; or if we make one of the benzenes a methylbenzene or a phenol — in any of these cases widths and positions of large numbers of lines will alter drastically. Therefore, for Johnson's suggestion to be accepted, we must imagine the wholesale and preferential production of χ in the interstellar medium, with the production of all its congeners and potential precursors somehow prohibited. This seems unlikely in the extreme. There is no reason to anticipate the UV stability of χ to be significantly greater than that of many of its congeners or precursors. Accordingly, either χ is not the unique molecule which explains all the unidentified diffuse interstellar lines, or there is a pervasive interstellar organism which preferentially synthesizes the molecule. For reasons to be discussed shortly, the latter seems inadmissable; thus, the former seems inescapable.

But the possibility that the 4430 Å line is a Soret band appears to be promising, and it seems not unlikely that at least some of the additional unidentified diffuse lines will be found to be due to polycyclic aromatic hydrocarbons, porphyrins, and comparable complex organic molecules. It is a remarkable fact that the interstellar absorption feature near 2000 Å, which is often attributed to graphite, is also characteristic of polycyclic aromatic hydrocarbons (Donn 1965). Indeed, with the large ratio [H]/[C] in the interstellar medium, it would be quite remarkable if the unsaturated ring structure of graphite could survive very long with respect to the near-saturated polycyclic aromatics; the mean free time between collisions of polycyclic aromatics with H atoms in diffuse clouds is $\ll$ the mean time to photoejection of a H atom from such a molecule.

XI. INTERSTELLAR BIOLOGY

The existence of moderately complex organic molecules in the interstellar medium may suggest that such molecules are in some way connected with biological processes somewhere. There are two principal possibilities: (1) that such molecules support or are the metabolic products of an interstellar biota; and (2) that such molecules, participating in planetary condensation from the interstellar medium, can make a significant contribution to the origin of planetary life. We consider these possibilities in turn.

There is no *a priori* objection to interstellar organisms. With the temperature difference between the diluted $10^{4°}$ K interstellar starlight radiation field and the $3°$ K black body background radiation, a heat engine with an efficiency $>$ 99.9 per cent can be driven; and the existence of interstellar masers and refrigerators shows that major departures from thermodynamic equilibrium, such as characterize living systems, are not uncommon in the interstellar environment. We consider a hypothetical interstellar organism with a $\sim 10^{-5}$ cm – about the size of a rabies virus, or the smallest known free living organism, PPLO. Suppose this organism resides in a typical diffuse cloud with a total number density ~ 1 cm^{-3}, and with gas kinetic velocities v ~ 1 km/sec. The organism must be made mostly of C, N, and O, or atoms of lower abundance, the density of which we assume is n $\sim 10^{-3}$ cm^{-3}. The mean free time between collisions of the putative organism with such "heavy" atoms, $t_c \approx (n \, \sigma_c v)^{-1}$, where σ_c is the *collision* cross section. Thus $t_c \approx 1$ yr. If the organism is assumed spherical, with a density ~ 1 g cm^{-3}, it contains N ≈ 2 x 10^8 heavy atoms. Assuming an accomodation coefficient of unity, so all incident atoms stick, these calculations imply a replication time scale $\tau \approx 2$ x 10^8 yrs. Thus there could only have been ~ 50 generations in the history of the Galaxy, far too little time to evolve, through natural selection, an organism able to function in so inclement an environment. It is difficult to imagine the organism smaller than we have assumed; a larger organism only compounds the difficulty – σ_c increase as a^2 but N increases as a^3. It is unlikely that v is in error by as much as a factor 10. The only flexibility is in n. If n is 10^5 times larger, so we are considering a moderately dense cloud, then τ is reduced to ~ 2 x 10^3 yrs. But the cloud lifetime is only $\sim 10^6$ yrs, giving us ~ 500 generations, which is still not enough. As we go to higher densities we slowly gain generations as $\sim n^{1/2}$.

There is a residual possibility that organisms might arise in very dense clouds, or on low mass planets or satellites, where the evaporation of the atmosphere permits gradual adaptation of indigenous organisms to interstellar conditions; they then might be spewed back into the interstellar medium by mechanisms similar to those described earlier for grains. However, the interstellar radiation field seems to make these possibilities untenable. The mean lethal dose, in the critical 2000 - 3000 Å region where the nucleic acids preferentially absorb, for the most radiation-resistant bacteria and bacterial spores known (Hollaender 1955) is $\sim 10^4$ ergs cm^{-2}. In free interstellar space this dose is accumulated in $\sim$ 1 yr; *i.e.,* the time to accrete one heavy atom is about the same as the time to be killed. Near a star the time to acquire such a dose is, of course, much shorter. Again the difficulties can be circumvented in a dense cloud with 10 to 15 mag UV extinction, but again the cloud has such a short lifetime that any surviving organisms will soon be ultraviolet-parboiled. The effects of low-energy cosmic rays and especially X-rays are additional serious impediments.

XII. INTERSTELLAR CONTRIBUTIONS TO PLANETARY BIOLOGY

The second case of interest is that interstellar organic molecules, entrapped in a condensing cloud, make a significant contribution to the origin of life on planets formed from the cloud. If such a contribution is to be numerically significant it must compare with the production of organic compounds in primitive planetary atmospheres. For example, laboratory measurements of UV quantum yields and the early evolution of the sun suggest (Sagan and Khare 1971) that ~ 200 kg cm^{-2} of organic molecules were synthesized in the first 10^9 yrs of Earth's history. Comparable quantities are expected from hypervelocity shocks (Bar-Nun *et al.*, 1970). We have calculated earlier that much larger quantities of organics are available in condensing clouds. But it is unlikely that such molecules will survive planet formation (Sagan 1965). The concentration of radioactive elements of large ionic radii near the lunar surface argues for early melting of the Moon (Ringwood 1970), as does its basaltic composition; and radioactive dating of material from the Fra Mauro Apollo 14 site indicates extensive lunar remelting 3.85 x 10^9 yrs ago (Wasserburg 1971). The energy for the initial melting of the Moon is likely to be the gravitational potential energy of accretion, particularly if the Moon accumulated rapidly, so that very large quantities of potential energy were not radiated away as heat. The later melting events are attributable to radioactive decay of elements concentrated by earlier melting events at the lunar surface. Similar events must have occurred on all larger objects. Studies of meteoritic irons indicate that melting events occurred in the parent bodies of these objects, which were almost certainly smaller than the Moon (although the survival of carbonaceous chondrite organics implies that *their* parent bodies remained cool). The specific gravitational potential energy of accretion is GM/R where G is the Newtonian gravitational constant, M the mass of the object and R its final radius. Assuming that all such objects have approximately the same density, as is true throughout the solar system, GM/R $\propto$ R^2. The maximum gravitational potential energy per gram is released in the final stages of accretion. More is released in larger objects. Thus maximum melting should occur near the end of planetary formation. Melting requires temperatures $\geqslant 1200^\circ$ C. Temperatures 1000° cooler are still adequate to denature, pyrolize or thoroughly dissociate all organic molecules of interest. Alternative proposals, *e.g.,* that the Moon accreted from already differentiated material (Gold 1971), do not enhance the survival of interstellar organics. Interstellar molecules were certainly accreted subsequently – say in the first 10^9 years after planetary formation – but the exogenous supply does not approach the indigenous supply from atmospheric chemistry. It also appears implausible that there is a specific molecule, required for the origin of life, which can only be synthesized under interstellar conditions.

Accordingly, the contribution of interstellar organic chemistry to problems in biology is not substantive but analogical. The interstellar medium reveals the operation of chemical processes which, on the Earth and perhaps on vast numbers of planets throughout the universe, led to the origin of life, but the actual molecules of the interstellar medium are unlikely to play any significant biological role.

This research has been supported in part by NASA Grant NGR 33-010-101. I am indebted to David Buhl, George Field, T. Gold, Bishun Khare, James Pollack, E. E. Salpeter, Joseph Veverka, and David Wallace for stimulating discussions on various aspects of this problem, and to Messrs. Field, Gold, Salpeter, Wallace, George Herbig, Fred Johnson and Charles Townes for reading and commenting on an early draft of this paper. Parts of this article appeared in *Nature*.

REFERENCES

Baldwin, B. and Schaeffer, Y. 1971, *J. G. R.,* **76,** 4653.

Ball, J. A.; Gottlieb, C. A.; Lilley, A. E. and Radford, H. E. 1971a, *IAU Circular No. 2350.*

————. 1971b, *Ap. J.,* **162,** L203.

Bar-Nun, A; Bar-Nun,N.; Bauer, S.; and Sagan, C. 1970, *Science,* **168,** 470.

Barrett, A. H. 1971, at the Conference on *Interstellar Molecules and Cosmochemistry,* New York.

Calvert, J. G. and Pitts, J. N. 1966, in *Photochemistry* (New York: John Wiley and Sons).

Calvin, M. 1969, in *Chemical Evolution* (Oxford: University Press).

Cameron, A. G. W. 1971, private communication.

Donn, B. 1965, *Ap. J.,* **152,** L129.

Ezer, D. and Cameron, A. G. W. 1967, *Can. J. Phys.,* **45,** 3429.

Gilra, D. P. 1971, *Nature,* **229,** 237.

Glasstone, R. M.; Laidler, K. H. and Eyring, H. 1941, in *The Theory of Rate Processes* (New York: McGraw-Hill).

Gold, T. 1971, *Proc. Am. Phil. Soc.,* **115,** 74.

Hackwell, J. A.; Gehrz, R. D. and Woolf, N. J. 1970, *Nature,* **227,** 822.

Herbig, G. H. 1970, *Proc. XVI Liege Symposium,* in press.

Hollaender, A., ed., 1955, *Radiation Biology. II. Ultraviolet and Related Radiations* (New York: McGraw-Hill).

Horowitz, N. H.; Drake, F. D.; Miller, S. L.; Orgel, L. E. and Sagan, C. 1970, in *Biology and the Future of Man,* ed. P. Handler (Oxford: University Press).

Hoyle, F.; Cameron, A. G. W. and Ruskol, E. 1971, in *Highlights of Astronomy* (Dordrecht: Reidel).

Hunter, C. E. and Donn, B. 1971, *Ap. J.,* **167,** 71.

Jastrow, R. and Cameron, A. G. W., eds., 1963, *Origin of the Solar System* (New York: Academic Press).

Johnson, F. M. 1967a, in *Interstellar Grains,* eds. J. M. Greenberg and T. P. Roark (Washington: NASA SP-140; U. S. Gov. Printing Office).

————. 1967b, in *Use of Space Systems for Planetary Geology and Geophysics, Am. Astronaut. Soc. Sci. Tech. Series,* **17,** 51.

————. 1971b, in *Interstellar Molecules and Cosmochemistry, Ann. N. Y. Acad. Sci.,* **194,** 3.

Kenyon, D. H. and Steinman, G. 1969, *Biochemical Predestination* (New York: McGraw-Hill).

Khare, B. N. and Sagan, C. 1971, this volume.

Knacke, R. F.; Cudaback, D. D. and Gaustad, J. E. 1969, *Ap. J.,* **158,** 151.

Lippincott, E. R.; Eck, R. V.; Dayhoff, M. O. and Sagan, C. 1967, *Ap. J.* **147,** 753.

Litvak, M. M. 1970, at the *Conference on Interstellar Molecules and the Origin of Life,* NASA Ames Research Center.

Oort, J. H. 1963, in *The Moon, Meteorites, and Comets,* eds. B. M. Middlehurst and G. P. Kuiper (Chicago: University of Chicago Press).

Opik, E. J. 1963, *Advances Astron. Ap.,* **2,** 219.

Ringwood, A. E. 1970, *J. G. R.,* **75,** 6453.

Sagan, C. 1957, *Evolution,* **11,** 40.

————. 1965, in *The Origins of Prebiological Systems,* ed. S. W. Fox (New York: Academic Press)

————. 1970, in *Highlights of Astronomy* (Dordrecht: Reidel).

Sagan, C.; Dayhoff, M. O.; Lippincott, E. R. and Eck, R. V. 1967, *Nature,* **213,** 273.

Sagan, C. and Khare, B. N. 1970, *B.A.A.S.,* **2,** 340.

————. 1971, *Science,* **173,** 417.

Sagan, C. and Miller, S. L. 1960, *Astron. J.,* **65,** 499.

Salpeter, E. E. 1971, in *Highlights of Astronomy* (Dordrecht: Reidel).

Snyder, L. E.; Buhl, D.; Zuckerman, B. and Palmer, P. 1969, *Phys. Rev. Letters,* **22,** 679.

Snyder, L. E. and Buhl, D. 1971, *Ap. J.,* **1963,** L47.

Solomon, P. M. 1971, at the Conference on *Interstellar Molecules and* Cosmochemistry, New York.

Solomon, P.M.; Jefferts, K. B.; Penzias, A. A. and Wilson, R. W. 1971 *Ap. J.,* **168,** L107.

Spitzer, L. 1968, in *Nebulae and Interstellar Matter,* ed. B. M. Middlehurst and L. H. Aller (Chicago: University of Chicago Press).

Stein, R. F. and Cameron, A. G. W., eds., 1966, *Stellar Evolution* (New York: Plenum).

Stein, R. F. and Gillett, F. C. 1969, *Ap. J. (Letters),* **155,** L197.

Studier, M. H.; Hayatsu, R. and Anders, E. 1968, *Geochim. Cosmochim. Acta.,* **32,** 151.

Townes, C. 1971, in *Highlights of Astronomy* (Dordrecht: Reidel).

Turner, B. E. 1971, *Ap. J.,* **163,** L35.

Urey, H. C. 1966, *M.N.R.A.S.,* **131,** 199.

van de Hulst, H. C. 1957, *Light Scattering by Small Particles* (New York: John Wiley and Sons).

Wasserburg, G. J. 1971, at *COSPAR Symposia,* Seattle.

Zuckerman, B.; Ball, J. A. and Gottlieb, C. A. 1971, *Ap. J.,* **163,** L41.

INDEX